T0189365

Lecture Notes in Computer Science 14560

Founding Editors

Gerhard Goos
Juris Hartmanis

The series Lecture Notes in Computer Science (LNCS), including its subseries Lecture Notes in Artificial Intelligence (LNAI) and Lecture Notes in Bioinformatics (LNBI), has established itself as a medium for the publication of new developments in computer science and information technology research, teaching, and education.

LNCS enjoys close cooperation with the computer science R & D community, the series counts many renowned academics among its volume editors and paper authors, and collaborates with prestigious societies. Its mission is to serve this international community by providing an invaluable service, mainly focused on the publication of conference and workshop proceedings and postproceedings. LNCS commenced publication in 1973.

Venanzio Capretta · Robbert Krebbers ·
Freek Wiedijk
Editors

Logics and Type Systems in Theory and Practice

Essays Dedicated to Herman Geuvers
on The Occasion of His 60th Birthday

 Springer

Editors
Venanzio Capretta 🆔
University of Nottingham
Nottingham, UK

Robbert Krebbers
Radboud University Nijmegen
Nijmegen, The Netherlands

Freek Wiedijk
Radboud University Nijmegen
Nijmegen, The Netherlands

ISSN 0302-9743   ISSN 1611-3349 (electronic)
Lecture Notes in Computer Science
ISBN 978-3-031-61715-7   ISBN 978-3-031-61716-4 (eBook)
https://doi.org/10.1007/978-3-031-61716-4

This Springer imprint is published by the registered company Springer Nature Switzerland AG
The registered company address is: Gewerbestrasse 11, 6330 Cham, Switzerland

If disposing of this product, please recycle the paper.

Herman Geuvers

Preface

This Festschrift is a collection of essays in celebration of the 60th birthday of Herman Geuvers. The essays span a wide range of subjects, as a tribute to Herman's diverse research pursuits: He is a passionate researcher with a keen interest in a variety of topics in the field of theoretical computer science. Herman is known for his openness to a new topic and his eagerness to collaborate on it, or advise a student on it.

In 1993 Herman defended his PhD thesis "Logics and Type Systems" at Radboud University Nijmegen, in which he studied the connection between logics and typed lambda calculi via the Curry-Howard isomorphism. This theme remained close to his heart throughout his career. Together with his group he worked on a spectrum of subjects: inductive/coinductive types in type theory, homotopy type theory, the Curry-Howard correspondence for classical logic, and more. In addition, Herman played a key role in the TYPES community, and co-authored multiple handbooks on type theory.

Coinduction and infinite structures are also important in Herman's research. In his early career, he formulated a seminal model of coinduction in polymorphic lambda calculus. In more recent years, he investigated the notion of bisimulations, and particularly its relationship with apartness—a concept that played a key role in his earlier work on constructive mathematics.

Another theme throughout Herman's career is proof assistants—particularly those based on type theory such as Coq—to formalize results in mathematics and computer science. Herman led the development of the Constructive Coq Repository Nijmegen (C-CORN) whose milestones include the first formalized constructive proof of the fundamental theorem of algebra and the fundamental theorem of calculus.

Herman aspires to make proof assistants more accessible to a broader group of users, not just experts in type theory and mathematical logic. With his group, he worked on a semantic web for formalized mathematics, a Wiki for formalized mathematics, and the use of machine learning to make proof assistants more scalable.

Not only is Herman an original, insightful, and innovative researcher, he also is a friendly, generous, and supportive advisor, group leader, director of education (2009–2013), and director of research (2001–2003 and 2015–2020). He is always available and has provided a wonderful research environment for many PhD students, postdocs, and assistant and associate professors over the years.

The 17 essays in this Festschrift were contributed by Herman's (former) PhD students, postdocs, colleagues, and collaborators. Each paper were reviewed with the help of 2–3 experts, who are listed below.

We thank Herman for being an inspiring researcher and a fantastic advisor, group leader, and collaborator. And we congratulate him on his 60th birthday!

March 2024

Venanzio Capretta
Robbert Krebbers
Freek Wiedijk

List of Expert Reviewers

- Jean-Paul Allouche
- Jos Baeten
- Steffen van Bakel
- Harsh Beohar
- Yves Bertot
- Jasmin Blanchette
- Thierry Coquand
- Luís Cruz-Filipe
- Ugo de'Liguoro
- Michele De Pascalis
- Michel Dekking
- Mariangiola Dezani-Ciancaglini
- Joerg Endrullis
- Martin Escardo
- Daniil Frumin
- Rob van Glabbeek
- Anton Golov
- Jan Friso Groote
- Sean Holden
- Tonny Hurkens
- Bart Jacobs
- Tom de Jong
- Cezary Kaliszyk
- Shin-Ya Katsumata
- Dexter Kozen
- Alberto Larrauri
- Antonio Lorenzin
- Damien Pous
- Narad Rampersad
- Jurriaan Rot
- Ana Sokolova
- Laurent Théry
- Frits Vaandrager
- Niels van der Weide
- Tim Willemse
- Hans Zantema

Contents

Sequential Value Passing Yields a Kleene Theorem for Processes

Jos C. M. Baeten[1(✉)] and Bas Luttik[2]

[1] CWI, Science Park 123, 1098 Amsterdam, The Netherlands
Jos.Baeten@cwi.nl
[2] Eindhoven University of Technology, Eindhoven, The Netherlands

Abstract. Communication with value passing has received ample attention in process theory. Value passing through a sequential composition has received much less attention. In recent work, we found that sequential value passing is the essential ingredient to prove the analogue of the classical theorem of the equivalence of pushdown automata and context-free grammars in a setting of interactive processes and bisimulation. Subsequently, we found that the treatment of sequential value passing in the process setting can be simplified considerably. We report on this simplification here, and find another application of sequential value passing, viz. a Kleene theorem for processes.

Keywords: sequential value passing · process theory · pushdown automaton · context-free grammar · bisimilarity · Kleene theorem

We dedicate this paper to our valued colleague Herman Geuvers on the occasion of his sixtieth birthday. We admire his meticulous and thorough style.

1 Introduction

This paper contributes to our ongoing project to integrate the theory of automata and formal languages on the one hand and concurrency theory on the other hand. We do not treat automata as language acceptors. Instead, we treat them as processes. That is, we view automata as transition systems, and consider them modulo bisimilarity.

It is well-known that Kleene's theorem [20], which states that a language is accepted by a finite automaton if and only if it is denoted by a regular expression, does not have a direct process-theoretic pendant modulo bisimilarity. Milner showed that there exist finite automata that are not bisimilar to the transition system associated with a regular expression [23]. We shall prove in this paper that it suffices to extend regular expressions with a simple notion of sequential value passing to obtain a process-theoretic variant of Kleene's theorem.

In [5], it was shown that a Kleene theorem can also be obtained in bisimulation semantics if the syntax of regular expressions is enriched with parallel composition, synchronisation and encapsulation. The Kleene theorem we establish here provides an alternative to that result.

© The Author(s), under exclusive license to Springer Nature Switzerland AG 2024
V. Capretta et al. (Eds.): *Logics and Type Systems in Theory and Practice*, LNCS 14560, pp. 1–16, 2024.
https://doi.org/10.1007/978-3-031-61716-4_1

In [4], we already looked at the classical theorem that a language is accepted by a pushdown automaton if and only if it is defined by a context-free grammar. In the process setting, a context-free grammar is a process algebra with actions, choice, sequencing and recursion. We proved that every process given by a finite guarded recursive specification over this algebra is also a process defined by a pushdown automaton, but not the other way around, the algebra is not sufficiently expressive to specify all processes defined by pushdown automata. Adding sequential value passing suffices: the set of processes given by a finite guarded recursive specification over the extended algebra coincides with the set of processes defined by pushdown automata. The variant of sequential value passing used in [4], however, is semantically significantly more involved than the variant that we propose here. Thus, we also present a simplification of the result in [4].

Communication with value passing has received ample attention in process theory (see, e.g., [19]). Value passing through a sequential composition has received much less attention (but see, e.g., [13,25]). Kleene algebra with tests, proposed in [21], gives an axiomatic treatment of regular expressions with conditionals. There are two important differences between this theory and the theory presented in this article. First, Kleene algebra with tests is catered towards language equivalence; indeed, it includes axioms that are not valid in bisimulation semantics. Second, it does not include a facility to specify sequential passing which, as we will show, is an essential ingredient to get a Kleene theorem in bisimulation semantics. In [18], a process algebra with guards is proposed, which yields a treatment of conditionals in bisimulation semantics. Also this work, however, lacks a facility to specify sequential value passing; state attributes are, instead, changed implicitly through an effect associated with actions. This approach makes it unsuitable for establishing the type of correspondence results we obtain in this article.

2 Preliminaries

As a common semantic framework we use the notion of a *transition system*.

Definition 1. *A* transition system space *is a quadruple* $(\mathcal{S}, \mathcal{A}, \rightarrow, \downarrow)$, *where*

1. $\mathcal{S}$ *is a set of* states;
2. $\mathcal{A}$ *is a set of* actions;
3. $\rightarrow \subseteq \mathcal{S} \times \mathcal{A} \times \mathcal{S}$ *is an* $\mathcal{A}$-labelled *transition relation; and*
4. $\downarrow \subseteq \mathcal{S}$ *is the set of final or accepting states.*

A transition system *is a transition system space with a special designated* root state *or* initial state $\uparrow$, *i.e., it is a quintuple* $(\mathcal{S}, \mathcal{A}, \rightarrow, \uparrow, \downarrow)$ *such that* $(\mathcal{S}, \mathcal{A}, \rightarrow, \downarrow)$ *is a transition system space, and* $\uparrow \in \mathcal{S}$.

Note that, by the requirement that there is a designated root state, a transition system has a non-empty set of states. It will be technically convenient to also consider the structure with an empty set of states, an empty transition relation, an empty set of accepting states and an undefined root state. This structure will be referred to as the inconsistent *transition system.*

We write $s \xrightarrow{a} s'$ for $(s, a, s') \in \rightarrow$ and $s{\downarrow}$ for $s \in {\downarrow}$. We say that s' is reachable from s if, for some $n \in \mathbb{N}$, there exist $s_0, \ldots, s_n \in \mathcal{S}$ and $a_1, \ldots, a_n \in \mathcal{A}$ such that $s = s_0$, $s_i \xrightarrow{a_{i+1}} s_{i+1}$ for all $0 \leq i < n$, and $s_n = s'$. A transition system that has finitely many states reachable from the root state (if it exists) and finitely many transitions between them is called a finite automaton.

By considering language equivalence classes of transition systems, we recover languages as a semantics, but we can also consider other equivalence relations. Notable among these is *bisimilarity*.

Definition 2. *Let $(\mathcal{S}, \mathcal{A}, \rightarrow, {\downarrow})$ be a transition system space. A symmetric binary relation R on $\mathcal{S}$ is a* bisimulation *if it satisfies the following conditions for every $s, t \in \mathcal{S}$ such that $s \, R \, t$ and for all $a \in \mathcal{A}$:*

1. if $s \xrightarrow{a} s'$ for some $s' \in \mathcal{S}$, then there is a $t' \in \mathcal{S}$ such that $t \xrightarrow{a} t'$ and $s' \, R \, t'$; and
2. if $s{\downarrow}$, then $t{\downarrow}$.

We write $s \leftrightarrow t$ if and only if there a bisimulation relating s and t, and say that s is bisimilar to t.

The results of this paper do not rely on abstraction from internal computations, so we do not consider the silent step τ here, and we can use the *strong* version of bisimilarity defined above, which does not give special treatment to τ-labelled transitions. But in general (in other work) we have to use a version of bisimilarity that accomodates for abstraction from internal activity; the finest such notion of bisimilarity is *divergence-preserving branching bisimilarity*, which was introduced in [14] (see also [22] for an overview of recent results).

We see that bisimilarity is an equivalence relation on a transition system space, so it divides a transition system space into a number of equivalence classes.

3 Regular Expressions

In [23], Robin Milner considered regular expressions in the process setting (we define these below), and found that not all finite automata can be defined by a regular expression (modulo bisimulation). He posed the question how the set of finite automata defined by a regular expression can be characterized, a question that was solved in [2]. Here, we answer the question which ingredient needs to be added to regular expressions to characterize all finite automata modulo bisimulation.

We present regular expressions as the closed terms in the theory TSP+IT of [8]. The syntax of this theory has the following elements:

- **0** is the inactive and not accepting process (deadlock), the one-state automaton without transitions where the state is not final;
- **1** is the inactive and accepting process, the one-state automaton without transitions where the state is final;
- for a given set $\mathcal{A}$ of actions we have the prefix operators $a.\_$ for each $a \in \mathcal{A}$;
- the binary operator $+$ is alternative composition or choice;

- the binary operator · is sequential composition;
- the unary operator _* is iteration or Kleene star.

We give the behaviour of terms over this algebra by means of structural operational semantics (see [1]): we define a unary acceptance or termination predicate ↓ (written postfix) and, for every $a \in \mathcal{A}$, a binary transition relation $\xrightarrow{a}$ (written infix), by means of the transition system specification in Table 1. The rules in Table 1 should be read as follows: if the premises above the line are satisfied for a certain substitution, then the conclusion(s) below the line are also valid for the same substitution. These rules turn the set of regular expressions into a transition system space. Each regular expression has a transition system in which every step and every termination is provable from this specification. Each regular expression has a transition system with finitely many reachable states and finitely many transitions, so is a finite automaton.

A *regular process* is a bisimulation equivalence class of finite automata.

Table 1. Operational semantics for regular expressions.

$$\frac{}{a.x \xrightarrow{a} x} \qquad \frac{x \xrightarrow{a} x'}{x + y \xrightarrow{a} x' \quad y + x \xrightarrow{a} x'}$$

$$\frac{}{1 \downarrow} \qquad \frac{x \downarrow}{x + y \downarrow \quad y + x \downarrow}$$

$$\frac{x \downarrow \quad y \downarrow}{x \cdot y \downarrow} \qquad \frac{x \xrightarrow{a} x'}{x \cdot y \xrightarrow{a} x' \cdot y} \qquad \frac{x \downarrow \quad y \xrightarrow{a} y'}{x \cdot y \xrightarrow{a} y'}$$

$$\frac{}{x^* \downarrow} \qquad \frac{x \xrightarrow{a} x'}{x^* \xrightarrow{a} x' \cdot x^*}$$

Since the rules in Table 1 are in *path* format (see [7]), bisimilarity is a congruence relation for the operators of regular expressions. Consequently, we can consider the equational theory of regular expressions. From [8, 10], we know that there is no finite axiomatization that is sound and ground-complete. For more information, see [16]. For all regular expressions x, we have $1 \cdot x \underset{\leftarrow}{\rightarrow} x \underset{\leftarrow}{\rightarrow} x \cdot 1$ and $0 \cdot x \underset{\leftarrow}{\rightarrow} 0$ (but not $x \cdot 0 \underset{\leftarrow}{\rightarrow} 0$!), which gives us the motivation to use the symbols 1 and 0.

We see alternative composition is commutative and associative, and sequential composition is associative but not commutative, and so we can leave out brackets as usual.

We further note that, modulo bisimilarity, sequential composition distributes from the right over choice $((x + y) \cdot z \underset{\leftarrow}{\rightarrow} x \cdot z + y \cdot z)$, but not from the left $(x \cdot (y + z) \not\underset{\leftarrow}{\rightarrow} x \cdot y + x \cdot z)$. If we consider the classical algorithm that finds a regular expression with the same language as a given finite automaton, we see that it makes essential use of the distributivity law that is not valid for bisimulation.

Each regular expression generates a finite automaton. The reverse direction is valid in language equivalence, but not in bisimilarity, as the following example shows (here, the state on the left is the root state, and both states are accepting).

Theorem 1. *There is no regular expression of which the transition system is bisimilar to the finite automaton in Fig. 1.*

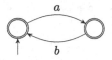

Fig. 1. A finite automaton not denoted by a regular expression.

Proof. Milner [23] proves this for a slightly more complicated example than the one in Fig. 1. His proof can be easily transposed to the present situation.

Thus, we look for an extension of TSP+IT in order to find expressions for all finite automata. In [5], extensions with various forms of parallel composition are studied. There, it is found that an extension with parallel composition and value passing communication yields all regular processes. Here, we present an extension that is more straightforward: just extending with sequential value passing suffices. In order to define sequential value passing, we use operators for signals and conditions, based on [3, 8].

4 Signals and Conditions

The key to the extension is that information is needed about the state which the process is in at a given time. In [4], we used expressions in propositional logic to express a property of the current state of a process. This gives considerable overhead, as we need to evaluate the expressions again at each step in a process, assigning truth values to the propositional variables that occur in an expression. Here, we just assume that the states of a process have an attribute that takes a unique value from a finite data set $\mathcal{D}$. We need that $\mathcal{D}$ is finite, since we will use in examples indexed summations over elements of $\mathcal{D}$, and we can use $\sum_{d \in \mathcal{D}}$ notation as an abbreviation. A difficulty is that we need to ensure that the attribute of a state has a unique value: it is impossible to have two different values of an attribute at the same time.

First, we introduce an operator, called the *root-signal emission operator* $^{\wedge\!\!\blacktriangle}$ in [3, 8], that exposes the value of the attribute. A term $d^{\wedge\!\!\blacktriangle}x$ represents the process x that starts out in a state in which the attribute has the value d. Only one value can be shown at a time, so we need to declare terms like $d^{\wedge\!\!\blacktriangle}1 + e^{\wedge\!\!\blacktriangle}1$, $(d^{\wedge\!\!\blacktriangle}1) \cdot (e^{\wedge\!\!\blacktriangle}1)$ or $d^{\wedge\!\!\blacktriangle}(e^{\wedge\!\!\blacktriangle}1)$ (if $d \neq e$) *inconsistent*. We associate the inconsistent transition system with these inconsistent terms. In the operational semantics, we add an extra predicate $\mathcal{C}$, indicating that the term is consistent, and giving the value if there is one.

Thus, for $d \in \mathcal{D}$, $x \, \mathcal{C}d$ means that (the initial state of) expression x is consistent, and has value d. On the other hand, $x \, \mathcal{C}\ominus$ means that (the initial state of) expression x is consistent, and no value is given. We write $\mathcal{D}_\ominus$ for $\mathcal{D} \cup \{\ominus\}$. We use $d, e \in \mathcal{D}$ and $\delta, \varepsilon \in \mathcal{D}_\ominus$. We have a very simple partial ordering on $\mathcal{D}_\ominus$: $\ominus$ is below all elements of $\mathcal{D}$, and elements of $\mathcal{D}$ are incomparable. We write $\delta \preceq \varepsilon$ if $\delta = \varepsilon$ or $\delta = \ominus$. We define the predicate $\mathcal{C}$ by the operational rules in Table 2 for regular expressions and the root-signal emission operator.

Table 2. Operational semantics of consistency and state values ($d \in \mathcal{D}, \delta, \varepsilon \in \mathcal{D}_\ominus$).

$$\overline{\mathbf{0}\ \mathcal{C}\ominus} \qquad \overline{\mathbf{1}\ \mathcal{C}\ominus} \qquad \overline{a.x\ \mathcal{C}\ominus}$$

$$\frac{x\ \mathcal{C}\delta \quad y\ \mathcal{C}\varepsilon \quad \delta \preceq \varepsilon}{x + y\ \mathcal{C}\varepsilon \quad y + x\ \mathcal{C}\varepsilon} \qquad \frac{x\ \mathcal{C}\delta \quad x\!\!\downarrow\!\!\!\!\!/}{x \cdot y\ \mathcal{C}\delta}$$

$$\frac{x\!\downarrow \quad x\ \mathcal{C}\delta \quad y\ \mathcal{C}\varepsilon \quad \delta \preceq \varepsilon}{x \cdot y\ \mathcal{C}\varepsilon} \qquad \frac{x\!\downarrow \quad x\ \mathcal{C}\varepsilon \quad y\ \mathcal{C}\delta \quad \delta \preceq \varepsilon}{x \cdot y\ \mathcal{C}\varepsilon}$$

$$\frac{x\ \mathcal{C}\delta}{x^*\ \mathcal{C}\delta} \qquad \frac{x\ \mathcal{C}\delta \quad \delta \preceq d}{d\ ^{\wedge}\!\!\!\!\!{\scriptstyle\blacktriangle} x\ \mathcal{C}d}$$

Table 3. Operational semantics for regular expressions and signal emission with consistency conditions ($\delta, \varepsilon \in \mathcal{D}_\ominus, d \in \mathcal{D}$).

$$\overline{\mathbf{1}\!\downarrow} \qquad \frac{x\ \mathcal{C}\delta}{a.x \xrightarrow{a} x} \qquad \frac{x \xrightarrow{a} x' \quad x + y\ \mathcal{C}\delta}{x + y \xrightarrow{a} x' \quad y + x \xrightarrow{a} x'}$$

$$\frac{x\!\downarrow \quad x + y\ \mathcal{C}\delta}{x + y\!\downarrow \quad y + x\!\downarrow} \qquad \frac{x\!\downarrow \quad y\!\downarrow \quad x \cdot y\ \mathcal{C}\delta}{x \cdot y\!\downarrow}$$

$$\frac{x \xrightarrow{a} x' \quad x \cdot y\ \mathcal{C}\delta \quad x' \cdot y\ \mathcal{C}\varepsilon}{x \cdot y \xrightarrow{a} x' \cdot y} \qquad \frac{x\!\downarrow \quad y \xrightarrow{a} y' \quad x \cdot y\ \mathcal{C}\delta}{x \cdot y \xrightarrow{a} y'}$$

$$\frac{x\ \mathcal{C}\delta}{x^*\!\downarrow} \qquad \frac{x \xrightarrow{a} x' \quad x' \cdot x^*\ \mathcal{C}\delta}{x^* \xrightarrow{a} x' \cdot x^*}$$

$$\frac{x\!\downarrow \quad d\ ^{\wedge}\!\!\!\!\!{\scriptstyle\blacktriangle} x\ \mathcal{C}d}{d^{\wedge}\!\!\!\!\!{\scriptstyle\blacktriangle} x\!\downarrow} \qquad \frac{x \xrightarrow{a} x' \quad d\ ^{\wedge}\!\!\!\!\!{\scriptstyle\blacktriangle} x\ \mathcal{C}d}{d^{\wedge}\!\!\!\!\!{\scriptstyle\blacktriangle} x \xrightarrow{a} x'}$$

In Table 3, we repeat the rules of Table 1, with extra conditions to ensure consistency. In addition, we give the operational rules for the signal emission operator.

Note that these rules ensure that a step can only occur between consistent terms, so if we can derive $x \xrightarrow{a} x'$, then terms x and x' are consistent. Also, if we can derive $x \downarrow$, then x is consistent.

Notice that the fifth rule in Table 2, the first rule for sequential composition, has a so-called *negative premise*: we can conclude $x \cdot y\ \mathcal{C}\delta$ provided $x\downarrow$ does not hold. It is well-known that transition system specifications with negative premises may not define a unique transition relation that agrees with provability from the transition system specification [12, 15, 17]. To show that the transition system specification presented here does indeed define a unique transition relation that agrees with provability, it suffices to define a *stratification* (see [17, Definition 2.11]). First note that, since the rules defining the predicates $\mathcal{C}\delta$ and $\downarrow$ do not have premises referring to $\xrightarrow{a}$, and the negative premise only occurs in a rule defining $\mathcal{C}\delta$, so we can ignore the rules with $\xrightarrow{a}$ in the conclusion. The mapping S from expressions of the form $x\ \mathcal{C}\delta$ and $x\downarrow$ to natural numbers defined by

$$S(0\ \mathcal{C}\delta) = S(1\ \mathcal{C}\delta) = S(a.x\ \mathcal{C}\delta) = 0$$
$$S(x + y\ \mathcal{C}\delta) = S(x \cdot y\ \mathcal{C}\delta) = S(x\ \mathcal{C}\delta) + S(y\ \mathcal{C}\delta) + 1, \quad \text{and}$$
$$S(d \wedge x\ \mathcal{C}\delta) = S(x^*\ \mathcal{C}\delta) = S(x\downarrow) = S(x\ \mathcal{C}\delta)$$

is a stratification. In [17] it is proved that whenever a stratification exists, then the transition system specification defines a unique transition relation that agrees with provability in the transition system specification.

The rules in Tables 2 and 3 turn the set of consistent expressions over the extended syntax into a transition system space, and thus Definition 2 yields a notion of bisimilarity on the consistent expressions. We will have no need to consider inconsistent expressions in this paper, or to do calculations on expressions that may be inconsistent, so we will not define bisimilarity on inconsistent expressions. By interpreting all inconsistent expressions as the inconsistent transition system, that is not hard to do, however.

Note that the set of consistent expressions is not closed under alternative composition: $d \wedge 1$ and $e \wedge 1$ are both consistent, but, if $d \neq e$, then $d \wedge 1 + e \wedge 1$ is not consistent. Further note that bisimilarity as defined in Definition 2 is not a congruence on the set of all expressions, as $d \wedge 1 \rightleftharpoons e \wedge 1$, but $d \wedge 1 + d \wedge 1$ is not bisimilar to $d \wedge 1 + e \wedge 1$. It is not difficult to define a variant of bisimulation on the set of all expressions that is a congruence. All that is required is that a term satisfying $\mathcal{C}\delta$ can only be related to a term also satisfying $\mathcal{C}\delta$. As all operational rules are in *panth* format, as defined in [24], the resulting bisimilarity is a congruence. We leave the precise formulation as further work, as we do not use this notion in the present paper.

An expression over this extended syntax that satisfies $\mathcal{C}\delta$ for some $\delta \in \mathcal{D}_\ominus$ is consistent, and exposes the attribute d of the state when it satisfies $\mathcal{C}d$ for some $d \in \mathcal{D}$. We can depict the attribute values in a state, as shown in Fig. 2 for the term $d \wedge (1 + a.(e \wedge 0))$, but we emphasize that these values are not part of the transition system, they just occur to help the reader.

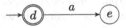

Fig. 2. Example showing attribute values.

Next, we define the *guarded command*. Given a attribute value d, we write $d :\rightarrow x$, with the intuitive meaning '*if the attribute of the current state has value d then x can be executed*'. Thus, d is a guard: x can only be executed in a state with attribute value d.

In the operational semantics, we use as additional relations the conditional steps $\xrightarrow{d,a}$ ($d \in \mathcal{D}, a \in \mathcal{A}$) and as additional predicates the conditional acceptance $^d\downarrow$ ($d \in \mathcal{D}$). In Table 4, we give the operational rules for guarded command.

In Table 5, we give the conditional steps and conditional acceptance for regular expressions and root signal emission.

Table 4. Operational semantics for guarded command ($\delta \in \mathcal{D}_\ominus, d \in \mathcal{D}$).

$$\frac{x \, C\delta \quad \delta \preceq d}{d :\rightarrow x \, C\ominus}$$

$$\frac{x \xrightarrow{a} x' \quad d :\rightarrow x \, C\ominus}{d :\rightarrow x \xrightarrow{d,a} x'} \qquad \frac{x \xrightarrow{d,a} x'}{d :\rightarrow x \xrightarrow{d,a} x'}$$

$$\frac{x \downarrow \quad d :\rightarrow x \, C\ominus}{d :\rightarrow x \, ^d\downarrow} \qquad \frac{x \, ^d\downarrow}{d :\rightarrow x \, ^d\downarrow}$$

Notice that if $d \neq e$, then the term $d :\rightarrow (e \wedge \mathbf{1})$ is inconsistent, whereas the term $d\wedge(e :\rightarrow \mathbf{1})$ is consistent, and bisimilar to $d\wedge\mathbf{0}$. Notice that a consistent term of the form $d\wedge x$ can do no conditional steps and no conditional termination: all steps $x \xrightarrow{e,a} x'$ and $x \, ^e\downarrow$ for $e \neq d$ disappear, and for all steps $x \xrightarrow{d,a} x'$ and $x \, ^d\downarrow$ the conditions are removed.

We can easily extend the stratification given above to include the guarded command operator by defining $S(d :\rightarrow x \, C\delta) = S(x \, ^d\downarrow) = S(x \, C\delta)$; as above, there is no need to consider rules with $\xrightarrow{d,a}$ in the conclusion. So the extended set of rules define a transition system space on the consistent expressions, and we also have a notion of bisimilarity on the extended set of consistent terms.

Now in order to get a notion of bisimilarity that is a congruence on the extended syntax, we also need to relate the conditional steps and conditional acceptance of bisimilar terms. We have no use for this bisimilarity in the present paper, so do not give the details here. We emphasize that the conditional steps and conditional acccceptance are only needed to generate the (unconditional) steps and acceptance of the transition system of a term. Thus, the conditional steps and conditional acceptance do not appear in the generated transition system space.

In order to illustrate the interplay of root signal emission and guarded command, and to show how nondeterminism can be dealt with, we give the following example.

Example 1. A coin toss can be described by the following term:

$$Toss = toss.(heads \wedge \mathbf{1}) + toss.(tails \wedge \mathbf{1}).$$

The behaviour of a player who wins one dollar when heads comes up and loses one dollar when tails comes up is specified by the term

$$Player = heads :\rightarrow win1\$ + tails :\rightarrow lose1\$.$$

Process $Toss \cdot Player$ shows how sequential value passing is achieved.

For a more involved example consider the process of tossing a coin until heads comes up:

$$Toss \cdot (tails :\rightarrow Toss)^* \cdot (heads :\rightarrow \mathbf{1}).$$

Table 5. Conditional steps and conditional termination for regular expressions and root signal emission ($d \in \mathcal{D}, \delta \in \mathcal{D}_\ominus$).

$$\frac{x \xrightarrow{d,a} x' \quad y\,\mathcal{C}d}{x+y \xrightarrow{a} x' \quad y+x \xrightarrow{a} x'} \qquad\qquad \frac{x \xrightarrow{d,a} x' \quad y\,\mathcal{C}\ominus}{x+y \xrightarrow{d,a} x' \quad y+x \xrightarrow{d,a} x'}$$

$$\frac{x\,{}^d\!\downarrow \quad y\,\mathcal{C}d}{x+y\downarrow \quad y+x\downarrow} \qquad\qquad \frac{x\,{}^d\!\downarrow \quad y\,\mathcal{C}\ominus}{x+y\,{}^d\!\downarrow \quad y+x\,{}^d\!\downarrow}$$

$$\frac{x\,{}^d\!\downarrow \quad y\,{}^d\!\downarrow}{x\cdot y\,{}^d\!\downarrow} \qquad \frac{x\,{}^d\!\downarrow \quad y\downarrow \quad y\,\mathcal{C}\delta \quad \delta \preceq d}{x\cdot y\,{}^d\!\downarrow}$$

$$\frac{x\downarrow \quad x\,\mathcal{C}d \quad y\,{}^d\!\downarrow}{x\cdot y\downarrow} \qquad\qquad \frac{x\downarrow \quad x\,\mathcal{C}\ominus \quad y\,{}^d\!\downarrow}{x\cdot y\,{}^d\!\downarrow}$$

$$\frac{x \xrightarrow{d,a} x' \quad x\cdot y\,\mathcal{C}d \quad x'\cdot y\,\mathcal{C}\delta}{x\cdot y \xrightarrow{a} x'\cdot y} \qquad\qquad \frac{x \xrightarrow{d,a} x' \quad x\cdot y\,\mathcal{C}\ominus \quad x'\cdot y\,\mathcal{C}\delta}{x\cdot y \xrightarrow{d,a} x'\cdot y}$$

$$\frac{x\downarrow \quad x\,\mathcal{C}d \quad y \xrightarrow{d,a} y'}{x\cdot y \xrightarrow{a} y'} \qquad\qquad \frac{x\downarrow \quad x\,\mathcal{C}\ominus \quad y \xrightarrow{d,a} y'}{x\cdot y \xrightarrow{d,a} y'}$$

$$\frac{x\,{}^d\!\downarrow \quad y \xrightarrow{d,a} y'}{x\cdot y \xrightarrow{d,a} y'} \qquad\qquad \frac{x\,{}^d\!\downarrow \quad y\,\mathcal{C}\delta \quad \delta \preceq d \quad y \xrightarrow{a} y'}{x\cdot y \xrightarrow{d,a} y'}$$

$$\frac{x \xrightarrow{d,a} x' \quad x'\cdot x^*\,\mathcal{C}\delta}{x^* \xrightarrow{d,a} x'\cdot x^*} \qquad\qquad \frac{x\,{}^d\!\downarrow}{d^{\wedge}x \downarrow} \qquad \frac{x \xrightarrow{d,a} x'}{d^{\wedge}x \xrightarrow{a} x'}$$

We show the transition system of this process in Fig. 3. It is a finite automaton. We have labelled two states with their attribute values to clarify the correspondence with the given expressions, but the attributes are not formally part of the transition system.

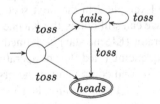

Fig. 3. The transition system of the coin toss.

Let us call the extended syntax TSP$^*_{sc}$, TSP with iteration, signals and conditions. Obviously, every consistent closed term over this syntax denotes a finite transition system. We use the conditional steps, conditional acceptance and signals in the operational rules to generate this transition system, but formally, they are not included in it. We now have all ingredients to prove the Kleene theorem.

Theorem 2. *Let t be a finite automaton. Then there is a consistent closed term over TSP^*_{sc} of which the transition system is isomorphic to t.*

Proof. Let t be a finite automaton. Assign a distinct value from $\mathcal{D}$ to each state, with i for the initial state. The term in TSP^*_{sc} consists of three parts, and has the form $Init \cdot Loop^* \cdot Exit$.

- The initial part $Init$ has a summand $a.(d^\wedge 1)$ whenever $i \xrightarrow{a} d$, and a summand 1 if $i \downarrow$;
- The looping part $Loop$ has a summand $d :\rightarrow a.(e^\wedge 1)$ for every step $d \xrightarrow{a} e$ $(d, e \in \mathcal{D})$;
- The exit part $Exit$ has a summand $d :\rightarrow 1$ whenever $d \downarrow$.

Notice that all three terms are consistent. The term $Init \cdot Loop^* \cdot Exit$ is also consistent; the transition system associated with it is isomorphic to t, so it is certainly bisimilar to t. Notice that the star height of this term is 1, i.e. there is no nesting of iterations. In fact, there is only a single iteration. As a corrollary, each consistent TSP^*_{sc} term is bisimilar to a term with only one iteration, a result well-known for while programs, see [21].

To illustrate the general procedure, we give a consistent TSP^*_{sc}-term for the regular process of Fig. 1. We label the initial state by i, and the other state by j.

$$1 + a.(j^\wedge 1) \cdot (i :\rightarrow a.(j^\wedge 1) + j :\rightarrow b.(i^\wedge 1))^* \cdot (i :\rightarrow 1 + j :\rightarrow 1)$$

The term starts with the options of the initial state. Next, there is the iteration of the behaviour of the process, with the final states encoded as the possible exits of the iteration.

5   Sequencing

In [4], sequential composition is not used but sequencing, as this fits better with stacks and pushdown automata (see Example 2 in the next section). It requires a number of adaptations in the theory, but we can still obtain the Kleene theorem. As the operational semantics of the iteration operator (Kleene star) is defined in terms of sequential composition, iteration based on sequencing is a different operator. When we use sequencing, we can only continue with the second component if the first component cannot do any further steps; e.g., the process $(a.1 + 1); b.1$ cannot do a b-step. Therefore, a loop cannot be exited by a term composed with sequencing, and we need to incorporate iteration as a *binary* operator (as, indeed, Kleene's original iteration operator, see [10]).

In this section we present TSP+IT with sequential composition $\cdot$ replaced by sequencing $;$, and Kleene star $\_^*$ replaced by $\_^{*;}\_$. We denote this theory by TSP;IT. Sequencing with its operational rules was first considered in [11], and studied extensively in [6,9]. The Kleene star based on sequencing is new here.

Table 6. Operational semantics for sequencing and binary Kleene star.

$$\frac{x \downarrow \quad y \downarrow}{x \, ; y \downarrow} \qquad \frac{x \xrightarrow{a} x'}{x \, ; y \xrightarrow{a} x' \, ; y} \qquad \frac{x \downarrow \quad x \nrightarrow \quad y \xrightarrow{a} y'}{x \, ; y \xrightarrow{a} y'}$$

$$\frac{x \xrightarrow{a} x'}{x^{*;}y \xrightarrow{a} x' \, ; x^{*;}y} \qquad \frac{y \downarrow}{x^{*;}y \downarrow} \qquad \frac{y \xrightarrow{a} y'}{x^{*;}y \xrightarrow{a} y'}$$

In Table 6, we give the operational rules for sequencing and the binary Kleene star. We write $x \xrightarrow{a}\!\!\!\!\!/\,$ for "there does not exist x' such that $x \xrightarrow{a} x'$" and $x \nrightarrow$ for "$x \xrightarrow{a}\!\!\!\!\!/\,$ for all $a \in \mathcal{A}$".

The crucial difference with the sequential composition operator is the third rule for sequencing: it is only allowed to continue with the second component if the first component cannot do any step. In the rules for binary Kleene star, we see again that there is no premise of the form $x\downarrow$, so termination of the body is irrelevant, and in the last rule we see that y can take a step irrespectively of the fact whether or not x can take a step.

Again, we see the occurrence of a negative premise here, so the rules may not define a unique transition relation that agrees with provability from the transition system specification. Since the negative premises refer to the transition relation, and due to the presence of (binary) Kleene star, defining a stratification in this case is considerably more involved than before, and we leave it to future work to spell it out in detail.

To define sequential value passing, we use signals and conditions to pass state information along a sequencing operator, similar to what we did in the previous section. First of all, we consider signal emission, and the $\mathcal{C}$ predicate, in Table 7. There, we write $x \rightarrow$ for "there exist a and x' such that $x \xrightarrow{a} x'$". Note that the term $(d^{\wedge}a.1)^{*;}(e^{\wedge}1)$ is inconsistent if $d \neq e$.

Table 7. Consistency and signals for sequencing and binary Kleene star ($d \in \mathcal{D}, \delta, \varepsilon \in \mathcal{D}_\ominus$).

$$\frac{x \, \mathcal{C}\delta \quad x\!\!\not\downarrow}{x \, ; y \, \mathcal{C}\delta} \qquad \frac{x\downarrow \quad x \rightarrow \quad x \, \mathcal{C}\delta \quad y\!\!\not\downarrow}{x \, ; y \, \mathcal{C}\delta}$$

$$\frac{x\downarrow \quad x \nrightarrow \quad x \, \mathcal{C}\delta \quad y \, \mathcal{C}\varepsilon \quad \delta \preceq \varepsilon}{x \, ; y \, \mathcal{C}\varepsilon} \qquad \frac{x\downarrow \quad x \nrightarrow \quad x \, \mathcal{C}\varepsilon \quad y \, \mathcal{C}\delta \quad \delta \preceq \varepsilon}{x \, ; y \, \mathcal{C}\varepsilon}$$

$$\frac{x\downarrow \quad y\downarrow \quad x \, \mathcal{C}\delta \quad y \, \mathcal{C}\varepsilon \quad \delta \preceq \varepsilon}{x \, ; y \, \mathcal{C}\varepsilon} \qquad \frac{x\downarrow \quad y\downarrow \quad x \, \mathcal{C}\varepsilon \quad y \, \mathcal{C}\delta \quad \delta \preceq \varepsilon}{x \, ; y \, \mathcal{C}\varepsilon}$$

$$\frac{x \, \mathcal{C}\delta \quad y \, \mathcal{C}\varepsilon \quad \delta \preceq \varepsilon}{x^{*;}y \, \mathcal{C}\varepsilon} \qquad \frac{x \, \mathcal{C}\varepsilon \quad y \, \mathcal{C}\delta \quad \delta \preceq \varepsilon}{x^{*;}y \, \mathcal{C}\varepsilon}$$

Next, we modify the rules of Table 6 by adding consistency conditions, in Table 8.

Table 8. Operational rules for sequencing and binary Kleene star with consistency conditions $(\delta, \varepsilon \in \mathcal{D}_\ominus)$.

$$\frac{x \downarrow \quad y \downarrow \quad x \,;y\, C\delta}{x\,;y \downarrow} \qquad \frac{x \xrightarrow{a} x' \quad x\,;y\, C\delta \quad x'\,;y\, C\varepsilon}{x\,;y \xrightarrow{a} x'\,;y}$$

$$\frac{x \downarrow \quad x \not\rightarrow \quad y \xrightarrow{a} y' \quad x\,;y\, C\delta}{x\,;y \xrightarrow{a} y'}$$

$$\frac{x^{*;}y\, C\delta \quad y\downarrow}{x^{*;}y\downarrow} \qquad \frac{x \xrightarrow{a} x' \quad x^{*;}y\, C\delta \quad x'\,;x^{*;}y\, C\varepsilon}{x^{*;}y \xrightarrow{a} x'\,;x^{*;}y}$$

$$\frac{y \xrightarrow{a} y' \quad x^{*;}y\, C\delta}{x^{*;}y \xrightarrow{a} y'}$$

Subsequently, we add the guarded command, and conditional steps and conditional termination. In Table 9, we use an extra abbreviation: we write $x \xrightarrow{d,} $ for "there does not exist x' and $a \in \mathcal{A}$ such that $x \xrightarrow{d,a} x'$".

Let us call the extended syntax TSP;$^*_{sc}$. Again, we have a Kleene theorem.

Theorem 3. *Let t be a finite automaton. Then there is a consistent closed term over TSP;$^*_{sc}$ of which the transition system is isomorphic to t.*

Proof. As before. The term now becomes $Init\,;Loop^{*;}Exit$.

6 Pushdown Processes

In [4], we considered the theory TSP;$_{sc}$ obtained by leaving out the binary Kleene star operator from the theory of the previous section, and considering recursion over this theory. Take $\mathcal{P}$ to be a finite set of *process identifiers*. The set $\mathcal{P}$ is a parameter of the theory.

A recursive specification over TSP;$_{sc}$ is a mapping Δ from $\mathcal{P}$ to the set of process expressions, that may contain elements of $\mathcal{P}$. The idea is that the process expression y associated with a process identifier $X \in \mathcal{P}$ by Δ *defines* the behaviour of X. We prefer to think of Δ as a collection of *defining equations* $X \stackrel{\text{def}}{=} y$, exactly one for every $X \in \mathcal{P}$.

In Table 10, we provide operational rules for recursion. As stated before, the occurring negative premises may not define a unique transition relation that agrees with provability from the transition system specification. This occurs in the defining equation $X \stackrel{\text{def}}{=} X\,;a.1 + 1$. For then, if $X \not\rightarrow$, according to the rules for sequencing and recursion we find that $X \xrightarrow{a} 1$, which is a contradiction. On the other hand, the transition $X \xrightarrow{a} 1$ is not provable from the transition system specification.

We remedy the situation by restricting our attention to *guarded* recursive specifications, i.e., we require that every occurrence of a process identifier in the definition of some (possibly different) process identifier occurs within the scope of an action prefix.

Table 9. Conditional steps and conditional termination for sequencing and binary Kleene star $(d \in \mathcal{D}, \delta \in \mathcal{D}_\ominus)$.

$$\frac{x \overset{d}{\downarrow} \quad y \overset{d}{\downarrow}}{x \,;\, y \overset{d}{\downarrow}} \qquad \frac{x \overset{d}{\downarrow} \quad y \downarrow \quad y\,C\delta \quad \delta \preceq d}{x \,;\, y \overset{d}{\downarrow}}$$

$$\frac{x \downarrow \quad x\,Cd \quad y \overset{d}{\downarrow}}{x \,;\, y \downarrow} \qquad \frac{x \downarrow \quad x\,C\ominus \quad y \overset{d}{\downarrow}}{x \,;\, y \overset{d}{\downarrow}}$$

$$\frac{x \overset{d,a}{\longrightarrow} x' \quad x\,;\,y\,Cd \quad x'\,;\,y\,C\delta}{x\,;\,y \overset{a}{\longrightarrow} x'\,;\,y} \qquad \frac{x \overset{d,a}{\longrightarrow} x' \quad x\,;\,y\,C\ominus \quad x'\,;\,y\,C\delta}{x\,;\,y \overset{d,a}{\longrightarrow} x'\,;\,y}$$

$$\frac{x \downarrow \quad x \nrightarrow \quad x\,Cd \quad y \overset{d,a}{\longrightarrow} y'}{x\,;\,y \overset{a}{\longrightarrow} y'} \qquad \frac{x \downarrow \quad x \nrightarrow \quad x \overset{d}{\nrightarrow} \quad x\,C\ominus \quad y \overset{d,a}{\longrightarrow} y'}{x\,;\,y \overset{d,a}{\longrightarrow} y'}$$

$$\frac{x \overset{d}{\downarrow} \quad x \nrightarrow \quad x \overset{d}{\nrightarrow} \quad y \overset{d,a}{\longrightarrow} y'}{x\,;\,y \overset{d,a}{\longrightarrow} y'} \qquad \frac{x \overset{d}{\downarrow} \quad x \nrightarrow \quad x \overset{d}{\nrightarrow} \quad y\,C\delta \quad \delta \preceq d \quad y \overset{a}{\longrightarrow} y'}{x\,;\,y \overset{d,a}{\longrightarrow} y'}$$

$$\frac{x \overset{d,a}{\longrightarrow} x' \quad y\,C\ominus \quad x'\,;\,x^{*i}y\,C\delta}{x^{*i}y \overset{d,a}{\longrightarrow} x'\,;\,x^{*i}y} \qquad \frac{x \overset{d,a}{\longrightarrow} x' \quad y\,Cd \quad x'\,;\,x^{*i}y\,C\delta}{x^{*i}y \overset{a}{\longrightarrow} x'\,;\,x^{*i}y}$$

$$\frac{y \overset{d}{\downarrow} \quad x\,C\ominus}{x^{*i}y \overset{d}{\downarrow}} \qquad \frac{y \overset{d}{\downarrow} \quad x\,Cd}{x^{*i}y \downarrow}$$

$$\frac{y \overset{d,a}{\longrightarrow} y' \quad x\,C\ominus}{x^{*i}y \overset{d,a}{\longrightarrow} y'} \qquad \frac{y \overset{d,a}{\longrightarrow} y' \quad x\,Cd}{x^{*i}y \overset{a}{\longrightarrow} y'}$$

If Δ is guarded, then it is straightforward to prove that the mapping S from process expressions to natural numbers inductively defined by $S(1) = S(0) = S(a.x) = 0$, $S(x_1 + x_2) = S(x_1 \,;\, x_2) = S(x_1) + S(x_2) + 1$, and $S(X) = S(y)$ if $(X \overset{\text{def}}{=} y) \in \Delta$ gives rise to a stratification S' from transitions to natural numbers defined by $S'(x \overset{a}{\longrightarrow} x') = S(x)$ for all $a \in \mathcal{A}$ and process expressions x and x'.

Example 2. Let us consider the stack S of unbounded capacity that is only accepting when it is empty. In [8, Section 6.6], we give the following guarded recursive specification.

$$S \overset{\text{def}}{=} 1 + \sum_{d \in \mathcal{D}} push(d).T_d \cdot S \qquad T_d \overset{\text{def}}{=} pop(d).1 + \sum_{e \in \mathcal{D}} push(e).T_e \cdot T_d$$

Now suppose that we want to define the stack that is always accepting, irrespective of the contents. We show the one-state pushdown automaton of this stack in Fig. 4. The transitions on the left show that we can execute $push(d)$ for some $d \in \mathcal{D}$ whenever the stack is empty (denoted by the empty string) or has some $e \in \mathcal{D}$ on top, and the transition on the right shows we can execute $pop(d)$ for some $d \in \mathcal{D}$ whenever the stack has d on top, replacing it by the empty string. Consider now the following specification.

$$S' \overset{\text{def}}{=} 1 + \sum_{d \in \mathcal{D}} push(d).T'_d \cdot S' \qquad T'_d \overset{\text{def}}{=} 1 + pop(d).1 + \sum_{e \in \mathcal{D}} push(e).T'_e \cdot T'_d$$

Table 10. Operational rules for recursion ($\delta \in \mathcal{D}_\ominus$).

$$\frac{y \xrightarrow{a} y' \quad X \stackrel{\text{def}}{=} y}{X \xrightarrow{a} y'} \qquad \frac{y\downarrow \quad X \stackrel{\text{def}}{=} y}{X\downarrow}$$

$$\frac{y\, C\delta \quad X \stackrel{\text{def}}{=} y}{X\, C\delta} \qquad \frac{y \xrightarrow{d,a} y' \quad X \stackrel{\text{def}}{=} y}{X \xrightarrow{d,a} y'} \qquad \frac{y \,^d\!\downarrow \quad X \stackrel{\text{def}}{=} y}{X \,^d\!\downarrow}$$

This specification will not define the always accepting stack, as it is forgetful (can lose part of the stack contents) and moreover, its transition system is unboundedly branching, see [4]. The following specification does give the right result.

$$S'' \stackrel{\text{def}}{=} 1 + \sum_{d \in \mathcal{D}} push(d).T''_d\, ; S'' \qquad T''_d \stackrel{\text{def}}{=} 1 + pop(d).1 + \sum_{e \in \mathcal{D}} push(e).T''_e\, ; T''_d$$

This provides the motivation to use sequencing rather than sequential composition for stacks and pushdown automata.

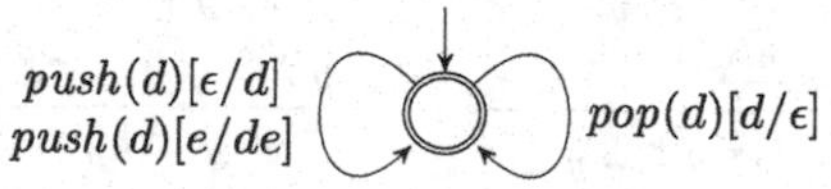

$$push(d)[\epsilon/d]$$
$$push(d)[e/de]$$
$$pop(d)[d/\epsilon]$$

Fig. 4. Pushdown automaton of the always terminating stack ($d, e \in \mathcal{D}$).

In [4], we proved that a process (i.e., a bisimulation equivalence class of transition systems) is defined by a pushdown automaton if and only if it can be specified by a finite guarded recursive specification over TSP with sequencing, propositional signals and conditions. The following theorem shows that the simpler theory of sequential value passing presented here can be used instead.

Theorem 4. *A process is defined by a pushdown automaton if and only if it is defined by a finite guarded recursive specification over TSP;$_{sc}$.*

The proof of this theorem is not so much different from the proof in [4]. The simplification is in the setup by means of the operational semantics and the two-sorted syntax.

7   Conclusion

We investigated sequential value passing. By taking an unstructured finite data set for signals and conditions instead of terms over propositional logic, the treatment is simplified considerably. We presented this simplification both in a setting with sequential composition and in a setting with sequencing.

In [3], propositional signals and conditions were introduced, and studied in a setting with parallel composition, in order to study signal observation between processes in a parallel or distributed setting. We think that there, the combination of different signals is more important, and our simplification does not work so well. This needs to be investigated further.

We proved a Kleene theorem for processes: extending regular expressions with sequential value passing suffices to denote all finite automata.

In this paper we focussed on the correspondence results. We leave as future work to investigate the equational theory of the theory presented here. We see that signal emission and guarded command can lead to inconsistency, in the equational theory we need to introduce the inaccessible process $\perp$ that can never be reached, see [3,4,8].

This paper contributes to our ongoing project to integrate automata theory and process theory. As a result, we can present the foundations of computer science using a computer model with explicit interaction (as opposed to viewing automata just as language acceptors). Such a computer model relates more closely to the computers we see all around us.

Acknowledgements. We are grateful to the two anonymous reviewers for their careful reviews; their comments and suggestions led to improvements in the presentation.

References

1. Aceto, L., Fokkink, W.J., Verhoef, C.: Structural operational semantics. In: Bergstra, J.A., Ponse, A., Smolka, S.A. (eds.) Handbook of Process Algebra, pp. 197–292. North-Holland/Elsevier (2001). https://doi.org/10.1016/b978-044482830-9/50021-7
2. Baeten, J.C.M., Corradini, F., Grabmayer, C.A.: A characterization of regular expressions under bisimulation. J. ACM **54**(2), 6–28 (2007). https://doi.org/10.1145/1219092.1219094
3. Baeten, J.C.M., Bergstra, J.A.: Process algebra with propositional signals. Theor. Comput. Sci. **177**(2), 381–405 (1997). https://doi.org/10.1016/S0304-3975(96)00253-8
4. Baeten, J.C.M., Carissimo, C., Luttik, B.: Pushdown automata and context-free grammars in bisimulation semantics. Logical Methods Comput. Sci. **19**, 15:1–15.32 (2023). https://doi.org/10.46298/LMCS-19(1:15)2023
5. Baeten, J.C.M., Luttik, B., Muller, T., van Tilburg, P.: Expressiveness modulo bisimilarity of regular expressions with parallel composition. Math. Struct. Comput. Sci. **26**(6), 933–968 (2016). https://doi.org/10.1017/S0960129514000309
6. Baeten, J.C.M., Luttik, B., Yang, F.: Sequential composition in the presence of intermediate termination (extended abstract). In: Peters, K., Tini, S. (eds.) Proceedings Combined 24th International Workshop on Expressiveness in Concurrency and 14th Workshop on Structural Operational Semantics and 14th Workshop on Structural Operational Semantics, EXPRESS/SOS 2017, Berlin, Germany, 4 September 2017. EPTCS, vol. 255, pp. 1–17 (2017). https://doi.org/10.4204/EPTCS.255.1. http://arxiv.org/abs/1709.00049
7. Baeten, J.C.M., Verhoef, C.: A congruence theorem for structured operational semantics with predicates. In: Best, E. (ed.) CONCUR '93, 4th International Conference on Concurrency Theory, Hildesheim, Germany, 23–26 August 1993, Proceedings. Lecture Notes in Computer Science, vol. 715, pp. 477–492. Springer (1993). https://doi.org/10.1007/3-540-57208-2_33
8. Baeten, J.C., Basten, T., Reniers, M.: Process Algebra: Equational Theories of Communicating Processes, vol. 50. Cambridge university press, Cambridge (2010). https://doi.org/10.1017/CBO9781139195003

9. Belder, A., Luttik, B., Baeten, J.: Sequencing and intermediate acceptance: axiomatisation and decidability of bisimilarity. In: Roggenbach, M., Sokolova, A. (eds.) 8th Conference on Algebra and Coalgebra in Computer Science, CALCO 2019. Leibniz International Proceedings in Informatics, LIPIcs, Schloss Dagstuhl - Leibniz-Zentrum für Informatik (2019). https://doi.org/10.4230/LIPIcs.CALCO.2019.11

10. Bergstra, J.A., Fokkink, W., Ponse, A.: Chapter 5 - process algebra with recursive operations. In: Bergstra, J., Ponse, A., Smolka, S. (eds.) Handbook of Process Algebra, pp. 333–389. Elsevier Science, Amsterdam (2001). https://doi.org/10.1016/B978-044482830-9/50023-0

11. Bloom, B.: When is partial trace equivalence adequate? Formal Aspects Comput. 6(3), 317–338 (1994). https://doi.org/10.1007/BF01215409

12. Bol, R.N., Groote, J.F.: The meaning of negative premises in transition system specifications. J. ACM 43(5), 863–914 (1996). https://doi.org/10.1145/234752.234756

13. Garavel, H.: Nested-unit petri nets: a structural means to increase efficiency and scalability of verification on elementary nets. In: Devillers, R.R., Valmari, A. (eds.) Application and Theory of Petri Nets and Concurrency - 36th International Conference, PETRI NETS 2015, Brussels, Belgium, 21–26 June 2015, Proceedings. Lecture Notes in Computer Science, vol. 9115, pp. 179–199. Springer, Heidelberg (2015). https://doi.org/10.1007/978-3-319-19488-2_9

14. van Glabbeek, R.J., Weijland, W.P.: Branching time and abstraction in bisimulation semantics. J. ACM 43(3), 555–600 (1996). https://doi.org/10.1145/233551.233556

15. van Glabbeek, R.J.: The meaning of negative premises in transition system specifications II. J. Log. Algebr. Program. 60–61, 229–258 (2004). https://doi.org/10.1016/j.jlap.2004.03.007

16. Grabmayer, C., Fokkink, W.J.: A complete proof system for 1-free regular expressions modulo bisimilarity. CoRR abs/2004.12740 (2020). https://arxiv.org/abs/2004.12740

17. Groote, J.F.: Transition system specifications with negative premises. Theor. Comput. Sci. 118(2), 263–299 (1993). https://doi.org/10.1016/0304-3975(93)90111-6

18. Groote, J.F., Ponse, A.: Process algebra with guards: combining hoare logic with process algebra. Formal Aspects Comput. 6(2), 115–164 (1994). https://doi.org/10.1007/BF01221097

19. Hennessy, M.: Value-passing in process algebras. In: Baeten, J.C.M., Klop, J.W. (eds.) CONCUR 1990. LNCS, vol. 458, pp. 31–31. Springer, Heidelberg (1990). https://doi.org/10.1007/BFb0039048

20. Kleene, S.C.: Representation of events in nerve nets and finite automata. In: Automata Studies, pp. 3–41 (1956)

21. Kozen, D.: Kleene algebra with tests. ACM Trans. Program. Lang. Syst. 19(3), 427–443 (1997). https://doi.org/10.1145/256167.256195

22. Luttik, B.: Divergence-preserving branching bisimilarity. In: Dardha, O., Rot, J. (eds.) Proceedings Combined 27th International Workshop on Expressiveness in Concurrency and 17th Workshop on Structural Operational Semantics, EXPRESS/SOS 2020, and 17th Workshop on Structural Operational Semantics, Online, 31 August 2020. EPTCS, vol. 322, pp. 3–11 (2020). https://doi.org/10.4204/EPTCS.322.2

23. Milner, R.: A complete inference system for a class of regular behaviours. J. Comput. Syst. Sci. 28(3), 439–466 (1984). https://doi.org/10.1016/0022-0000(84)90023-0

24. Verhoef, C.: A congruence theorem for structured operational semantics with predicates and negative premises. Nord. J. Comput. 2(2), 274–302 (1995)

25. Visser, E., Benaissa, Z.: A core language for rewriting. In: Kirchner, C., Kirchner, H. (eds.) 1998 International Workshop on Rewriting Logic and its Applications, WRLA 1998, Abbaye des Prémontrés at Pont-à-Mousson, France, September 1998. Electronic Notes in Theoretical Computer Science, vol. 15, pp. 422–441. Elsevier (1998). https://doi.org/10.1016/S1571-0661(05)80027-1

YACC: Yet Another Church Calculus

A Birthday Present for Herman Inspired by His Supervisor Activity

Franco Barbanera[1], Mariangiola Dezani-Ciancaglini[2(✉)], Ugo de'Liguoro[2], and Betti Venneri[3]

[1] Dipartimento di Matematica e Informatica, Università di Catania, Catania, Italy
[2] Dipartimento di Informatica, Università di Torino, Turin, Italy
dezani@di.unito.it
[3] Dipartimento di Statistica, Informatica, Applicazioni, Università di Firenze, Florence, Italy

Abstract. A novel typed λ-calculus à la Church with intersection types is proposed. The novelty is the presence of three type constructors representing different roles of the standard intersection type constructor. The main properties are Subject Reduction and the characterisation of typed λ-terms reducing to head normal forms.

Keywords: Lambda Calculus · Intersection Types · Typing à la Church

1 Introduction

There are essentially two approaches to type theory for λ-calculus, introduced by Curry [18] and Church [14], respectively. Barendregt [8] dubbed them *à la Curry* and *à la Church*. Types are built out of a denumerable set of variables using the arrow type constructor, i.e. they are defined by the grammar:

$$T ::= \varphi \mid T \to T$$

where φ is a type variable and $T \to U$ is the type of functions which applied to arguments of type T return results of type U. These types are called *simple types*. In the typed λ-calculus à la Curry types are assigned to pure λ-terms by using syntax-driven type inference rules. For example, the identity function $\lambda x.x$ has all types $T \to T$, where T is arbitrary. Each λ-term can receive either an infinite number of types or no type. In the typed λ-calculus à la Church, instead,

F. Barbanera—Partially supported by Project "National Center for HPC, Big Data e Quantum Computing", Programma M4C2, Investimento 1.3.
U. de' Liguoro—Partially supported by Project INDAM-GNCS "Fondamenti di Informatica e Sistemi Informatici".

V. Capretta et al. (Eds.): *Logics and Type Systems in Theory and Practice*, LNCS 14560, pp. 17–35, 2024.
https://doi.org/10.1007/978-3-031-61716-4_2

each variable is "decorated" with its type, and then each typed term has exactly one type which is part of its structure. For example, there are infinitely many identity functions $\lambda x : T.x$ of type $T \to T$, one for each type T. By comparing the Church and Curry approaches, the typed term $\lambda x : T.x$ is simply a compact code for the type deduction (à la Curry) assigning $T \to T$ to $\lambda x.x$. Namely, $\lambda x.x$ is the code of the type deduction tree, while T marks the type assumption of x.

Intersection types [15] extend simple types by adding the intersection type constructor $\cap$ and the universal type ω. They are defined by

$$\mu ::= \varphi \mid \omega \mid \mu \to \mu \mid \mu \cap \mu \tag{1}$$

with the assumption that $\cap$ takes precedence over $\to$. Intersection types were born as types à la Curry with the following rule for the introduction of intersection

$$\frac{\Gamma \vdash_\cap P : \mu \quad \Gamma \vdash_\cap P : \nu}{\Gamma \vdash_\cap P : \mu \cap \nu} \; [\cap I]$$

where P is a λ-term and Γ is a typing environment associating term variables to types

$$\Gamma ::= \emptyset \mid \Gamma, x : \mu$$

For instance, we can get $\vdash_\cap \lambda x.x : (\phi \to \phi) \cap (\psi \to \psi)$. This rule, which allows to derive, for a λ-term, the intersection of two types, is *proof-functional* (i.e., roughly, having a conclusion depending not only on the premises, but on their derivations as well), as already noticed in [24] and further investigated in [2,4]. This implies that, if we wish to encode an à la Curry typing into an à la Church term, we need to somewhat internalise, inside the "Church types", some information about derivations. In fact, on page 18 of the master thesis [10], supervised by Herman Geuvers, one can find the question:

"what can we replace to *?* in $\vdash_\cap \lambda x : ?.x : (\phi \to \phi) \cap (\psi \to \psi)$?"

In Sect. 6 various answers proposed in the literature related to this question will be discussed. For Combinatory Logic a simple proposal can be found in [26]. There it is shown that the actual problem of formulating intersection types à la Church does not concern application, but annotations of λ-abstractions[1]. For example, we need to find differently annotated λ-terms à la Church encoding the following type derivations à la Curry:

$$\vdash_\cap \lambda x.\lambda y.xy : (\phi \cap \psi \to \phi) \to \phi \cap \psi \to \phi$$

and

$$\vdash_\cap \lambda x.\lambda y.xy : ((\phi \to \phi) \to \phi \to \phi) \cap ((\psi \to \psi) \to \psi \to \psi)$$

Our Typed Intersection λ-calculus (dubbed TIC) annotates the λ-abstractions with different type constructors. More precisely the TIC terms encoding these derivations are

---

[1] We say the type σ decorates the variable x in x^σ and the type κ annotates the λ-abstraction λx in $\lambda x : \kappa$.

$$\lambda x : \phi \wedge \psi \to \phi.\lambda y : \phi \wedge \psi.x^{\phi \wedge \psi \to \phi} y^{\phi \wedge \psi}$$

and

$$\lambda x : (\phi \to \phi) \sqcap (\psi \to \psi).\lambda y : \phi \sqcap \psi.x^{(\phi \to \phi) \wedge (\psi \to \psi)} y^{\phi \wedge \psi}$$

where the type constructor $\wedge$ connects types in variable decorations and in annotations of λ-abstractions, while the type constructor $\sqcap$ signals the use of an intersection introduction rule on λ-abstractions. The types of these TIC terms are $(\phi \wedge \psi \to \phi) \to \phi \wedge \psi \to \phi$ and $((\phi \to \phi) \to \phi \to \phi) \wedge ((\psi \to \psi) \to \psi \to \psi)$, respectively. Note that the type constructor $\sqcap$ only appears in annotations of λ-abstractions.

We also want to distinguish when two types refer to the same or two different occurrences of a variable. As an example consider

$$\vdash_\cap \lambda x.\lambda y.yx : (\phi \to \psi) \cap \phi \to ((\phi \to \psi) \cap \phi \to \psi) \to \psi$$

and

$$\vdash_\cap \lambda x.xx : (\phi \to \psi) \cap \phi \to \psi$$

We realise this by introducing a third (and last) type constructor, i.e. the TIC terms representing these derivations are

$$\lambda x : (\phi \to \psi) \wedge \phi.\lambda y : (\phi \to \psi) \wedge \phi \to \psi.y^{(\phi \to \psi) \wedge \phi \to \psi} x^{(\phi \to \psi) \wedge \phi}$$

and

$$\lambda x : (\phi \to \psi) \,\&\, \phi.x^{\phi \to \psi} x^\phi$$

where the type constructor $\&$ connects the types of the different occurrences of a variable. The types of these TIC terms are

$$(\phi \to \psi) \wedge \phi \to ((\phi \to \psi) \wedge \phi \to \psi) \to \psi$$

and

$$(\phi \to \psi) \,\&\, \phi \to \psi$$

Note that the type constructor $\&$ appears in annotations of λ-abstractions and on the left of the arrow constructor.

To sum up, TIC has three different constructors we can annotate λ-abstractions with. They represent different uses of intersection: the constructor $\wedge$ means that an occurrence of a variable has an intersection type; the constructor $\&$ connects types of different occurrences of the same variable; the constructor $\sqcap$ connects types of the same occurrence of a variable in different derivations, which are put together by an application of the intersection introduction rule.

It is commonly held that typed λ-calculi à la Church are the theoretical basis of programming languages with types and higher-order functions. In fact programming languages require decidable type checking, which is not the case of intersection types in general. However today, although with some restrictions, intersection types are present in various programming languages. Some pointers to them are given in [11]. Our long term aim is that TIC could stimulate spreading intersection types in programming.

Outline. The typed λ-terms of TIC are defined in Sect. 2. The type checking algorithm is presented in Sect. 3. Section 4 contains the reduction rules together with the proofs of Subject Reduction and a comparison with the β-reduction of pure λ-terms. The relations between TIC and the type assignment system of [6] are discussed in Sect. 5 where, as a byproduct, we show the characterisation of TIC terms having a head normal form. Section 6 concludes with some pointers to related works and some remarks.

2   Syntax of TIC

As usual, types are build from a denumerable set of type variables, ranged over by $\varphi, \phi, \psi, \dots$ and from the constant type ω. The type ω can only appear alone as type of the constant TIC term Ω or on the left of arrow types. The types are *strict* according to [3], i.e., on the right of arrow types one can find no intersections but only type variables and arrow types. As explained in the Introduction, TIC has three constructors representing different uses of the intersection. The constructor $\wedge$ can connect types in superscripts of variables, in annotations of λ-abstractions and in term typings. Then $\wedge$ appears in typing judgements either at top level or on the left of arrow types. The constructor $\&$ connects types in annotations of λ-abstractions, when these types decorate different occurrences of the same variable. Then $\&$ appears on the left of arrow types. Also the constructor $\sqcap$ connects types in annotations of λ-abstractions, but when these types decorate the same occurrence in terms "glued" by applying the intersection introduction rule. Then $\sqcap$ cannot appear inside arrow types. Hence, types are formally defined as follows.

Definition 1 (Types).

(base types)	$\alpha ::= \varphi \mid \theta \to \alpha$	*(main types)*	$\sigma, \tau ::= \alpha \mid \sigma \wedge \sigma$
(relevant left types)	$\vartheta ::= \sigma \mid \vartheta \,\&\, \vartheta$	*(left types)*	$\theta ::= \omega \mid \vartheta$
	(conjunctive types)	$\kappa ::= \theta \mid \kappa \sqcap \kappa$	

The base types are the types of variables in the initial assumptions, see Axiom [VAR] in Fig. 1. The main types occur as decorations of variables, annotations of λ-abstractions and types of TIC terms different from Ω (the only type of Ω is ω). The left types occur on the left of arrows and as annotations of λ-abstractions; the relevant ones do not include ω. The conjunctive types are used only as annotations of λ-abstractions.

   The constructors $\wedge$ and $\sqcap$ are neither idempotent nor commutative nor associative, while the constructor $\&$ is neither idempotent nor commutative, but it is associative[2]. The precedence between type constructors is: $\wedge$, $\&$, $\sqcap$ and $\to$.

---

[2] Non idempotent intersection types have been used to define principal typings [16] and for characterising complexity classes of λ-terms [9,13]. In [19] non associative intersection types model polynomial time computations.

We define an equivalence relation on main types and, in terms of that, a preorder relation between relevant left types and main types. The latter relation is used in the typing Rule $[\to E]$ in order to equate $\&$ and $\wedge$ type constructors.

Definition 2 ($\simeq$ and $\ltimes$). *Let $\simeq$ be the equivalence relation induced by considering $\wedge$ associative. We define $\vartheta \ltimes \sigma$ if $\vartheta = \&_{i \in I} \tau_i$ and $\sigma \simeq \wedge_{i \in I} \tau_i$.*

Note that in the definition of $\ltimes$ none of the τ_i can have $\&$ as top level constructor, being τ_i a main type.

As pseudo-terms we allow variables decorated with main types, λ-abstractions annotated with conjunctive types (and hence, possibly, by left types), applications and the constant TIC term Ω. We assume Barendregt's convention, i.e. that the names of free and bound variables are always different.

Definition 3 (Pseudo-terms).

$$M ::= x^\sigma \mid \lambda x : \kappa.M \mid MM \mid \Omega$$

Out of the set of pseudo-terms, we identify the TIC terms by means of a type system. The notations used in the type system are specified in the informal description of the rules.

Definition 4 (TIC terms). *A TIC term is a pseudo-term which can be typed in the type system shown in Fig. 1.*

TIC terms encode their typing derivations, as shown in Sect. 3.

The axiom [VAR] allows to derive only base types for variables, the advantage being that we decorate variables by types we actually "use".

Rule $[\omega]$ types the constant Ω by ω.

In Rule $[\to I]$ we use $\iota(M, x)$ which, roughly, collects in a left type the decorations of all the free occurrences of x in M, if any, and returns ω otherwise. Namely, we define[3]

$$\iota(x^\sigma, x) = \sigma \quad \iota(y^\sigma, x) = \omega \text{ if } y \neq x \quad \iota(\lambda y : \sigma.M, x) = \iota(M, x) \quad \iota(\Omega, x) = \omega$$

$$\iota(MN, x) = \begin{cases} \iota(M, x) \,\&\, \iota(N, x) & \text{if } \iota(M, x) \neq \omega \text{ and } \iota(N, x) \neq \omega, \\ \iota(M, x) & \text{if } \iota(N, x) = \omega \\ \iota(N, x) & \text{if } \iota(M, x) = \omega \end{cases}$$

Notice that either $\iota(M, x) = \&_{i \in I} \sigma_i$ or $\iota(M, x) = \omega$. If $\iota(M, x) = \&_{1 \leq i \leq n} \sigma_i$, then x occurs free in M and the occurrences of x in M are $x^{\sigma_1}, \ldots, x^{\sigma_n}$ from left to right. If $\iota(M, x) = \omega$, then x does not occur free in M. Therefore Rule $[\to I]$ is *relevant* according to [1].

Rule $[\to E\omega]$ allows to apply a TIC term typed by $\omega \to \alpha$, for some α, to Ω.

Rule $[\to E]$ ensures that M (typed by $\vartheta \to \alpha$) can be considered a function taking N (typed by σ) as argument proviso $\vartheta \ltimes \sigma$ according to Definition 2.

---

[3] By Barendregt's convention, $x \neq y$ in the third clause.

$$\frac{}{\vdash x^\alpha : \alpha} \, [\text{VAR}] \qquad \frac{}{\vdash \Omega : \omega} \, [\omega] \qquad \frac{\vdash M : \alpha \quad \iota(M, x) = \theta}{\vdash \lambda x : \theta.M : \theta \to \alpha} \, [\to I] \qquad \frac{\vdash M : \omega \to \alpha}{\vdash M\Omega : \alpha} \, [\to E\omega]$$

$$\frac{\vdash M : \vartheta \to \alpha \quad \vdash N : \sigma \quad \vartheta \Bowtie \sigma}{\vdash MN : \alpha} \, [\to E] \qquad \frac{\vdash M : \sigma \quad \vdash N : \tau}{\vdash M \bigwedge N : \sigma \wedge \tau} \, [\wedge I]$$

Fig. 1. TIC typing rules.

$$\frac{\dfrac{\vdash x^\beta : \beta \quad \vdash x^\alpha : \alpha}{\vdash x^\beta x^\alpha : \omega \to \alpha} \quad \vdash \Omega : \omega}{\dfrac{\vdash x^\beta x^\alpha \Omega : \alpha}{\vdash M : \beta \& \alpha \to \alpha}} \qquad \frac{\dfrac{\vdash y^\alpha : \alpha}{\vdash \lambda z : \omega.y^\alpha : \omega \to \alpha}}{\vdash N_1 : \beta} \qquad \frac{\dfrac{\vdash y^\phi : \phi}{\vdash \lambda z : \omega.y^\phi : \omega \to \phi}}{\vdash N_2 : \alpha}$$

$$\frac{\qquad\qquad\qquad \vdash N : \beta \wedge \alpha}{\vdash MN : \alpha}$$

where

$$\alpha = \phi \to \omega \to \phi \qquad\qquad \beta = \alpha \to \omega \to \alpha$$
$$N_1 = \lambda y : \alpha.\lambda z : \omega.y^\alpha \qquad N_2 = \lambda y : \phi.\lambda z : \omega.y^\phi$$
$$M = \lambda x : \beta \& \alpha.x^\beta x^\alpha \Omega \qquad N = \lambda y : \alpha \sqcap \phi.\lambda z : \omega \sqcap \omega.y^{\alpha \wedge \phi}$$

Fig. 2. Example of typing.

Rule $[\wedge I]$ uses the (partial) operator $\bigwedge$ which, applied to two TIC terms, builds a TIC term, if any, typed by the $\wedge$ of the types of the two TIC terms. The subjects in the premises of this rule cannot be Ω, since they have main types. But they can contain Ω as subterm. The subject in the conclusion of this rule is a TIC term. The operator $\bigwedge$ generates conjunctive types as type annotations of λ-abstractions, see the second clause below. We define:

$$
\begin{aligned}
x^\sigma \bigwedge x^\tau &= x^{\sigma \wedge \tau} \\
(\lambda x : \kappa.M) \bigwedge (\lambda x : \kappa'.N) &= \lambda x : \kappa \sqcap \kappa'.M \bigwedge N \\
(MN) \bigwedge (M'N') &= (M \bigwedge M')(N \bigwedge N') \\
\Omega \bigwedge \Omega &= \Omega
\end{aligned}
$$

The operator $\bigwedge$ is neither idempotent nor commutative nor associative, being $\wedge$ and $\sqcap$ not so.

Remark that only Rule $[\wedge I]$ derives a type which is an intersection (built by using $\wedge$) of two main types, while Axiom [VAR], Rules $[\to I]$, $[\to E\omega]$ and $[\to E]$ derive base types. For example, we can derive a type for the TIC term representing an auto-application function with the first projection function as argument:

$$\vdash (\lambda x : (\alpha \to \omega \to \alpha) \& \alpha).x^{\alpha \to \omega \to \alpha} x^\alpha \Omega)(\lambda y : \alpha \sqcap \phi.\lambda z : \omega \sqcap \omega.y^{\alpha \wedge \phi}) : \alpha$$

as shown in Fig. 2.

3 Type Reconstruction

The mandatory feature of typed calculi à la Church is the possibility of reading from a typed term a derivation which justifies this term. Indeed, type reconstruction must be decidable and each typed term must have a unique type. We dub $type(M)$ the unique type of the TIC term M. Let ζ range over main types and ω. We define below the function der which, applied to a pseudo-term M, either returns a derivation with conclusion $\vdash M : \zeta$, for some ζ, or fails. The correctness of der easily follows from the following Inversion Lemma. To show this lemma it is handy to split a TIC term, built using $\bigwedge$, into its components. This is done by means of projections, the partial functions on TIC terms defined as follows.

Definition 5 (Projections). *Let $j = 1, 2$. We define*

$$
\begin{aligned}
\Pi_j(x^{\sigma_1 \wedge \sigma_2}) &= x^{\sigma_j} \\
\Pi_j(\lambda x : \kappa_1 \sqcap \kappa_2.M) &= \lambda x : \kappa_j.\Pi_j(M) \\
\Pi_j(MN) &= \Pi_j(M)\Pi_j(N) \\
\Pi_j(\Omega) &= \Omega
\end{aligned}
$$

We immediately get the following from the above definition.

Fact 1. $\Pi_1(M \bigwedge N) = M$ *and* $\Pi_2(M \bigwedge N) = N$.

Lemma 1 (Inversion Lemma).

1. *If $\vdash x^\sigma : \zeta$, then $\zeta = \sigma$.*
2. *If $\vdash \lambda x : \theta.M : \zeta$, then $\zeta = \theta \to \alpha$ and $\vdash M : \alpha$ and $\iota(M, x) = \theta$.*
3. *If $\vdash \lambda x : \kappa_1 \sqcap \kappa_2.M : \zeta$, then $\zeta = \sigma_1 \wedge \sigma_2$ and $\vdash \lambda x : \kappa_j.\Pi_j(M) : \sigma_j$ with $j = 1, 2$.*
4. *If $\vdash MN : \zeta$, then one and only one of the following cases holds:*
   (a) $\zeta = \alpha$ and $\vdash M : \omega \to \alpha$ and $N = \Omega$;
   (b) $\zeta = \alpha$ and $\vdash M : \vartheta \to \alpha$ and $\vdash N : \sigma$ and $\vartheta \ltimes \sigma$;
   (c) $\zeta = \tau_1 \wedge \tau_2$ and $\vdash \Pi_j(MN) : \tau_j$ with $j = 1, 2$.
5. *If $\vdash \Omega : \zeta$, then $\zeta = \omega$.*
6. *If $\vdash M_1 \bigwedge M_2 : \zeta$, then $\zeta = \tau_1 \wedge \tau_2$ and $\vdash M_j : \tau_j$ with $j = 1, 2$.*

Proof. All points are shown by induction on derivations.

(1). If the last applied rule is [VAR], it is trivial. If the last applied rule is [$\wedge I$], we get $\zeta = \tau_1 \wedge \tau_2$ and $\dfrac{\vdash x^{\sigma_1} : \tau_1 \quad \vdash x^{\sigma_2} : \tau_2}{\vdash x^{\sigma_1 \wedge \sigma_2} : \tau_1 \wedge \tau_2}$. By induction, $\sigma_j = \tau_j$ for $j = 1, 2$.

(2). Easy since the last applied rule must be [$\to I$]. Rule [$\wedge I$] cannot be the last applied rule, since otherwise the abstracted variable would be annotated with a conjunctive type.

(3). The last applied rule must be [$\wedge I$], i.e. $\zeta = \sigma_1 \wedge \sigma_2$ and, by Fact 1, we also have $\vdash \lambda x : \kappa_j.\Pi_j(M) : \sigma_j$ with $j = 1, 2$. The last applied rule cannot be [$\to I$], since the abstracted variable is annotated with a conjunctive type.

Variable

$$\text{der}(x^\sigma) = \begin{cases} \vdash x^\sigma : \sigma & \text{if } \sigma \text{ is a base type} \\[2ex] \dfrac{\text{der}(x^{\tau_1}) \quad \text{der}(x^{\tau_2})}{\vdash x^\sigma : \sigma} \, [\wedge I] & \text{if } \sigma = \tau_1 \wedge \tau_2 \end{cases}$$

λ-abstraction

$$\text{der}(\lambda x : \kappa.M) = \begin{cases} \dfrac{\text{der}(M) \quad \iota(M,x) = \kappa}{\vdash \lambda x : \kappa.M : \kappa \to \alpha} \, [\to I] \\ \qquad\qquad \text{if } \iota(M,x) = \kappa \text{ and } \text{type}(M) = \alpha \\[3ex] \dfrac{\text{der}(\lambda x : \kappa_1.\Pi_1(M)) \quad \text{der}(\lambda x : \kappa_2.\Pi_2(M))}{\vdash \lambda x : \kappa.M : \tau_1 \wedge \tau_2} \, [\wedge I] \\ \text{if } \kappa = \kappa_1 \sqcap \kappa_2 \text{ and } \text{type}(\lambda x : \kappa_i.\Pi_i(M)) = \tau_i \text{ for } i = 1, 2 \end{cases}$$

Application

$$\text{der}(MN) = \begin{cases} \dfrac{\text{der}(M)}{\vdash MN : \alpha} \, [\to E\omega] \\ \qquad\qquad \text{if } \text{type}(M) = \omega \to \alpha \text{ and } N = \Omega \\[3ex] \dfrac{\text{der}(M) \quad \text{der}(N) \quad \vartheta \ltimes \sigma}{\vdash MN : \tau} \, [\to E] \\ \text{if } \text{type}(M) = \vartheta \to \alpha \text{ and } \text{type}(N) = \sigma \text{ and } \vartheta \ltimes \sigma \\[3ex] \dfrac{\text{der}(\Pi_1(MN)) \quad \text{der}(\Pi_2(MN))}{\vdash MN : \tau_1 \wedge \tau_2} \, [\wedge I] \\ \qquad\qquad \text{if } \text{type}(\Pi_i(MN)) = \tau_i \text{ for } i = 1, 2 \end{cases}$$

Ω

$$\text{der}(\Omega) = \dfrac{}{\vdash \Omega : \omega} \, [\omega]$$

Fig. 3. Definition of the function *der*.

(4). If the last applied rule is $[\to E\omega]$ we get $\zeta = \alpha$ and $\vdash M : \omega \to \alpha$ and $N = \Omega$. If the last applied rule is $[\to E]$ we get $\zeta = \alpha$ and $\vdash M : \vartheta \to \alpha$ and $\vdash N : \sigma$ and $\vartheta \ltimes \sigma$. If the last applied rule is $[\wedge I]$ we get $\zeta = \tau_1 \wedge \tau_2$ and, by Fact 1, we also have $\vdash \Pi_j(MN) : \tau_j$ for $j = 1, 2$.

(5). Easy since last applied rule must be $[\omega]$.

(6). If the last applied rule is $[\wedge I]$ it is immediate. The last applied rule cannot be $[\to I]$, since $[\to I]$ annotates the λ-abstraction with a left type, while $\wedge$ requires to annotate the λ-abstraction with a conjunctive type which is not a left type. The last applied rule cannot be $[\to E]$, since this would require the TIC term in functional position to have an arrow type. This is impossible, since TIC terms obtained by applying $\wedge$ have intersection types. □

The function *der* is defined by structural induction on pseudo-terms in Fig. 3. For a λ-abstraction we apply the first clause if κ is a left type and the second clause otherwise. For an application, by Lemma 1(4), we can decide which clause to apply first by looking at the argument: if it is Ω, then we use the first clause. Otherwise we compute the type of M: if it is an arrow type we use the second clause, otherwise we use the third one.

For example the application of *der* to

$$(\lambda x : (\alpha \to \omega \to \alpha) \& \alpha).x^{\alpha \to \omega \to \alpha}x^{\alpha}\Omega)(\lambda y : \alpha \sqcap \phi.\lambda z : \omega \sqcap \omega.y^{\alpha \wedge \phi})$$

returns the derivation shown in Fig. 2.

It is important to observe that type reconstruction for a given pseudo-term is univocally driven by the syntactical structure and the type annotations and decorations of the pseudo-term. Thus the type of a pseudo-term, if any, is unique[4].

4   Operational Semantics of TIC

In this section we give the reduction rules of TIC. To this aim, the following lemma shows that, if a TIC term is build by repeated applications of Rule $[\wedge I]$, then we can extract, by means of projection compositions, the TIC terms which are in the premises of the applications of Rule $[\wedge I]$. We use $\widetilde{\Pi}$ to range over compositions of projections. The equivalence $\simeq$ is given in Definition 2.

Lemma 2. *1. If* $\vdash M : \tau_1 \wedge \tau_2$, *then* $\vdash \Pi_j(M) : \tau_j$ *for* $j = 1, 2$.
2. If $\vdash M : \tau$ *and* $\tau \simeq \tau_1 \wedge \ldots \wedge \tau_n$ *with* $n \geq 2$, *then there is* $\widetilde{\Pi_i^\tau}$ *such that*
$\vdash \widetilde{\Pi_i^\tau}(M) : \tau_i$ *for each* i ($1 \leq i \leq n$).

Proof. (1). The proof is by cases on the last applied rule in the derivation of $\vdash M : \tau_1 \wedge \tau_2$. For Rule $[\wedge I]$ it follows from Fact 1. The last rule cannot be $[\text{VAR}]$, $[\to I]$, $[\to E\omega]$ or $[\to E]$, since the type in the conclusion of all these rules is a base type. The last rule cannot be $[\omega]$, since the type in the conclusion of this rule is ω.

(2). The proof is by induction on n. Point (1) gives the proof for $n = 2$. If $n > 2$, then $\tau = \sigma_1 \wedge \sigma_2$ and $\sigma_1 \simeq \tau_1 \wedge \ldots \wedge \tau_m$ and $\sigma_2 \simeq \tau_{m+1} \wedge \ldots \wedge \tau_n$ for some $m < n$. By Point (1) $\vdash \Pi_j(M) : \sigma_j$ for $j = 1, 2$. By induction, if $i \leq m$ there is $\widetilde{\Pi_i^{\sigma_1}}$ such that $\vdash \widetilde{\Pi_i^{\sigma_1}}(\Pi_1(M)) : \tau_i$, otherwise there is $\widetilde{\Pi_{i-m}^{\sigma_2}}$ such that $\vdash \widetilde{\Pi_{i-m}^{\sigma_2}}(\Pi_2(M)) : \tau_i$. So, in the first case we can choose $\widetilde{\Pi_i^\tau} = \widetilde{\Pi_i^{\sigma_1}} \circ \Pi_1$, and $\widetilde{\Pi_i^\tau} = \widetilde{\Pi_{i-m}^{\sigma_2}} \circ \Pi_2$ in the second one. $\square$

---

[4] This somewhat sounds as a circular argument, since we are using type reconstruction (which uses *type(_)*) to justify uniqueness of types. As a matter of fact, we implicitly assumed that uniqueness of typing is simultaneously proven together with the definition of *der*.

Before providing the reduction rules of TIC, let us define

$$M[x^{\sigma_i} := N_i]_{i \in I}$$

as the pseudo-term obtained by simultaneously substituting (using the standard notion of substitution for λ-calculus), for each $i \in I$, the occurrence of x^{σ_i} in M with N_i.

TIC has three reduction rules which can be applied on TIC terms inside any suitably typed context. The first two rules are used when the λ-abstraction is typed by Rule $[\rightarrow I]$, so the annotation of this λ-abstraction is a left type. These rules just replace the abstracted variable with the argument by respecting types. For this reason we distinguish the case in which the λ-abstraction is annotated with $\&_{i \in I} \sigma_i$ (Rule $[\beta\&]$) or with ω (Rule $[\beta\omega]$). Rule $[\beta\&]$ also depends on the type of the argument. Rule $[\beta\sqcap]$ is used when the application has been typed by Rule $[\wedge I]$, so the annotation of this λ-abstraction is a conjunctive type. In this case we reduce the first projection of the TIC term in function position applied to the first projection of the TIC term in argument position and the second projection of the TIC term in function position applied to the second projection of the TIC term in argument position. Then the two obtained results become arguments of the operator $\bigwedge$. The typability of the redex (see Theorem 2) ensures that this application of $\bigwedge$ is defined.

Definition 6 (Reduction Rules).

$[\beta\&]$ $(\lambda x : \&_{i \in I} \sigma_i.M)N \longrightarrow M[x^{\sigma_i} := N_i]_{i \in I}$
          where $\sigma = \text{type}(N) \simeq \wedge_{i \in I} \sigma_i$ *and* $N_i = \widetilde{\Pi_i^\sigma}(N)$ *for each* $i \in I$

$[\beta\omega]$ $(\lambda x : \omega.M)\Omega \longrightarrow M$

$[\beta\sqcap]$ $\dfrac{(\lambda x : \kappa_j.M_j)N_j \longrightarrow M_j' \quad j = 1, 2}{(\lambda x : \kappa_1 \sqcap \kappa_2.M_1 \bigwedge M_2)(N_1 \bigwedge N_2) \longrightarrow M_1' \bigwedge M_2'}$

For example, by taking the TIC term of Fig. 2,

$$(\lambda x : (\alpha \rightarrow \omega \rightarrow \alpha) \& \alpha).x^{\alpha \rightarrow \omega \rightarrow \alpha} x^\alpha \Omega)(\lambda y : \alpha \sqcap \phi.\lambda z : \omega \sqcap \omega.y^{\alpha \wedge \phi})$$
$$\longrightarrow_{\beta\&} (\lambda y : \alpha.\lambda z : \omega.y^\alpha)(\lambda y : \phi.\lambda z : \omega.y^\phi)\Omega$$
$$\longrightarrow_{\beta\&} (\lambda z : \omega.\lambda y : \phi.\lambda z : \omega.y^\phi)\Omega$$
$$\longrightarrow_{\beta\omega} \lambda y : \phi.\lambda z : \omega.y^\phi$$

where $\alpha = \phi \rightarrow \omega \rightarrow \phi$ and the arrows are decorated with the applied rule. The following is, instead, a simple application of Rule $[\beta\sqcap]$

$$\dfrac{(\lambda x : \phi.x^\phi)y^\phi \longrightarrow y^\phi \quad (\lambda x : \psi.x^\psi)y^\psi \longrightarrow y^\psi}{(\lambda x : \phi \sqcap \psi.x^{\phi \wedge \psi})y^{\phi \wedge \psi} \longrightarrow y^{\phi \wedge \psi}} \ [\beta\sqcap]$$

We prove now the Subject Reduction Theorem, namely that the reduction rules applied to TIC terms produce TIC terms having the same type.

Theorem 2 (Subject Reduction). *If* $\vdash M : \tau$ *and* $M \longrightarrow N$, *then* $\vdash N : \tau$.

Proof. The proof is by induction on $\longrightarrow$.

Rule $[\beta\&]$. Let $M = (\lambda x : \vartheta.M')N'$, the judgement $\vdash (\lambda x : \vartheta.M')N' : \tau$ implies $\tau = \alpha$ and $\vdash \lambda x : \vartheta.M' : \vartheta \to \alpha$ and $\vdash N' : \sigma$ and $\vartheta \ltimes \sigma$ by Lemma 1(4b). Let $\vartheta = \&_{i \in I}\sigma_i$, then $\vartheta \ltimes \sigma$ gives $\sigma \simeq \wedge_{i \in I}\sigma_i$. From $\vdash \lambda x : \vartheta.M' : \vartheta \to \alpha$ we get $\vdash M' : \alpha$ and $\iota(M', x) = \vartheta$ by Lemma 1(2). Let $\widehat{\Pi_i^\sigma}$ be defined as in Lemma 2(2). We can now obtain $\vdash M'[x^{\sigma_i} := N_i]_{i \in I} : \alpha$, where $N_i = \widehat{\Pi_i^\sigma}(N')$ for $i \in I$, by substituting (a derivation of) $\vdash x^{\sigma_i} : \sigma_i$ with (a derivation of) $\vdash N_i : \sigma_i$ for all $i \in I$ in the proof of $\vdash M' : \alpha$ (capture of free variables is prevented by Barendregt's convention).

Rule $[\beta\omega]$. Let $M = (\lambda x : \omega.M')\Omega$. The judgement $\vdash (\lambda x : \omega.M')\Omega : \tau$ implies $\vdash \lambda x : \omega.M' : \omega \to \tau$ and $\vdash \Omega : \omega$ by Lemma 1(4a). By Lemma 1(2), from $\vdash \lambda x : \omega.M' : \omega \to \tau$ we get $\vdash M' : \tau$.

Rule $[\beta\sqcap]$. We have
$$\frac{(\lambda x : \kappa_j.M_j)N_j \longrightarrow M_j' \quad j = 1, 2}{(\lambda x : \kappa_1 \sqcap \kappa_2.M_1 \bigwedge M_2)(N_1 \bigwedge N_2) \longrightarrow M_1' \bigwedge M_2'}$$
and $\vdash (\lambda x : \kappa_1 \sqcap \kappa_2.M_1 \bigwedge M_2)(N_1 \bigwedge N_2) : \tau$.
By Lemma 1(6) the judgement $\vdash (\lambda x : \kappa_1 \sqcap \kappa_2.M_1 \bigwedge M_2)(N_1 \bigwedge N_2) : \tau$ implies $\tau = \tau_1 \wedge \tau_2$ and $\vdash (\lambda x : \kappa_j.M_j)N_j : \tau_j$ with $j = 1, 2$. By induction, $\vdash M_j' : \tau_j$ with $j=1,2$, so we conclude using Rule $[\wedge I]$. □

In the remaining of this section we show some relations between the TIC reduction rules and the β-reduction rule for pure λ-terms. We denote by $\longrightarrow_\beta$ one step of β-reduction between pure λ-terms. We define $\|M\|$ as the pure λ-term obtained from M by erasing all types and replacing Ω with $(\lambda x.xx)(\lambda x.xx)$.

Definition 7. *The mapping $\| \ \|$ from pseudo-terms to pure λ-terms is defined by:*
$$\|x^\sigma\| = x \quad \|\lambda x : \kappa.M\| = \lambda x.\|M\| \quad \|MN\| = \|M\|\|N\| \quad \|\Omega\| = (\lambda x.xx)(\lambda x.xx)$$

It is useful and easy to verify that $\bigwedge$ can be applied only to TIC terms that are mapped by $\| \ \|$ to the same pure λ-term, which is also the image through $\| \ \|$ of the resulting TIC term.

Fact 3. $M \bigwedge N$ *is defined iff* $\|M\| = \|N\|$. *Moreover* $\|M\| = \|N\| = \|M \bigwedge N\|$.

We show now that $\| \ \|$ relates $\longrightarrow$ to $\longrightarrow_\beta$.

Theorem 4. *If* $M \longrightarrow N$, *then* $\|M\| \longrightarrow_\beta \|N\|$.

Proof. The proof is by induction on $\longrightarrow$.

The cases of Rules $[\beta\&]$ and $[\beta\omega]$ are easy.

For Rule $[\beta\sqcap]$ we get $M = (\lambda x : \kappa_1 \sqcap \kappa_2.M_1 \bigwedge M_2)(N_1 \bigwedge N_2)$ and $N = M_1' \bigwedge M_2'$ and $(\lambda x : \kappa_j.M_j)N_j \longrightarrow M_j'$ with $j = 1, 2$. By induction, we have $\|(\lambda x : \kappa_j.M_j)N_j\| \longrightarrow_\beta \|M_j'\|$ with $j = 1, 2$. We conclude since, by Fact 3, $\|M\| = \|(\lambda x : \kappa_1 \sqcap \kappa_2.M_1 \bigwedge M_2)(N_1 \bigwedge N_2)\| = \|(\lambda x : \kappa_j.M_j)N_j\|$ and $\|N\| = \|M_1' \bigwedge M_2'\| = \|M_j'\|$ with $j = 1, 2$. □

Theorem 5. *If* $\|M\| \longrightarrow_\beta P$, *then there is a TIC term N such that* $\|N\| = P$ *and either* $M \longrightarrow N$ *or* $N = M$.

Proof. If we reduce $(\lambda x.xx)(\lambda x.xx) \longrightarrow_\beta (\lambda x.xx)(\lambda x.xx)$ inside $\|M\|$, then $\|M\| = P$ and we can take $N = M$. Otherwise, it is enough to consider M to be a redex, so let $M = (\lambda x : \kappa.M')M''$ where $\|M'\| = P'$ and $\|M''\| = Q$ and $P = P'[x := Q]$. The proof is by induction on the number of $\sqcap$ in κ.

If $\kappa = \omega$, then necessarily $M'' = \Omega$ and we can choose $N = M'$.

If $\kappa = \&_{i \in I}\tau_i$, then $\vdash M'' : \sigma$ with $\sigma \simeq \wedge_{i \in I}\tau_i$. Let $\widetilde{\Pi_i^\sigma}$ be defined as in Lemma 2(2). Now we can take

$$N = M'[x^{\tau_i} := N_i]_{i \in I}$$

where $N_i = \widetilde{\Pi_i^\sigma}(M'')$ for $i \in I$.

If $\kappa = \kappa_1 \sqcap \kappa_2$, then $M' = M_1 \wedge M_2$ and $M'' = M_1' \wedge M_2'$. We get

$$\|(\lambda x : \kappa_j.M_j)M_j'\| \longrightarrow_\beta P$$

since $\|M\| = \|(\lambda x : \kappa_j.M_j)M_j'\|$ for $j = 1, 2$. Then by induction there are N_j such that $\|N_j\| = P$ and $(\lambda x : \kappa_j.M_j)M_j' \longrightarrow N_j$ for $j = 1, 2$. We can take $N = N_1 \wedge N_2$, which is defined by Fact 3. $\square$

For example, if $\alpha = \phi \to \omega \to \phi$ and

$$M = (\lambda x : (\alpha \to \omega \to \alpha) \& \alpha.x^{\alpha \to \omega \to \alpha}x^\alpha \Omega)(\lambda y : \alpha \sqcap \phi.\lambda z : \omega \sqcap \omega.y^{\alpha \wedge \phi})$$

then $\|M\| = (\lambda x.xx((\lambda t.tt)(\lambda t.tt)))(\lambda y.\lambda z.y)$ and $\|M\| \longrightarrow_\beta P$, where

$$P = (\lambda y.\lambda z.y)(\lambda y.\lambda z.y)((\lambda t.tt)(\lambda t.tt))$$

We can take

$$N = (\lambda y : \alpha.\lambda z : \omega.y^\alpha)(\lambda y : \phi.\lambda z : \omega.y^\phi)\Omega$$

since $\|N\| = P$ and $M \longrightarrow N$.

5   Relations with Intersection Type Assignment Systems

There is a family of intersection type assignment systems for untyped λ-calculus, see [7, Part III]. Here we only consider the most popular one, originally defined

$$\frac{}{\Gamma, x : \mu \vdash_\cap x : \mu} \; [\cap \text{VAR}] \qquad \frac{}{\Gamma \vdash_\cap P : \omega} \; [\cap \omega] \qquad \frac{\Gamma, x : \mu \vdash_\cap P : \nu}{\Gamma \vdash_\cap \lambda x.P : \mu \to \nu} \; [\cap \to I]$$

$$\frac{\Gamma \vdash_\cap P : \mu \to \nu \quad \Gamma \vdash_\cap Q : \mu}{\Gamma \vdash_\cap PQ : \nu} \; [\cap \to E] \qquad \frac{\Gamma \vdash_\cap P : \mu \quad \mu \leq \nu}{\Gamma \vdash_\cap P : \nu} \; [\cap \leq]$$

$$\frac{\Gamma \vdash_\cap P : \mu \quad \vdash_\cap P : \nu}{\Gamma \vdash_\cap P : \mu \cap \nu} \; [\cap I]$$

Fig. 4. Type assignment system of [6].

in [6] and recalled in Fig. 4, where the types are as in (1) on page 1. The type constructor $\cap$ is idempotent, commutative and associative and it takes precedence over $\to$. The relation $\leq$ is the reflexive and transitive closure of the preorder relation induced by the following axioms and rules

$$\mu \leq \omega \qquad \omega \leq \omega \to \omega \qquad \mu \cap \nu \leq \mu \qquad (\mu \to \nu) \cap (\mu \to \nu') \leq \mu \to \nu \cap \nu'$$

$$\frac{\mu \leq \nu \quad \mu \leq \nu'}{\mu \leq \nu \cap \nu'} \qquad \frac{\mu' \leq \mu \quad \nu \leq \nu'}{\mu \to \nu \leq \mu' \to \nu'}$$

Let $\widehat{\zeta}$ be the type obtained from ζ by replacing $\wedge$ and $\&$ with $\cap$. It is easy to verify that the pure λ-term obtained from a TIC term by the mapping given in Definition 7 is typable in the system of [6]. Moreover, the type of the pure λ-term is obtained by applying the mapping $\widehat{}$ to the type of the TIC term and the environment is determined by the function ι. Notice that $\{x : \widehat{\vartheta} \mid \iota(M, x) = \vartheta\}$ is finite since x must occur in M by definitions of ι and ϑ.

Theorem 6. *If* $\vdash M : \zeta$*, then* $\{x : \widehat{\vartheta} \mid \iota(M, x) = \vartheta\} \vdash_\cap \|M\| : \widehat{\zeta}$.

The proof of this theorem, by induction on the derivation of $\vdash M : \zeta$ and by cases on the last applied rule, is standard. The replacement of Ω with $(\lambda x.xx)(\lambda x.xx)$ agrees with the fact that each unsolvable pure λ-term has only types μ with $\mu \geq \omega$ in the system $\vdash_\cap$, see [17]. The vice versa does not holds, there are no TIC terms representing derivations involving the subsumption rule or proving intersection types without using the intersection introduction rule.

An important property of the system $\vdash_\cap$ is the Approximation Theorem. We recall that the set of *approximants* is defined by

$$A ::= \perp \mid H \qquad\qquad H ::= \lambda x.H \mid x A_1 \ldots A_m$$

where $m \geq 0$, see [5, Section 14.3]. The *direct approximant* of a pure λ-term P (notation $\flat(P)$) is defined by

$$\flat(\lambda x_1 \ldots \lambda x_n.y P_1 \ldots P_m) = \lambda x_1 \ldots \lambda x_n.y \flat(P_1) \ldots \flat(P_m)$$
$$\flat(\lambda x_1 \ldots \lambda x_n.(\lambda y.P')Q P_1 \ldots P_m) = \perp$$

where $m, n \geq 0$.

The *set of approximants* of a pure λ-term P (notation $\mathcal{A}(P)$) is obtained by considering the direct approximants of the pure λ-terms which can be obtained by β-reducing P

$$\mathcal{A}(P) = \{\flat(P') \mid P \longrightarrow_\beta^* P'\}$$

According to the Approximation Theorem below, the types of a pure λ-term are all and only the types of its approximants, when the system $\vdash_\cap$ allows approximants as subjects.

Theorem 7 (Approximation Theorem [20]). $\Gamma \vdash_\cap P : \mu$ *if and only if there is* $A \in \mathcal{A}(P)$ *such that* $\Gamma \vdash_\cap A : \mu$.

The interest of this theorem lies in the fact that approximants have *principal typings*. This means that for each approximant A there is a pair $\langle \Gamma; \mu \rangle$ such that, if $\Gamma' \vdash_\cap A : \mu'$, then $\langle \Gamma'; \mu' \rangle$ can be obtained from $\langle \Gamma; \mu \rangle$ using suitable transformations, see [16].

The principal typing of approximants are as in Definition 8 below, where the operator $\uplus$ on typing environments is defined by

$$\Gamma \uplus \Gamma' = \{x : \mu \cap \mu' \mid x : \mu \in \Gamma \; x : \mu' \in \Gamma'\} \cup$$
$$\{x : \mu \mid x : \mu \in \Gamma \; x \notin \Gamma'\} \cup \{x : \mu' \mid x \notin \Gamma \; x : \mu' \in \Gamma'\}$$

and where we write $x \notin \Gamma$ if x does not occur as subject in Γ.

Definition 8 (Principal typing [16]).

- *The principal typing of $\bot$ is $\langle \emptyset; \omega \rangle$.*
- *If $\langle \Gamma, x : \nu; \mu \rangle$ is the principal typing of H, then $\langle \Gamma; \nu \to \mu \rangle$ is the principal typing of $\lambda x.H$.*
- *If $\langle \Gamma; \mu \rangle$ is the principal typing of H and $x \notin \Gamma$, then $\langle \Gamma; \omega \to \mu \rangle$ is the principal typing of $\lambda x.H$.*
- *If $\langle \Gamma_i; \mu_i \rangle$ is the principal typing of A_i for $1 \leq i \leq m$, then*

$$\langle \{x : \mu_1 \to \dots \mu_m \to \varphi\} \uplus \biguplus_{1 \leq i \leq m} \Gamma_i; \varphi \rangle$$

(where φ is fresh) is the principal typing of $x A_1 \dots A_m$.

We can now show that, for each principal typing $\langle \Gamma; \mu \rangle$ of an approximant A, there is a TIC term M such that A is a direct approximant of $\|M\|$ and $\widehat{type(M)} = \mu$ and $\Gamma = \{x : \widehat{\vartheta} \mid \iota(M, x) = \vartheta\}$.

Theorem 8. *If $\langle \Gamma; \mu \rangle$ is the principal typing of A, then there is a TIC term M such that $A = \flat(\|M\|)$ and $\vdash M : \zeta$ with $\widehat{\zeta} = \mu$ and $\Gamma = \{x : \widehat{\vartheta} \mid \iota(M, x) = \vartheta\}$.*

Proof. The proof is by structural induction on A. If $\langle \Gamma; \mu \rangle$ is the principal typing of A we dub μ the principal type of A.

$A = \bot$. We can take $M = \Omega$.

$A = \lambda x.H$. There are two cases. The principal type of H cannot be ω.
Case 1: the principal typing of A is $\langle \Gamma; \nu \to \mu' \rangle$ and $\langle \Gamma, x : \nu; \mu' \rangle$ is the principal typing of H. By induction there is M' such that $H = \flat(\|M'\|)$ and $\vdash M' : \zeta'$ with $\widehat{\zeta'} = \mu'$ and $\Gamma' = \{y : \widehat{\vartheta} \mid \iota(M', y) = \vartheta\}$, where $\Gamma' = \Gamma, x : \nu$. Since $\mu' \neq \omega$, we actually have $\zeta' \neq \omega$. Let $\zeta' = \tau$. Moreover if $\iota(M', x) = \vartheta'$, then $\widehat{\vartheta'} = \nu$. We can choose $M = \lambda x : \vartheta'.M'$. Starting from $\vdash M' : \tau$ we can derive $\vdash M : \vartheta' \to \tau$ by Rule $[\to I]$. Lastly $\Gamma' = \{y : \widehat{\vartheta} \mid \iota(M', y) = \vartheta\}$ implies $\Gamma = \{y : \widehat{\vartheta} \mid \iota(M, y) = \vartheta\}$.
Case 2: the principal typing of A is $\langle \Gamma; \omega \to \mu' \rangle$ and $\langle \Gamma; \mu' \rangle$ is the principal typing of H. By induction there is M' such that $H = \flat(\|M'\|)$ and $\vdash M' : \zeta'$ with $\widehat{\zeta'} = \mu'$ and $\Gamma = \{y : \widehat{\vartheta} \mid \iota(M', y) = \vartheta\}$. As in the previous case, $\zeta' = \tau$. We can then choose $M = \lambda x : \omega.M'$ and, using $\vdash M' : \tau$, we can derive $\vdash M : \omega \to \tau$ by Rule

$[\to I]$. Lastly $\Gamma = \{y : \hat{\vartheta} \mid \iota(M', y) = \vartheta\}$ implies $\Gamma = \{y : \hat{\vartheta} \mid \iota(M, y) = \vartheta\}$. In both cases $H = \flat(\|M'\|)$ implies $A = \flat(\|M\|)$.

$A = xA_1 \dots A_m$. By definition, the principal typing of A is

$$\langle\{x : \mu_1 \to \dots \mu_m \to \varphi\} \uplus \biguplus\nolimits_{1 \leq i \leq m} \Gamma_i; \varphi\rangle$$

where $\langle\Gamma_i; \mu_i\rangle$ is the principal typing of A_i for $1 \leq i \leq m$. By induction, there is M_i such that $A_i = \flat(\|M_i\|)$ and $\vdash M_i : \zeta_i$ with $\hat{\zeta}_i = \mu_i$ and

$$\Gamma_i = \{y : \hat{\vartheta} \mid \iota(M_i, y) = \vartheta\} \text{ for } 1 \leq i \leq m$$

We can choose $M = x^{\zeta_1 \to \dots \zeta_m \to \varphi} M_1 \dots M_m$. We can derive $\vdash M : \varphi$ by applying m times Rule $[\to E]$ or Rule $[\to E\omega]$ as required. Let $I_y = \{i \mid \iota(M_i, y) \neq \omega\}$. We get

$$\iota(M, x) = (\zeta_1 \to \dots \zeta_m \to \varphi) \,\&\, (\&_{i \in I_x} \iota(M_i, x))$$

and $\iota(M, y) = \&_{i \in I_y} \iota(M_i, y)$ for $y \neq x$. Since

$$\Gamma = \{x : \mu_1 \to \dots \mu_m \to \varphi\} \uplus \biguplus\nolimits_{1 \leq i \leq m} \Gamma_i$$

we have $\Gamma = \{y : \hat{\vartheta} \mid \iota(M, y) = \vartheta\}$. From $A_i = \flat(\|M_i\|)$ for $1 \leq i \leq m$ we get $A = \flat(\|M\|)$. $\qquad\square$

We say that $\langle\Gamma; \mu\rangle$ is a principal typing if it is the principal typing of some approximant. It should be clear that the most interesting typings in the system $\vdash_\cap$ are the principal typings. We prove that infinitely many TIC terms have types related to principal typings, also if there are TIC terms with typings which are not related to principal typings, see the examples in the Introduction. First we show that reduction preserves ι when we do not distinguish between $\wedge$ and $\&$ and we consider them idempotent, commutative and associative.

Lemma 3. *Let M be a TIC term. If $M \longrightarrow N$, then $\widehat{\iota(M, x)} = \widehat{\iota(N, x)}$ for all free variables x.*

Proof. The proof is by cases and by induction on the reduction rules.

Rule $[\beta\&]$. Then $M = (\lambda y : \&_{i \in I}\sigma_i.M')N'$ and $N = M'[y^{\sigma_i} := N_i]_{i \in I}$, where $type(N') = \sigma$ and $\sigma \simeq \wedge_{i \in I}\sigma_i$ and $N_i = \widehat{\Pi_i^\sigma}(N')$ for all $i \in I$. We get $\iota(M, x) = \iota(M', x) \,\&\, \iota(N', x)$, while in $\iota(N, x)$ we have the same main types (possibly repeated) connected by $\&$ and in an order which depends on where y occurs in M'. Anyway $\widehat{\iota(M, x)} = \widehat{\iota(N, x)}$ holds since $\frown$ maps $\wedge$ and $\&$ to $\cap$, and $\cap$ is idempotent, commutative and associative.

Rule $[\beta\omega]$. This case is immediate since $M = (\lambda x : \omega.M')$ and $N = M'$.

Rule $[\beta\cap]$. Then $M = (\lambda x : \kappa_1 \cap \kappa_2.M_1 \wedge M_2)(N_1 \wedge N_2)$ and $N = M_1' \wedge M_2'$ and $(\lambda x : \kappa_j.M_j)N_j \longrightarrow M_j'$ for $j = 1, 2$. By induction, $\iota((\lambda x : \kappa_j.M_j)N_j, x) = \widehat{\iota(M_j', x)}$ for $j = 1, 2$, which implies $\widehat{\iota(M, x)} = \widehat{\iota(N, x)}$. $\qquad\square$

As observed in the above proof, $M \longrightarrow N$ does not imply $\iota(M,x) = \iota(N,x)$. An example is $M = (\lambda y : (\phi \to \psi) \& \phi.y^{\phi \to \psi} y^{\phi}) x^{(\phi \to \psi) \wedge \phi}$ and $N = x^{\phi \to \psi} x^{\phi}$ since $\iota(M,x) = (\phi \to \psi) \wedge \phi$ and $\iota(N,x) = (\phi \to \psi) \& \phi$. However, $\widehat{\iota(M,x)} = \widehat{\iota(N,x)} = (\phi \to \psi) \cap \phi$, as prescribed by the previous lemma.

Theorem 9. *Given a principal typing $\langle \Gamma; \mu \rangle$ there are infinitely many TIC terms M such that $\widehat{type(M)} = \mu$ and $\Gamma = \{x : \widehat{\vartheta} \mid \iota(M,x) = \vartheta\}$.*

Proof. Theorem 8 builds a TIC term N which satisfies the requirements. Since reduction preserves types and the type of a term is unique, all TIC terms which reduce to N are typed by $type(N)$. Lemma 3 ensures the relation between ι and Γ for all TIC terms which reduce to N. Let $type(N) = \zeta$, then $\widehat{\zeta} = \mu$ implies either $\zeta = \alpha$ or $\zeta = \omega$ by Definition 8. Let $\mathbf{I}_\zeta = \lambda x : \alpha.x^\alpha$ if $\zeta = \alpha$ and $\mathbf{I}_\zeta = \lambda x : \omega.\Omega$ if $\zeta = \omega$. Then $\underbrace{\mathbf{I}_\zeta(...(\mathbf{I}_\zeta N)...)}_{n}$ reduces to N for all $n > 0$. $\square$

For example $\langle \emptyset; (\phi \to \psi) \cap \phi \to \psi \rangle$ is the principal typing of the auto-application function $\lambda x.xx$ and all TIC terms which reduce to

$$\lambda x : (\phi \to \psi) \& \phi.x^{\phi \to \psi} x^{\phi}$$

have type $(\phi \to \psi) \wedge \phi \to \psi$.

As last result we show that TIC characterises the set of λ-terms having head normal forms. This is not surprising, since the type assignment system of [6] characterises the pure λ-terms having head normal forms. As expected, a TIC term is in head normal form if it is of the shape

$$\lambda x_1 : \kappa_1 \ldots \lambda x_n : \kappa_n.y^\sigma M_1 \ldots M_m$$

where $n, m \geq 0$ and either $y = x_i$ for some i ($1 \leq i \leq n$) or y is a free variable.

Theorem 10. *A TIC term M has a head normal form iff $type(M) \neq \omega$.*

Proof. The left to right implication is immediate: since only Ω has type ω, all other TIC terms have types different from ω.

Vice versa if $\vdash M : \zeta$, then Theorem 6 implies $\Gamma \vdash_\cap \|M\| : \widehat{\zeta}$ with

$$\Gamma = \{x : \widehat{\vartheta} \mid \iota(M,x) = \vartheta\}$$

Since $\zeta \neq \omega$ gives $\omega \not\leq \widehat{\zeta}$, by the Approximation Theorem (Theorem 7) the pure λ-term $\|M\|$ reduces to an head normal form, let it be P. By repeated applications of Theorem 5 we can find a TIC term N such that M reduces to N and $\|N\| = P$. Then N is in head normal form. $\square$

6   Related Works and Conclusion

In the literature there are many proposals for typed λ-calculi à la Church with intersection types. We only recall here some of the most significant ones.

The calculus of [27] has branching types and types with quantification over type selection parameters. There $\vdash_\cap \lambda x.x : (\phi \to \phi) \cap (\psi \to \psi)$ is represented by $\Lambda(\text{join}\{i = \star, j = \star\}).\lambda x^{\{i=\phi, j=\psi\}}.x^{\{i=\phi, j=\psi\}}$, where $\text{join}\{i = \star, j = \star\}$ is a branching type. Branching types avoid duplication, since they "squash together" the premises of the intersection introduction typing rule.

In the calculus of [23] typing depends on an "imperative-like" formulation of context, assigning types to term-variables at a given mark/location, and on a new notion of store, that remembers, trough modalities, the associations between marks and types. For example, $\vdash_\cap \lambda x.x : (\phi \to \phi) \cap (\psi \to \psi)$ can be written as the term $(\lambda x : 0.x)@(\lambda 0 : \phi.0) \cap (\lambda 0 : \psi.0)$ where 0 is a mark, and the modality for the subterm $\lambda x : 0.x$ is $(\lambda 0 : \phi.0) \cap (\lambda 0 : \psi.0)$.

A parallel term constructor | representing the intersection is instead introduced in [12]. This allows to obtain, for any type derivation in the system $\vdash_\cap$, a corresponding type decorated term. For example, the term corresponding to $\vdash_\cap \lambda x.x : (\phi \to \phi) \cap (\psi \to \psi)$ is $\lambda x^\phi.x^\phi | \lambda y^\psi.y^\psi$.

A new and interesting solution is represented by *dimensional intersection type calculi* [21, 22]. The typing judgements are of the shape $\Gamma \vdash P : \mu$, where P is an *elaboration*, i.e., a λ-term where each sub-term is decorated with the set of types assigned to it. These decorations are enclosed between angle brackets. For example, the judgement in dimensional intersection type calculi corresponding to $\vdash_\cap \lambda x.x : (\phi \to \phi) \cap (\psi \to \psi)$ is

$$\vdash_{<>} (\lambda x.x\langle\phi, \psi\rangle)\langle\phi \to \phi, \psi \to \psi\rangle : (\phi \to \phi) \cap (\psi \to \psi)$$

Our proposal is definitely worthy of inclusion in the above scenario. Differently from the mentioned calculi, TIC has more type constructors and a less permissive type syntax than the type assignment system in [6]. Besides, it lacks the subsumption rule, as well as weakening and contraction. This implies that we do not have the isomorphism between typing à la Curry and à la Church, which is a feature of the calculi in [12, 21–23, 27]. However, due to the presence of the universal type ω, we characterise the λ-terms having head normal forms, while the calculi in [12, 21–23, 27] characterise strongly normalising λ-terms.

We split the intersection into distinct constructors taking inspiration from [25], where two distinct conjunction constructors reflect two possible shapes of derivations. The synchronous conjunction can be used only among equivalent deductions, while the asynchronous conjunction can be used among arbitrary derivations. This is the basis for building a logical system which is a proof theoretical justification of intersection type assignment systems.

The use of different constructors for abstracted variables and terms in our type system is reminiscent of the different linear logic conjunction operators used both on the left and on the right hand side of the linear logic entailment. This could be the starting point of an investigation on the notion of strong conjunction for linear logic and hence, via realisability, on a notion of "intersection linear types".

Acknowledgements. We are grateful to the anonymous referees for their comments and suggestions to improve the readability of this paper.

References

1. Anderson, A.R., Belnap, N.: Entailment: The Logic of Relevance and Necessity, vol. I. Princeton University Press, Princeton (1975)
2. Alessi, F., Barbanera, F.: Strong conjunction and intersection types. In: Tarlecki, A. (ed.) MFCS. LNCS, vol. 520, pp. 64–73. Springer, Heidelberg (1991). https://doi.org/10.1007/3-540-54345-7_49
3. van Bakel, S.: Strict intersection types for the lambda calculus. ACM Comput. Surv. **43**(3), 20:1–20:49 (2011). https://doi.org/10.1145/1922649.1922657
4. Barbanera, F., Martini, S.: Proof-functional connectives and realizability. Arch. Math. Logic **33**(3), 189–211 (1994). https://doi.org/10.1007/BF01203032
5. Barendregt, H.: The lambda calculus, its syntax and semantics. In: Studies in logic and foundations of mathematics, vol. 103. North-Holland (1981)
6. Barendregt, H., Coppo, M., Dezani-Ciancaglini, M.: A filter lambda model and the completeness of type assignment. J. Symb. Logic **48**(4), 931–940 (1983). https://doi.org/10.2307/2273659
7. Barendregt, H., Dekkers, W., Statman, R.: Lambda Calculus with Types. Cambridge University Press, Cambridge (2013). https://doi.org/10.1017/CBO9781139032636
8. Barendregt, H.: Lambda calculi with types. In: Abramsky, S., Gabbay, D.M., Maibaum, T. (eds.) Handbook of Logic in Computer Science, vol. 2: Background: Computational Structures, pp. 117–309. Oxford University Press, Oxford (1992). https://doi.org/10.1093/oso/9780198537618.003.0002
9. Bernadet, A., Lengrand, S.: Complexity of strongly normalising λ-terms via non-idempotent intersection types. In: Hofmann, M. (ed.) FOSSACS. LNCS, vol. 6604, pp. 88–107. Springer, Heidelberg (2011). https://doi.org/10.1007/978-3-642-19805-2_7
10. Blaauwbroek, L.: On the interaction between unrestricted union and intersection types and computational effects. Master's thesis, Eindhoven University of Technology (2017). https://pure.tue.nl/ws/portalfiles/portal/88387896/Masters_Thesis_Lasse_Blaauwbroek.pdf
11. Bono, V., Dezani-Ciancaglini, M.: A tale of intersection types. In: Hermanns, H., Zhang, L., Kobayashi, N., Miller, D. (eds.) LICS, pp. 7–20. ACM (2020). https://doi.org/10.1145/3373718.3394733
12. Bono, V., Venneri, B., Bettini, L.: A typed lambda calculus with intersection types. Theor. Comput. Sci. **398**(1–3), 95–113 (2008). https://doi.org/10.1016/j.tcs.2008.01.046
13. de Carvalho, D.: Execution time of lambda-terms via denotational semantics and intersection types. CoRR **abs/0905.4251** (2009). http://arxiv.org/abs/0905.4251
14. Church, A.: A formulation of the simple theory of types. J. Symb. Logic **5**(2), 56–68 (1940). https://doi.org/10.2307/2266170
15. Coppo, M., Dezani-Ciancaglini, M.: An extension of the basic functionality theory for the λ-calculus. Notre Dame J. Formal Logic **21**(4), 685–693 (1980). https://doi.org/10.1305/ndjfl/1093883253

16. Coppo, M., Dezani-Ciancaglini, M., Venneri, B.: Principal type schemes and lambda-calculus semantics. In: Hindley, R., Seldin, J.P. (eds.) To H.B.Curry: Essays on Combinatory Logic, Lambda-calculus and Formalism, pp. 535–560. Academic Press (1980)

17. Coppo, M., Dezani-Ciancaglini, M., Venneri, B.: Functional characters of solvable terms. Zeitschrift für Mathematishche Logik und Grundlagen der Mathematik 27(1), 45–58 (1981). https://doi.org/10.1137/0205036

18. Curry, H.B.: Functionality in combinatory logic. Proc. Natl. Acad. Sci. USA 20, 584–590 (1934). https://doi.org/10.1073/pnas.20.11.584

19. De Benedetti, E., Ronchi Della Rocca, S.: A type assignment for lambda-calculus complete both for FPTIME and strong normalization. CoRR abs/1410.6298 (2014). http://arxiv.org/abs/1410.6298

20. Dezani-Ciancaglini, M., Honsell, F., Motohama, Y.: Approximation theorems for intersection type systems. J. Log. Comput. 11(3), 395–417 (2001). https://doi.org/10.1093/logcom/11.3.395

21. Dudenhefner, A., Rehof, J.: Intersection type calculi of bounded dimension. In: Castagna, G., Gordon, A.D. (eds.) POPL, pp. 653–665. ACM (2017). https://doi.org/10.1145/3009837.3009862

22. Dudenhefner, A., Rehof, J.: Typability in bounded dimension. In: Ouaknine, J. (ed.) LICS, pp. 1–12. IEEE Computer Society (2017). https://doi.org/10.1109/LICS.2017.8005127

23. Liquori, L., Della Rocca, S.R.: Intersection-types à la Church. Inf. Comput. 205(9), 1371–1386 (2007). https://doi.org/10.1016/j.ic.2007.03.005

24. Lopez-Escobar, E.G.K.: Proof functional connectives. In: Di Prisco, C.A. (ed.) Methods in Mathematical Logic, pp. 208–221. Springer, Heidelberg (1985). https://doi.org/10.1007/BFb0075313

25. Pimentel, E., Ronchi Della Rocca, S., Roversi, L.: Intersection types from a proof-theoretic perspective. Fund. Inf. 121(1–4), 253–274 (2012). https://doi.org/10.3233/FI-2012-778

26. Venneri, B.: Intersection types as logical formulae. J. Log. Comput. 4(2), 109–124 (1994). https://doi.org/10.1093/logcom/4.2.109

27. Wells, J.B., Haack, C.: Branching types. In: Métayer, D.L. (ed.) ESOP. LNCS, vol. 2305, pp. 115–132. Springer, Heidelberg (2002). https://doi.org/10.1007/3-540-45927-8_9

Safe Smooth Paths Between Straight Line Obstacles

Yves Bertot(✉) 🆔

Inria Université Côte d'Azur, Sophia Antipolis, France
yves.bertot@inria.fr

Abstract. We describe a collections of algorithm to compute smooth trajectories between obstacles, with the objective that the obtained trajectories should be smooth. In particular, we use a vertical cell decomposition algorithm to avoid the obstacles, a best first search algorithm to obtain trajectory sketches and Bézier curves to implement smoothness. We also provide some insights into the correctness arguments for these algorithms. These correctness arguments are intended for use in a formal proof. While we have running implementations of the full program, the formal proofs of correctness are still incomplete.

1 Introduction

One benefit of formal verification is the removal of errors at early stages in the design of complex artifacts. This is being used extensively for software with great success. We wish to extend this to domains where the distance between formal models and the real application is bigger. We choose to work on robotics. For this field, there is a large distance between models and real implementations, because our knowledge of physics is incomplete, the world needs to be represented through geometrical abstractions, and there are limitations to what computers can compute accurately and what sensors and actuators can do in terms of precision.

To begin with we will look only at a question involving geometry. We wish to present a comprehensive Coq program to compute trajectories for a robot modeled as a point between obstacles that are given by straight line segments. We aim for smooth trajectories, since such trajectories can be traveled by wheeled robots without a need to stop to change direction. As an illustration of a robot able to exploit such a collection of trajectories, one may think of a gyropod (a two wheeled robot standing upward by active control, where the two wheels can rotate a different speeds). Such a robot has no restriction on rotation radius, but of course the curves must be taken more and more slowly as the curve is tight. For a sharp angle, such a robot can stop and change direction on the spot by having the two wheels rotate in opposite directions.

The program we wish to describe formally receives as inputs descriptions of problems like this drawing, where the bottom and top straight lines delimit the workspace, the other straight lines represent obstacles, and the cross marks represent the starting and final extremities of the required trajectory:

© The Author(s), under exclusive license to Springer Nature Switzerland AG 2024
V. Capretta et al. (Eds.): *Logics and Type Systems in Theory and Practice*, LNCS 14560, pp. 36–53, 2024.
https://doi.org/10.1007/978-3-031-61716-4_3

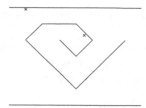

The output of the program should be smooth trajectories like the curved line that appears in this picture, when such trajectories are possible:

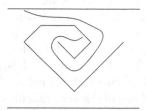

Using straight line segments for obstacles may seem unrealistic at first thought, but these segments can be connected together to form polygons and these polygons can in turn be used to safely over approximate obstacles with more complex geometry.

For trajectories, the most general way to describe them is to give a representation of the position of a point in time, as an interval of time $[0, T]$ and a continuous function $f : \mathbb{R} \to \mathbb{R}^2$, such that $f(0)$ is one of the given extremities, $f(T)$ is the other given extremity. the smoothness property could be expressed by the fact that the derivative is continuous and never 0 in the open interval $(0, T)$.

For the purpose of this paper, we will rather describe trajectories as the concatenation of elementary fragments, where each fragment is connected to the next one by an extremity. The elementary fragments we use here are either straight line segments or Bézier curves. Obviously, each elementary segment can be traveled by a function from interval $[0, 1]$ to $\mathbb{R}^2$, and thanks to a translation and scaling on inputs, a sequence of elementary fragments can be transformed into a trajectory represented by a single interval and continuous function.

By combining this function f with a function $g : \mathbb{R} \to \mathbb{R}$, we could guarantee more properties about the robot behavior, like limiting accelerations imposed on the payload, for instance.

The main property that we want to prove is that the trajectory does not collide with the obstacles. Since the obstacles are given as straight line segments, this main property is only about ensuring that there is no intersection between the set $f([0, T])$ and the union of these segments, viewed as sets of points. Making that the function f has indeed a continuous and nonzero speed in the open interval $(0, T)$ is an important guideline for our work, but obtaining a formal proof of this fact is beyond the scope of this article. We will however discuss some approaches to this property as we study the various parts of the trajectory computation.

Also, because of our limited time frame, we cannot address questions of optimality. We wish to describe a program that produces trajectories with the stated properties, but we have to accept any algorithm that gives reasonable solutions.

2   A Combination of Algorithms

The Coq program contains four phases, where each corresponds to a different algorithm. For the first two phases, the basic blocks are concerned with converting the geometry of the problem into a discrete problem, using sorting and a recursive treatment of the obtained sorted list. The result of the first two phases consists in a collection of cells, where each cell has a single higher edge and a single lower edge, connected by two vertical sides. The higher and lower edges are portions of obstacles, and one of the vertical sides may be reduced to a point. So in the general case the cells have four sides, but in some cases they may have only three. When not reduced to a point, the vertical sides are safe boundaries that can be crossed from one cell to its neighbors without collision with the obstacles. A convex polygon connected to neighbors by safe boundaries that can be crossed without collision with the obstacles.

For our running example, the result of these first two phases is described in the following picture, where safe boundaries are shown as dashed lines and cell numbers correspond to the order in which closed cells are added to the result. There are 14 cells, numbered from 0 to 13.

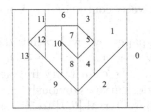

The problem becomes one of moving between cells. The collection of cells and safe boundaries is used to construct an abstract view of the trajectory, viewed simply as a path between the cross marks and cell boundaries. For this phase, a simple graph algorithm is used. A sketch of the trajectory is built as a broken line from the cross marks to centers of safe boundaries of the cells that contain them, and then between safe boundaries. This broken line already represents a possible path, and we hope to prove that this path is safe thanks to the fact that cells are convex and the segments are each included in one cell at a time, except for their extremities which are members of a cell boundary or equal to one of the cross marks.

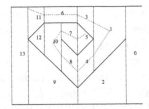

However, such a broken line is unsatisfactory as a robot trajectory, because each corner would impose that the robot stops to rotate and change direction. A last phase is added to find smooth curved lines in the vicinity of the broken line, and extra proofs are required to show that the smooth trajectories still avoid collisions with the obstacles. This phase can itself be decomposed in two parts, where one proposes a Bézier curve for testing, and the other checks that the given Bézier curve is safe. When the test check fails, a curve that is closer to the initial broken line is proposed. In the case where the repetitive process does not converge fast enough, it is possible to use the initial broken line as a fallback solution (but in this case, smoothness is not obtained).

As a summary, the input to the program is composed of three parts:

1. a top and a bottom straight line segments forming a box in which all other obstacles have to be included,
2. a sequence of segments describing the obstacles,
3. and a pair of points describing the extremities that have to be joined by some trajectory.

Given such an input, it may be that there exists no trajectory between the mandated extremities, for instance when an extremity is inside a closed polygon and the other extremity is outside.

When a trajectory exists, the output is a sequence of trajectory fragments, where each trajectory fragment is either a straight line segment or a bounded quadratic Bézier curve given by three control points. This sequence of fragments connects the two extremities (in the sense that one of the extremities of the first fragment is one of the input extremities, and one of the extremities of the last fragment is the other input extremity) and the fragments are connected, in the sense that two successive fragments of the sequence share an extremity. Moreover, the connection is smooth, in the sense that the connected fragments have the same tangent at the point of connection, and this tangent is traveled in the same direction.

The program that is described in this paper is fully implemented in the functional language provided inside Coq. This implementation has to relinquish on one of the constraints. In cases where making the trajectory smooth requires recursive computations whose termination is too hard to prove, a fallback solution constructed with adjacent segments that are not aligned is chosen. The Coq program can be extracted and processed for running in other contexts as will be illustrated later in the paper.

This project is currently in progress in the sense that the program computing the trajectories is fully implemented in Coq, but most of the formal proofs of correctness are currently being developed.

For the property of obstacle avoidance, we plan to use the fact that each trajectory fragment and each straight line segment can be viewed as the image set of a function of type $K \to K^2$, where K is the field used to represent coordinates, for the interval $[0, 1]$. So if an obstacle is represented by the function f_o and a trajectory fragment is represented by the function f_t, we can state that this trajectory fragment avoids this obstacle with the following mathematical formula:

$$\forall xy \in [0, 1], f_o(x) \neq f_t(y)$$

We can then state a general safety property by repeating this avoidance criterion for all obstacles and all trajectory fragments.

3   Producing a Sequence of Events

The first objective is to produce a collection of cells with the following properties:

1. The interior of each cell is free of obstacles
2. The parts of the boundaries of these cells that can be crossed safely to reach other cells are known precisely

To achieve this objective, we took inspiration from an algorithm known as *vertical cell decomposition* [9]. The intuitive operational model of this algorithm is that a vertical line is sweeping the working space from left to right. When the extremity of an obstacle segment or the intersection between two obstacle segments is encountered, some internal view of the cells is updated, ultimately producing a collection of closed cells. To represent this sweeping operations, we chose to implement a sorting algorithm that takes as input the list of segments and produces as output a list of *events*. Each event corresponds to an obstacle extremity and is annotated with the sequence of segments that have this event as leftmost extremity.

Here is an example with three segments and the corresponding list of events. For this example, we write $(a, [s_1; s_2])$ for an event whose location is given by point a and whose sequences contains s_1 and s_2.

$$(a, [s_2; s_1]); (d, [s_3]); (b, []); (c, []); (e, [])$$

We chose to simplify the problem by assuming that segments only intersect at their extremities. This assumption is not hard to satisfy, as it is possible to take a collection of segments which do not satisfy this assumption and produce a new collection of segments that cover exactly the same points of the plane but satisfy the assumption. We implemented a naive program that performs this cleaning step by visiting all pairs of edges and, upon detecting an intersection, replacing the two edges either by three edges or four edges depending on whether the extremity of one edge is in the middle of the other or whether they cross in their middle.

Another simplification to our presentation of the problem is that vertical segments are not allowed in our input. Again, we think that it is possible to work around this limitation: If one wishes to add vertical obstacles, only their extremities should be added as events without outgoing edges in the sequence of events. We expect the cells to be created where a safe boundary appears between the two events. It is enough to remove this safe boundary from the cells to avoid collisions with the corresponding vertical obstacle. This approach is not implemented in our current version of the algorithm.

The sorting algorithm works as follows: every time a new segment is processed, two events are considered for addition in the sequence of events. The first event is for the leftmost extremity of the segment. If an event already exists at the same location, then the current segment is simply added to the annotation of the existing event. If no event exists with that location, the event is simply added, again with the current segment in the event annotation. The second event for the rightmost extremity of the segment is added similarly, except that nothing is done for event annotations.

For instance in our example, segments s_1, s_2, and s_3 are processed in this order. The sequence of events after processing s_1 is

$$(a, [s_1]); (b, []).$$

After processing s_2, the sequence is

$$(a, [s_2; s_1]); (b, []); (c, []).$$

We see that in this case, no new event is added for the leftmost extremity of s_2, but s_2 is added to the sequence of segments annotating the event at location a.

We assume we are working in some abstract ordered field to represent the coordinates of points. For the algorithm itself, we shall only use the field operations and comparison functions. For the proofs we need to use some properties of the order. The data types that we need to create to describe our algorithm are as follows:

- A datatype of points, which are ordered pairs of coordinates in the field,
- A datatype of edges (for the obstacles), which are ordered pairs of points, with the invariant that the first coordinate of the first point is strictly smaller than the first coordinate of the second point,
- A datatype of events, which have two components, the first is a point and the second is sequence of edges, all of which have this point as first extremity.

For this algorithm to be correct, we need to guarantee the following properties:

- The set of segments obtained by collecting all outgoing edges of all events in the output must be the initial collection of obstacles.
- All segments attached to a given event have this event as first extremity.
- All right points of outgoing edges in the output must be present as locations of events in the output.
- The sequence of events is strictly sorted lexicographically, as a consequence each edge appears as outgoing edge of a single event.

Most of these properties have already proved formally in our development.

Processing the sorted list of events then implements the idea of the vertical line sweeping from left to right. In our example, this means that the next phase of the algorithm will be processing events located at a, d, b, c, e in this order.

4   Producing Cells

We add an extra constraint on the data: the whole working space is enclosed between a bottom and a top edge that do not cross and are long enough so that they define a quadrangle containing all obstacles, with no contact between the bottom and top edges and the obstacles.

To understand how cells are produced, we can keep the mental model of a sweeping line, which is the vertical line passing through the location of the current event. When producing cells, we distinguish between closed and open cells. The closed cells lie to the left of the sweeping line, they are complete and have well defined left and right sides. The open cells intersect the sweeping line. They have a well defined left side, but the right side is unknown, as it will be fixed when processing later events. The left and right sides of cells are vertical. Cells also have a top side and a bottom, which are fragments of edges (obstacles).

The data structure for a cell thus has four components: a list of points for the left side, a list of points for the right side, a low edge and a high edge. For the algorithms to work correctly, we guarantee the invariants that the high edge is above the low edge, points on each side are vertically aligned and ordered with respect to their vertical coordinate and the top (resp. low) point on each side belongs to the high (resp. low) edge of the cell. The preservation of this invariant has already been proved in our development.

The points that belong to each of the sides list correspond to unsafe points on the side of the cell. The doors into the cell are the vertical segments between two successive points in one of the sides. It often happens that the high and low edge meet on one side of the cell. In that case the side is restricted to only one point.

At a given position of the sweeping line, there is an ordered sequence of open cells. Two successive cells in this sequence are adjacent in the sense that the high edge of a cell is the low edge of the next cell. This sequence of cells is also complete, in the sense that it covers all the vertical space between the bottom edge and the top edge. All the open cells also have their left side well defined and lying to the left of the sweeping line, and sometimes the left side of a cell coincides with the position of the sweeping line.

The processing of a new event works as follows:

1. The sequence of open cells is decomposed into three parts: a first sequence of untouched cells, a second sequence of cells in contact with the given event, and a third sequence of untouched cells
2. All cells in contact are transformed into closed cells: their right side is described and they are added to the collection of closed cells
3. If the given event has n outgoing edges, then $n+1$ new open cells are created. The outgoing edges are first sorted vertically. The first cell has for its low edge a portion of the same obstacle that was used for the low edge of the first cell in the contact cells, and as high edge the first of the outgoing edges (after sorting). The last new cell is similar but with the last outgoing edge and a portion of the same obstacle as the high edge of the last contact cell. All other newly created cells have a left side that is reduced to a single point, because the low and high edges meet at the event that is being processed.

Special care must be taken to handle the case where events are vertically aligned. When this happens, a naive version of our algorithm produces cells that have the left side and the right side on the same vertical line. These cells are flat and are unfit for our later needs.

We modified our algorithm to memorize the position of the last processed event, together with the last opened cell and the last closed cell. When the current event is vertically aligned with the last processed event, we can avoid creating a flat closed cell by updating the last opened cell and the last closed cell. For the last closed cell, only the unsafe points on the right side need to be updated. However, this adds complexity in the proofs about the algorithm.

Thanks to Coq computation capability, this algorithm can be run on examples of small size. However, a full formal proof of this program has not been achieved yet. In the currently ongoing formal development, we proceed by showing that a collection of properties are invariant through the processing of events. The properties that we have proved invariant so far are as follows:

- The sequence of open cells is sorted vertically, the cells are adjacent, and they cover the whole space between the bottom and top edges,
- All low and high edges of the current open cells contain a point that is vertically aligned with the currently processed event,
- All sets composed of the interior of a cell and its high edge are disjoint,
- The obstacles are covered by the high edges of existing cells, closed or open and the outgoing edges of events that have not been processed yet.

We have also been able to show the following main results formally, but only in the case where the first coordinate of two successive events are different.

- There is no intersection between the inside of closed cells and the obstacles,
- The only intersection between obtacles and left or right sides of cells are the points listed in these cells' sides.
- At the end of processing, there remains only one open cell whose higher edge is the top edge and the lower edge is the bottom edge.

Future work will include providing formal proofs of these invariants and main results also for the case when the vertical alignment between successive events is allowed.

In what follows, we will describe algorithms that construct trajectory fragments that always intersect only two cells and a door between these cells. The absence of collision with obstacles is direct consequence of these main results.

5   Making a Broken Line Trajectory Between Two Points

With the graph of cells and doors between these cells, we can already construct safe trajectories between two given extremities, which we shall the source and target points for clarity. We only need the source and target to be either inside the same polygon made of obstacles, or inside the work space and outside of any polygon.

The first step is to find the two cells that contain the source and the target. If the source and target are in the same cell, it is enough to draw a straight line segment from the source to the target.

If the two cells are different, we can explore the cell graph to check for the existence of a path in this graph from the source cell to the target cell. This path is composed of edges of the graph and each of the graph edges is actually a door between two cells. To obtain a broken line trajectory, It is enough to construct a sequence of smaller trajectories between the middle points of each door that is an element of this path.

If two successive doors are not on the same vertical line, these two doors are on both sides of a same cell. A straight line segment from the middle of the first door to the middle of the second door is a safe trajectory in the interior of the cell (this property would not be satisfied if the cell had the same coordinate for the left vertical side and the right vertical side, hence our insistence on producing nonempty cells). This illustrated in the following drawing, where obstacles are given by solid lines, doors by vertical dotted lines, and the trajectory fragment is given by the dashed line.

Choosing to move from middle of doors to middle of doors is arbitrary and it may seem more efficient to choose a path that moves closer to one of the door extremities. This is true and any point inside the door could be chosen instead, we only need this point to be chosen far enough from the door extremities to give enough space for the later smoothing process to succeed, because the smooth curves does not meet the door at the same point.

When two successive doors are on the same vertical line, a straight line between the center of one door and the center of the other is most probably unsafe, because it will touch the lowest bound of the highest of the two doors, and this lowest bound is deemed unsafe. To avoid this collision, we build a trajectory composed of two straight line segments. The first segment goes from the middle of the first door to the center of the cell. The second segment goes from the center of the cell to the middle of the second door. Such a two segment trajectory is safe because the two extremities are safe, and the interiors of these segments lie inside the interior of the cell, which is safe.

Here again, choosing to move to the center of the cell is arbitrary, any other point inside the cell could be chosen.

An example of safe trajectory between two doors on the same side of a cell is given in the following figure. In this figure, solid lines represent obstacles, vertical dotted lines represent doors between cells, and the slanted dashed lines represent a broken line trajectory between the centers of two doors that are on the same vertical.

For the rest of this article, we assume that this trajectory is correct. The trajectory is described by a sequence of segments, where the two extremities of each segment are inside the doors of a cell or inside the cell itself, two successive segments share an extremity, and the first and last extremities are the requested source and target.

6   Making a Smooth Trajectory

The trajectory described in the previous section is given by a sequence of segments, where each segment shares an extremity with its predecessor and the other extremity with its successor. We shall call these extremities the *corners* of the trajectory. We wish to replace each corner with a smooth curve and to connect each curve fragment in a smooth fashion, in other words, so that a point can travel the curve with continuous nonzero speed.

We expect that the smooth connection constraint can also be rephrased with the following sufficient criteria: the curve fragments are connected, their tangents at the connection are parallel, and these curves are locally on opposite sides of a straight line that crosses these tangents. This is illustrated in the following figures, each where two curves connect at the point where several straight lines cross. These straight lines are the two tangents of the curves at the point of intersection (two distinct dashed lines) and the line crossing the tangents that separates locally the two curves (dotted line). In the first figure the two connected curves are on both sides of the tangent. In the second figure, the two connected curves are on the same side of the tangent.

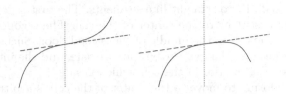

Such a property is easy to achieve with Bézier curves. In our case, it is enough to use quadratic Bézier curves. A quadratic Bézier curve is given by 3 points (in the same way that a polynomial of degree 2 is given by three coefficients). Given three points A, B, C a point on the corresponding Bézier curve is given by choosing a parameter t between 0 and 1 and computing three more points: D is the barycenter of A with weight $1 - t$ and B with weight t, E is the barycenter of B with weight $1 - t$ and C with weight t, and F is the barycenter of D with weight $1 - t$ and E with weight t. The point F is the point on the Bézier curve for the parameter t. When t varies between 0 and 1, the point moves on the Bézier curve from A to C, generally without passing through B. The curve is tangent to the segment (A, B) in A and tangent to the segment (B, C) in C.

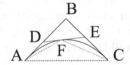

This construction has two advantages. First it makes it clear that the Bézier curve can be traveled by a continuous function from interval $[0, 1]$ to $\mathbb{R}^2$. Second, it makes it obvious that the convex hull of the triangle A, B, C contains the points of the trajectory. In other words, this convex hull is an over approximation of the trajectory. To check the absence of collision, it is enough to check that this convex hull is included in the interior of cells or doors.

A broken line A B C D can be transformed into a sequence of smoothly connected Bézier curves by adding a point M on the segment between B and C and replacing the three segments (A, B), (B, C), and (C, D) by the two Bézier curves given by points A B M on the one hand and M C D on the other hand. These two curves connect in M so that their tangents are respectively parallel to the line BM and the line MC, which are the same line. Moreover, any line that crosses BC in M has B on one side and C on the other side. Locally, the points of the first curve close to M are on the same side as B and the points of the second curve close to M are on the same side as C. We hope to be able to prove this thanks to the properties of Bernstein polynomials, which are useful in describing Bézier curves.

Our Bézier curve construction starts with a broken line that is safe, but the curves stray away from the corners, and in doing so they may collide with the obstacles. We designed two more algorithms, one to check that a Bézier curve stays in safe territory, and the other to repair trajectories whose Bézier curve do collide with the obstacles.

The first algorithm is based on a strong property of Bézier curves, which we shall call the *dichotomy property*. In the description above where we showed how to construct the point F which belongs to the Bézier curve, it turns out that the triplets $A\ D\ F$ and $F\ E\ C$ define two new Bézier curves that decompose the original Bézier curve in two complementary subsets. The triplet $A\ D\ F$ defines a curves that covers all the points of the curve defined by $A\ B\ C$ for a parameter $t' \leq t$, while the triplet $F\ E\ C$ defines a curves that covers all the points of the curve defined by $A\ B\ C$ for a parameter $t' \geq t$. The two convex hulls for $A\ D\ F$ and $F\ E\ C$ give over approximations of the Bézier curve that are much better than $A\ B\ C$ and the process can be repeated recursively.

Our checking algorithm repeats this dichotomy operation until it gets enough information to conclude that there is a guarantee that the Bézier curve is included in the union of the interior of two cells and the door between these cells, or that there is a guarantee that the Bézier curve steps out of this union, in which case we have proved that the Bézier curve is unsafe. A similar dichotomy argument is used to isolate the roots of a polynomial using Bernstein coefficient in [1] and there exists a proof that that dichotomy process will eventually reach a conclusive answer, this proof was already implemented formally by Zsidó [12]. It is still unsure whether this proof can be used to prove that the dichotomy process converges in our use case. As an alternative which does not require a complex proof, we may also wish to stop checking after a fixed number of dichotomy steps and declare the trajectory to be unsafe, even if we have not detected an actual collision.

When we have an unsafe Bézier curve, we can repair the situation by replacing the Bézier curve $A\ B\ C$ by the composite path obtained by adding two new points M and N which are the centers of (A, B) and (B, C) respectively, and taking the segment (A, M), the Bézier curve $M\ B\ N$, and finally the segment (N, C). This composite path is smooth, because the junctions at M and N have the same tangent and direction, and it is also closer to the initial broken line trajectory. This new trajectory needs to be checked again, and the replacement process can be repeated if need be. Ultimately, if the replacement process were repeated indefinitely the limit of the smooth path would be the broken line path.

Here again, there is an arbitrary choice in placing M and N at the centers of (A, B) and (B, C) respectively, any other choice of points in the interior of these segments will make the Bézier curve get closer to the initial trajectory sketch. The only constraint is that repeating the process will eventually produce a satisfactory curve.

The following figure illustrates the case where the faulty Bézier curve drawn as a dotted line is repaired into the trajectory composed of two straight line segments and a smaller Bézier curve. The vertical dashed line represents a door and the solid straight line below represents an obstacle.

In our program, we first produce a candidate Bézier curve for each corner, and we then check whether this curve is safe. If not, we produce new Bézier curves that are closer to the broken line and repeatedly check their safety. To obtain a program that is executable in Coq, the number of iterations is limited. When the limit is exceeded, the program falls back to the initial broken line corner. We only wish to prove that the checking procedure is correct. For this program, we will not be able to prove that the resulting trajectory is smooth.

7   Exploiting the Program and Visual Feedback

This program can be run inside Coq, using rational numbers for the number system. To compute vertical cells, most computations are ring operations, except when we want to know the vertical projection of a point on an edge, where we use a division operation, so the exact division operation provided for rational numbers suffices.

When defining Bézier curves, we only need to construct midpoints of existing points to define the new control points for the successive Bézier curves that we consider. Working with rational numbers again suffices to have exact computation.

The output can be displayed by simply traversing the output data structure and generating simple postscript commands for each of the straight line or Bézier curve segment. Our program uses quadratic Bézier curves, while postscript supports cubic Bézier curves, but converting from one to the other is easy. In the end, Postscript uses limited precision numbers instead of rational numbers. Approximations are needed, but these approximations are only useful for display purposes.

The Coq system also provides an extraction facility, so the algorithmic part of our program can be translated to a conventional programming language. We used this to produce OCaml code, which was later translated to Javascript, and we developed a hosting web page for this Javascript code, visible at the following link:

https://stamp.gitlabpages.inria.fr/trajectories.html.

The minimal user interface provided on that page gives access to a window with a board of 40 by 40 square tiles and two modes of operations, in the first mode of operation the user can add and remove obstacles, in the second mode of operation the user can add the source and target points. Two modes of display are also also provided, one where the square tiles are visible (this is more practical for reproducing experiments), and one where the cells obtained after vertical cell decomposition are visible.

The following illustration shows the result of running this program for a situation close to that used in our running example, where the display mode shows the cells.

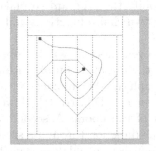

8   Future Work

We are planning several stages of future work concerning this program. In the short term, we wish to have more proofs concerning the formal model. At the time of writing these lines, the formal proof of the vertical cell decomposition algorithm is not complete. For instance, we have already proved that the vertical cell decomposition algorithm produces cells whose interior is safe, free of obstacles, but only for the case where events have different horizontal coordinates. Some of the invariants for the case where events are vertically aligned have already been proved, but some invariants are still work in progress.

For the other parts of the program, this article sketches the formal proof that we intend to develop, relying on already formalized knowledge that is scattered in previous work. We shall describe some of this existing knowledge in the next section.

Other future work relates to improvement of the program. The current algorithm finds a safe solution and we work to equip it with a proof of safeness, but the solution is not very efficient. An obvious reason for inefficiency is that the broken line path is only optimized with respect to the number of cells being traversed, irrespective of the length of the trajectory incurred. It will happen that a long trajectory will be preferred to a short one because the long trajectory traverses less cells. This raises the question of whether the program could be changed to obtain *optimal* trajectories for some criterion, like length or speed of traversal under some constraints like limited acceleration or curvature radius or both.

We do not intend to address this problem of optimality, because we believe it is not practically relevant, and for the objective of applying formal methods to problems originated in real life, it is even counterproductive. Finding a trajectory is already difficult, and as we see in this program four phases are composed, where several phases make arbitrary choices that may impact whether the found trajectory is optimal or not. For instance, the first two phases implement vertical cell decomposition. There is an arbitrary choice of a vertical direction here. If we were using a different sweeping direction, a different graph would be obtained, yielding different broken line trajectory sketches. Other examples of arbitrary choices concern the points in doors used for the broken line trajectory sketch and the points used for repairing faulty Bézier curves. Rather than studying optimality, we intend to modify the program to make it easily configurable so that users can provide their own choice functions and inherit correctness from a generic proof.

Still, there are a few improvements that will be considered for implementation and proof. First, a study of the literature on algorithms shows that the current best first search in the cell graph could be replaced by a weighted best first search algorithm, like the one proposed by Dijkstra [5]. However, it is not obvious how to assign a weight to each of the doors. A second idea would be to replace trajectories going in the same direction through a sequence of doors in a broken line with a straight line, as soon as the final door is visible from the first one, thus avoiding a wavy effect that is already visible in our working example. However, optimally choosing which sequences of broken line patterns should be replaced by straight lines is a difficult problem.

Another improvement that we intend to address is the question of progressive curvature. Railway and road designers know that curves should be designed in a way that avoids lateral shocks for the payload of the vehicles. To avoid such shocks, it is customary to use clothoids, curves where the radius of curvature is inversely proportional to the distance traveled along the curve. On such curves, the acceleration felt by the payload only evolves in a continuous manner. Replacing Bézier curves with clothoids is a practically relevant objective for applications in industrial contexts, but the mathematics behind this are more complex, because computing collisions between a clothoid and an obstacle is less elementary.

Another direction of future work is to connect this Coq program to an implementation in a conventional programming language such as C, and its use as embedded code in a physical toy robot. One of the questions raised is whether we can prove anything about the computations once they are done with floating point numbers, implying approximations at different stages of the algorithm.

9   Related Work

In this section, we mostly concentrate on related work that entails formal verification.

The computation of subareas of the plane is already studied in work on convex hulls [10] and triangulations [2,6]. The algorithm we use to decompose the workspace into cells relies extensively on the orientation predicate taken from the work of Knuth [8].

We argued in this article that the program we designed only uses rational computations. Up to the problem of reasoning about collisions between the broken line trajectories and the obstacles, rational numbers are sufficient. However, smooth curves based on Bézier curve pose a new problem. The coordinates of potential collision points are real algebraic numbers. So while the program description can rely on rational numbers, the proofs concerning collision avoidance require working in a real closed field. Work on the decidability of equality between algebraic numbers is related to the work done in CoRN on the fundamental theorem of Algebra [7]. The Mathematical Components library also contains descriptions of procedures to decide problems with polynomials [4].

Bézier curves can actually be viewed as parametric curves, where the polynomials giving each coordinate are easily described from the control points using Bernstein polynomials as a basis of the vector space of polynomials, so this work is also related to previous work on Bernstein polynomials [1,3,12].

A piece of work whose motivation has many similarities with ours is an experiment combining a motion planner programmed in Matlab with a formal verification performed by Isabelle [11]. This experiments describes a system that is more powerful than ours, as it relies on analysis, especially tools to reasons about solutions to differential equations, to check solutions that have been elaborated by a Matlab program. However, our objective is to describe a program that is eventually independent from the interactive theorem prover being used (in our case Coq, in their case Isabelle). To our understanding, their approach is to have Matlab and Isabelle being used at runtime to produce trajectories that are formally verified. On the other hand, our work contains a formal verification of the computations that happen at the time of elaborating the trajectories, while their work receives the results of these computations as if they were provided by an oracle.

10   Conclusion

It is unfortunate that this presentation is more a description of work in progress than a statement of achievement. At the time of writing these lines, the formal verification is focusing on proving the correctness of the vertical cell decomposition algorithm. The naive solution seems rather easy to verify, but the constraint of obtaining nonempty cells adds a lot of complexity in the program, as we need to pay attention to the case where two successive events are vertically aligned.

The full program can be seen on the public repository

https://github.com/math-comp/trajectories.

The incomplete proofs of the first two phases can be seen on the public repository

https://github.com/ybertot/VerticalCells.

In our section about future work, we showed a few directions that could be addressed as a continuation of this work. This is a project of formal verification and one could be tempted to prove formally everything that could be said about the program, like smoothness and even optimality of the found trajectories.

However, this case study is also an effort to make formal verification penetrate fields where this technique is seldom used. From this perspective, one should avoid *proving too much*. If formal methods are to be adopted widely, we need to show that there is a gain in productivity when using these techniques. This gain of productivity cannot be achieved if we devote too much time proving properties that are not really required by the end user. If absence of collision is required and smoothness is desired, we should prove absence of collision for an algorithm that provides smooth trajectories, but not necessarily prove smoothness, because doing the latter increases the cost of the project, and this increased cost may shed a bad light on the whole technique of formal verification.

In this respect, proving that our program produces trajectories which are optimal in any well defined sense is both extremely difficult and counterproductive for our advocacy effort.

Acknowledgments. The initial work on the vertical cell decomposition algorithms was done by Thomas Portet. Studies of potential collisions between Bézier curves and straight line segments were done by Quentin Vermande. Laurent Théry added the possibility to visualize the results on a web page.

References

1. Basu, S., Pollack, R., Roy M.F.: Algorithms in Real Algebraic Geometry, volume 10 of Algorithms and Computation in Mathematics, 2nd edn. Springer, Heidelberg (2006). https://doi.org/10.1007/3-540-33099-2
2. Bertot, Y.: Formal Verification of a geometry algorithm: a quest for abstract views and symmetry in coq proofs. In: Fischer, B., Uustalu, T. (eds.) ICTAC 2018, vol. 11187, pp. 3–10. Springer, Heidelberg (2018). https://doi.org/10.1007/978-3-030-02508-3_1
3. Bertot, Y., Guilhot, F., Mahboubi, A.: A formal study of Bernstein coefficients and polynomials. Math. Struct. Comput. Sci. **21**(04), 731–761 (2011)
4. Cohen, C., Mahboubi, A.: Formal proofs in real algebraic geometry: from ordered fields to quantifier elimination. Logical Methods Comput. Sci. **8**(102), 1–40 (2012)
5. Dijkstra, E.W.: A note on two problems in connexion with graphs. Numer. Math. **1**(1), 269–271 (1959)
6. Dufourd, J.F., Bertot, Y.: Formal study of plane Delaunay triangulation. In: Paulson, L., Kaufmann, M. (eds.) ITP 2010, vol. 6172, pp. 211–226. Springer, Heidelberg (2010). https://doi.org/10.1007/978-3-642-14052-5_16
7. Geuvers, H., Wiedijk, F., Zwanenburg, J.: A constructive proof of the fundamental theorem of algebra without using the rationals. In: Callaghan, P., Luo, Z., McKinna, J., Pollack, R. (eds.) TYPES 2000. LNCS, vol. 2277, pp. 96–111. Springer, Heidelberg (2000). https://doi.org/10.1007/3-540-45842-5_7
8. Knuth, D.: Axioms and Hulls. Number 606 in Lecture Notes in Computer Science. Springer-Verlag, Heidelberg (1991). DOI: https://doi.org/10.1007/3-540-55611-7
9. Latombe, J.-C.: Robot Motion Planning. Kluwer Academic Publishers, Norwell (1991)

10. Pichardie, D., Bertot, Y.: Formalizing convex hull algorithms. In: Boulton, R.J.,
    Jackson, P.B. (eds.) TPHOLs 2001. LNCS, vol. 2152, pp. 346–361. Springer, Hei-
    delberg (2001). https://doi.org/10.1007/3-540-44755-5_24
11. Rizaldi, A., Immler, F., Schürmann, B., Althoff, M.: A Formally Verified Motion
    Planner for Autonomous Vehicles. In: Lahiri, S.K., Wang, C. (eds.) ATVA 2018.
    LNCS, vol. 11138, pp. 75–90. Springer, Heidelberg (2018). https://doi.org/10.1007/
    978-3-030-01090-4_5
12. Zsidó, J.: Theorem of Three Circles in Coq. J. Autom. Reason. 53(2), 105–127
    (2014)

Learning Guided Automated Reasoning: A Brief Survey

Lasse Blaauwbroek[1,2], David M. Cerna[3], Thibault Gauthier[1], Jan Jakubův[1,4], Cezary Kaliszyk[4], Martin Suda[1], and Josef Urban[1(✉)]

[1] Czech Technical University in Prague, Prague, Czechia
Josef.Urban@gmail.com
[2] Radboud University Nijmegen, Nijmegen, Netherlands
[3] Czech Academy of Sciences Institute of Computer Science, Prague, Czechia
[4] University of Innsbruck, Innsbruck, Austria

Abstract. Automated theorem provers and formal proof assistants are general reasoning systems that are in theory capable of proving arbitrarily hard theorems, thus solving arbitrary problems reducible to mathematics and logical reasoning. In practice, such systems however face large combinatorial explosion, and therefore include many heuristics and choice points that considerably influence their performance. This is an opportunity for trained machine learning predictors, which can guide the work of such reasoning systems. Conversely, deductive search supported by the notion of logically valid proof allows one to train machine learning systems on large reasoning corpora. Such bodies of proof are usually correct by construction and when combined with more and more precise trained guidance they can be boostrapped into very large corpora, with increasingly long reasoning chains and possibly novel proof ideas.

In this paper we provide an overview of several automated reasoning and theorem proving domains and the learning and AI methods that have been so far developed for them. These include premise selection, proof guidance in several settings, AI systems and feedback loops iterating between reasoning and learning, and symbolic classification problems.

1 Introduction

No one shall drive us from the semantic AI paradise of computer understandable math and science!

– AGI'18 [154]

Automated Reasoning (AR) [126] and Automated Theorem Proving (ATP) systems are general AI systems that are in principle capable of solving arbitrary mathematical and reasoning problems. Their theoretical *completeness* means that *any* solvable problem, regardless of its difficulty, will be eventually solved.

In practice, today's AR and ATP systems however soon encounter combinatorial explosion, typically preventing them from solving hard open problems in a reasonable time. Regardless of which proof calculi they implement, they have

V. Capretta et al. (Eds.): *Logics and Type Systems in Theory and Practice*, LNCS 14560, pp. 54–83, 2024.
https://doi.org/10.1007/978-3-031-61716-4_4

to make many decisions concerning which theories and lemmas to use, which inference steps or tactics to choose, which instantiations to apply, how to divide the problems, split the clauses, propose useful lemmas, etc. These choices do not typically influence the theoretical correctness and completeness of the proof mechanisms but rather the systems' efficiency and practical performance.

AR/ATP practitioners have often *manually designed and optimized* a spectrum of heuristics, efficiency and calculus improvements based on their practical and theoretical insights. Such manual designs can sometimes be very successful, as witnessed, e.g., by the large performance boosts brought by using the right term orderings [61], literal selection function [53] and age-weight ratios [98] in ATPs, sophisticated indexing methods for ATPs [121] and SAT solvers [102], as well as by human-designed improvements of the underlying calculi such as CDCL [69,143] in SAT solving, ordering-based constraints [105], and AVATAR-style splitting [124,159] in saturation-based ATP.

On the other hand, general mathematics is undecidable and arbitrarily complicated, and it seems increasingly hard to manually design more complex heuristics for more complex domains and problems. At the same time, *automated design and optimization* of parameters, heuristics, functions and algorithms has been a major topic in AI since its beginnings. Especially in the last decades the field of machine learning (ML) has produced a number of interesting *data-driven* [142] methods that can be used in AR and ATP. Perhaps the most interesting area of such research is how to combine and interleave the AR and ML methods, creating feedback loops and meta-systems that can continuously improve their skills and keep finding—for long time—harder and harder proofs and explanations.

In this survey, we cover the development of such methods starting from the early AI/TP systems and tasks such as high-level knowledge selection, to the today's spectrum of architectures and tasks that include guidance at various levels, a variety of learning-based methods with various speed/accuracy trade-offs, and a variety of combinations of the ML and AR methods. Automated Reasoning is however a large field and this brief survey does not make claims to be exhaustive. We mainly focus here on the fields of Automated and Interactive Theorem Proving. For some related AR fields similar surveys and overviews have been written recently. In particular, we recommend the recent exhaustive overview [57] discussing ML methods in the context of SAT and QSAT solving.

2   Early History

According to Davis [26], in the beginning of AR and ATP, two research directions emerged: (i) *heuristic*/AI emulation of human thought processes, such as Newell's and Simon's Logic Theorist [104] and later Bledsoe's work [14], and (ii) design of crisp algorithms based on logic transformations and calculi, such as the early proof procedures by Davis and Putnam [27], Gilmore [46] and Robinson [125]. The latter (*logicist*) approach largely prevailed in ATP in the first decades, thanks to major advances such as resolution, DPLL and paramodulation. Interesting examples of the heuristic/AI systems at that time include Lenat's Automated Mathematician (AM) [93], Langley's Bacon [92] and Colton's HR [22], which propose concepts and conjectures, thus also qualifying as symbolic ML.

First interesting combinations of state-of-the-art ATPs with ML methods (both statistical and symbolic) were designed in the 90 s by the Munich AR group [30,31]. The methods and systems included, for example, the invention of *tree (recursive) neural networks (TNNs)* for classifying logical expressions by Goller and Kuchler [50], and the development of the E prover [137] by Schulz, which allowed proof guidance by symbolic patterns abstracted from related proof searches [135]. Related to that was the *hints* proof guidance in *Otter* developed for attacking open conjectures by Veroff [158], again based on symbolic abstraction (subsumption) of lemmas in related proofs.

At the same time, large ITP (*interactive theorem proving*) libraries started to appear, with the *Mizar* project [6] producing over 700 formal mathematical articles by 2001. Such large libraries have since become a natural target for combining ML and AR. Urban's 1998 MSc thesis [148] was perhaps the first attempt to learn symbolic heuristics over such formal libraries with the use of *inductive logic programming* (ILP – a symbolic ML approach). Already before that, the ILF project [25] started to work on translations between Mizar and ATPs. This was continued by the MPTP project [149,150] which in 2003 released a dataset of about 30000 Mizar ATP problems, and reported about the first large ATP and ML experiments over it. This included training a *naive Bayes* (Sect. 4.2) predictor over the Mizar library (Mizar Proof Advisor) to select suitable library facts (premises) for proving the next Mizar problems. In 2004, the Isabelle/Sledgehammer was developed [99], including the first heuristic (non-learning) premise selection methods (MePo) [100]. Such *hammer* systems connecting ITPs with ATPs have become an AR topic of its own, see [12].

Since efficient premise selection is a critical task in hammers, a lot of the initial AI/TP work focused on it. The 2007 MPTP Challenge[1] followed by the 2008 CASC LTB (large theory batch) competition introduced benchmarks of large-theory problems and suitable settings for AI/TP system training and evaluation. This quickly led to fast non-learning methods such as SInE [54] integrated in ATPs, as well as larger meta-systems such as MaLARea [152,156] that interleave proving with learning of premise selection.

The early non-learning systems typically focus on symbols and symbolic features in the formulas, with MePo and SInE using metrics such as symbolic overlap (Jaccard index) and symbol rarity, starting with the conjecture symbols and recursively adding the most related formulas until a certain size of the set of premises is reached. The Mizar Proof Advisor and *MaLARea* instead initially trained naive Bayes to associate the conjecture symbols (features) with the premise names, later adding more complicated syntactic features such as term walks and also semantic model-based features (Sect. 3.3), obtained by evaluating both conjectures and premises in a growing set of finite models.

These first approaches turned out quite successful, with MePo, SInE and MaLARea considerably increasing the performance of the underlying ATPs in

---

[1] https://www.tptp.org/MPTPChallenge/.

large theories. This has (re-)opened several research topics, such as (i) how to suitably characterize mathematical formulas and objects, (ii) what are the suitable ML methods, (iii) on what level should the ML guidance be applied, and (iv) how to construct larger feedback loops and meta-systems combining ML and AR. In the next sections we discuss some of these topics.

3   Characterization of Mathematical Knowledge

How knowledge is represented is often essential to understanding and gleaning deeper insights about the subject encompassing it. To take a prominent example, consider Fermat's Last Theorem. The theorem statement concerns what most would classify as number theory, yet the proof lives in the world of *elliptic curves*. There may be an elementary number-theoretic proof of the theorem. However, given the state of mathematics at the time, the shortest path was through a, at least to the mathematics dilettante, seemingly unrelated area. When developing methods for classifying mathematical expressions, analogously, proper representation is essential to extracting the necessary semantic notions.

3.1   Syntactic Features

As mentioned above, the work on premise selection started with syntactic characterizations of conjectures and premises. The early methods extend the extraction of symbols (already mentioned in the discussion of SInE and MePo in the previous section) by using subterms as additional features, applying various normalizations to them to increase the feature matching between the formulas, and using syntactic representation of types and their connections. The aim of such early investigations was to provide HOL Light with automation (a hammer) for proof development in the Flyspeck project. For example, [72] included features based on the subterms normalized by replacing variables by their types, their de Bruijn numbers, or merging all variables. As an example, the HOL theorem DISCRETE_IMP_CLOSED with the HOL Light statement:

$\forall$s:real^N$\rightarrow$bool e.
    &0 < e $\land$ ($\forall$x y. x IN s $\land$ y IN s $\land$ norm(y − x) < e $\implies$ y = x)
    $\implies$ closed s

is characterized by the following set of strings that represent unique features.

"real", "num", "fun", "cart", "bool", "vector_sub", "vector_norm",
"real_of_num", "real_lt", "closed", "_0", "NUMERAL", "IN", "=", "&0",
"&0 < Areal", "0", "Areal", "Areal^A", "Areal^A - Areal^A",
"Areal^A IN Areal^A->bool", " Areal^A->bool", "_0", "closed Areal^A->bool",
"norm (Areal^A - Areal^A)", "norm (Areal^A - Areal^A) < Areal"

The above approach resulted in many unique features, for which various feature weighting schemes were explored for efficient use with the premise selection predictors. The most efficient schemes were based on the linguistic TF-IDF [71], raising considerably the performance of the best k-nearest neighbor predictors. In TF-IDF, a term t, present in a collection of documents D, is weighted by the logarithm of the inverse of the term's frequency within D, that is:

$$\text{IDF}(t, D) = \log \frac{|D|}{|\{d \in D : t \in d\}|}$$

A further improvement to the syntactic characterization of terms was introduced in the work on machine learning for Sledgehammer [86], where walks through term graphs were considered. Adding such features again allows the abstraction of the global term structure and better sharing of such automatically created concepts/features between the statements.

3.2   ENIGMA Syntactic Features

Syntactic features are also heavily used by ENIGMA systems, where clauses are represented by finite numeric *feature vectors*. ENIGMA (*Efficient Learning-Based Inference Guiding Machine*) is a state-of-the-art machine learning guidance system for the ATP E [138]. In the first ENIGMA [62], the feature vector is constructed by traversing the clause syntax tree and collecting all top-down oriented symbol paths of length 3. For example, given the unit clause $plus(X, nul) = X$, we obtain triples $(\oplus, =, plus)$, $(\oplus, =, X)$, $(=, plus, X)$, and $(=, plus, nul)$, where $\oplus$ signifies the root node of the syntax tree of positive literals. Additionally, to abstract from variable names and to deal with possible collisions of Skolem symbols, all variables are replaced by a special name $\odot$ and all Skolem symbols by $\circledast$. After this renaming, the triples contain only the symbols from the problem signature Σ and 4 special symbols $\{\oplus, \ominus, \odot, \circledast\}$, where $\ominus$ is used as the root node of negative literals. This allows exhaustive enumeration of all possible triples, assigning each triple a unique number smaller than $(|\Sigma| + 4)^3$. This number is used as an index in the feature vector, and the vector value specifies the number of occurrences of the corresponding triple in the clause.

While the first version of ENIGMA yielded encouraging results, it was not yet ready to scale to benchmarks with larger signatures. This led to the second ENIGMA [63] with enhanced feature vectors. Instead of an exhaustive enumeration of all possible symbol triples, only the triples appearing in the training data were enumerated. This significantly reduced the vector length as many triples do not appear in the provided training data, and many can not appear at all, for example, $(=, =, =)$. This enumeration must be stored together with the trained model. The second ENIGMA additionally introduced the following additional clause features.

Count Features extend the feature vector with the clause length, and the counts of positive/negative literals. Moreover, for each symbol f we added the number of occurrences of f in positive/negative literals, together with the maximal depth of f in positive/negative literals. Count features allowed us to drop the clause length multiplier γ and to use the model prediction directly as the clause weight.

Horizontal Features provide a more accurate representation of clauses by feature vectors. For every term $f(t_1, \ldots, t_n)$, a new feature $f(s_1, \ldots, s_n)$ was introduced, where s_i is the top-level symbol of t_i. The number of occurrences of each horizontal feature is stored in the feature vector. Again, only the horizontal features that appear in the training data are considered. For example, the unit clause $plus(X, nul) = X$, yields horizontal features $= (plus, \circledast)$ and $plus(\circledast, nul)$, each occurring once in the clause.

Conjecture Features embed the conjecture to be proved in the feature vector. The first ENIGMA simply recommended the same clauses independently of the conjecture being proved. In the second ENIGMA, conjecture features were appended to the vector, making the vector twice the size. Thusly, the second ENIGMA was able to provide goal specific predictions, which was essential for experiments on Mizar problems which are much more heterogeneous than AIM benchmarks.

Another important step towards large data support in ENIGMA, was the implementation of *feature hashing*. This significantly reduced the feature vector size. ENIGMA uses a generic purpose string hashing function *sdbm*.[2] Each feature is represented by a unique string identifier, for example, $(=, plus, X)$ becomes "|=|plus|*|", and $plus(\circledast, nul)$ becomes ".plus.*.nul.". This string is passed through the hashing function, computed with a fixed-length data type representation (64 bit unsigned). The string hash modulo the selected *hash base* is used as the feature index. The hash base is intended to directly limit the vector size, at the price of occasional feature collisions.

In order to abstract from specific symbol names also in the context of features, the next ENIGMA [60] introduced a very simple method of *symbol anonymization*. During the extraction of clause features, all symbol names are replaced by symbol arities, keeping only the information whether the symbol is a function or a predicate. For example, the binary function symbol *plus* becomes simply *f2*, the ternary predicate symbol *ite* becomes *p3*, and so on. In this way, a decision tree classifier does not depend on symbol names, for the price of symbol collisions. While this rather trivial symbol anonymization was initially implemented mainly as a baseline for the graph neural networks, it performed surprisingly well in practice and it became a useful ENIGMA option. The symbol collisions reduce the size of training data, which is a favorable side effect of name anonymization.

3.3   More Semantic Features

The syntactic structure of mathematical statements does not always easily capture their intended meaning. For examples, the syntactic features of two complex

---

[2] Given the (code of the) i-th character s_i: $h_i = s_i + (h_{i-1} \ll 6) + (h_{i-1} \ll 16) - h_{i-1}$ with $h_0 = 0$.

formulas ϕ and $\neg\phi$ differ only very little, making them very close in various feature metrics. On the other hand, in the semantic Tarski-Lindenbaum algebra, these formulas are each other's complements, i.e., in some sense they are as distant as possible. To accommodate this, mechanisms for extracting features that are based on the meaning of formulae have been proposed.

The most common way of specifying the semantics of logical expressions in first-order logic is by considering which *models satisfy the expressions*. However, a formula may have many satisfying models. Thus, only *interesting* models that distinguish between formulas, are useful for characterization purposes. MaLARea-SG1 [156] used finite model finders to continuously search for counter-models for problems with too few recommended premises. The notion of model's interestingness is thus defined dynamically, based on the current state of the learned knowledge (the trained premise selector), with which it co-evolves. This led to considerable improvement of MaLARea's performance.

Another approach to obtain more semantic features considers the *unifiability of two formulas*. For most first-order automated reasoning calculi, the unifiability of formulas correlates with their use as principal formulas of an inference step. This means that first-order indexing structures [78] can be used as another source of more semantic features. The nodes of indexing structures, such as a substitution tree or discrimination tree, correspond to sets of unifiable formulas present in the given reasoning problem, while paths correspond to similar term structures. Thus, features corresponding to these nodes can characterize whether the given formulas can be combined for inferencing within a given calculus.

Latent semantic analysis (LSA) [29] is a method that creates low-rank vectors (*embeddings*) characterizing terms and documents based on the terms' co-occurrence in the documents. LSA-based features were used with some success in premise selection systems [77] and hammers [75]. LSA-based embeddings precede related more recent methods, which use the same idea of characterizing words by their context, such as Word2Vec [101] and neural embeddings.

3.4   Characterization Using Neural Networks

Already the early tree neural networks by Goller and Kuchler attempt to learn the representation of the symbols as neural sub-networks. More recently, starting with Word2Vec (a shallow neural network), a number of neural approaches have been experimented with for obtaining useful sentence and term embeddings.

For example, encoder-decoder neural frameworks attempt to summarize an input text (e.g., in English) in a vector that can be then decoded into suitable output (e.g., French text with corresponding meaning). A recent encoder-decoder approach was developed by Sutton et al. [3]. It focused on developing an embedding capturing semantic equivalence, i.e., formulas that are negations of each other are adequately distinguished. This research direction is connected to the work outlined in Sect. 8, where approximate reasoning and semantically rich encoding are used to build systems for various synthesis problems. Purgał [118] later developed an autoencoder for predicate logic formulas by training a network to decode a given formula's top symbol and its children. Another example is [165], which builds a representation of tactics by their meanings, i.e., how the tactic transforms a given state into several subsequent states.

An alternative to vectorization of symbolic expression as simple sequences of symbols is to use neural architectures that capture the structure of the expressions. This includes the pioneering work on Tree NNs by [51], capturing the recursive nature and tree-like structure of logical expressions. Related investigation is done in [35] for recognizing propositional logical entailment through evaluation of formulas in 'possible worlds', which is also related to the earlier model-based features in MaLARea. A related investigation [18] models expressions using a similar network architecture but instead aims to recognize properties such as *validity* through a top-down evaluation of the expression. The model thus tries to approximate decompositional reasoning.

An illustrative example of Tree NNs modeling symbolic expressions for the purpose of guiding an automated reasoner is *ENIGMA-NG* [20]. The authors provide a vector embedding that associates each predicate (function) symbol with a learned function $R^n \times \cdots \times R^n \to R^n$ where the number of arguments matches the symbol's arity. Thus, the term structure is encoded by the composition of these functions. Certain expressions whose semantic content cannot be directly extracted from the structure of the expression are grouped together. For example, all variables share the same encoding function. Clauses, rather than being considered as a composition of several instances of a binary *or* function, are handled by a separate RNN model capturing their set-like nature.

While Tree NNs capture the structure of symbolic expressions, a few issues remain. For example, the encoding used by *ENIGMA-NG* [20] requires the construction of a learnable function for each *symbol*, while all variables use a single learnable function. Graph neural networks (GNNs) [131] can provide an architecture able to abstract away the names used for the representation of symbolic expressions and, to some extent, distinguish variables and their occurrences. Early uses of GNNs for premise selection, such as FORMULANET [160], did not improve these deficiencies but provided direction towards a more semantic-preserving architecture. Olsak [106] then proposed a GNN-based name-invariant embedding using a hypergraph of clauses and terms, where none of the names appear and a symbol's meaning can only be inferred from the subgraph of its properties. While some loss of structure may still occur, for example, $f(t_1, t_2, t_1)$ and $f(t_2, t_1, t_2)$ are identically encoded, the embedding is a vast improvement over previous approaches, naturally treating also clauses as sets.

Such embeddings allow the neural architecture to draw analogies between different mathematical domains. Many algebraic operators are associative, commutative, and/or distributive, yet they have different names. The framework presented in [106] would recognize the local graph structure as analogous regardless of the different names used. While such neural architectures provide better generalization and improved cross-domain predictions, compared to simpler methods such as decision trees their predictions may be too confident. This problem was observed when integrating the above-mentioned GNN architecture into the inference selection mechanism of *plcop* [167]. The authors introduced *entropy regularization* to normalize the model's confidence and improve accuracy.

4   Premise Selection

The success of modern ATPs is partially due to their ability to select a small number of facts that are relevant to the conjecture. The act of selecting these relevant facts is referred to as *premise selection* [1], which can also be seen as a step in a more general abstraction-refinement framework [95]. Without good premise selection, ATPs can be easily overwhelmed by a number of possible deductions, which holds even for relatively simple conjectures. Non-learning ATP approaches include heuristics such as MePo and SInE, mentioned in Sect. 2. In this section, we cover the typically more precise learning approaches.

4.1   k-Nearest Neighbors (k-NN)

This approach to selection requires a measure of distance, more appropriately referred to as a *similarity relation*, between two facts. This similarity relation, in simple cases, is defined by the distance between two facts computed from a set of binary features. To increase precision, weights and a scaling factor are added to the features. Thus, the similarity relation between two facts is defined as follows:

$$s(a,b) = \sum_{f \in F(a) \cap F(b)} w(f)^{\tau_1}$$

where F is the feature vector, w is the weight vector and τ_1 is the scaling factor.

We then realize that if a proof relies on more dependencies, each of them is less valuable, so we divide by the number of dependencies. We additionally add a factor for the dependencies themselves. Given N the set of the k nearest neighbours, the relevance of fact a for goal g is:

$$\left(\tau_2 \sum_{b \in N | a \in D(b)} \frac{s(b,g)}{|D(b)|} \right) + \begin{cases} s(a,g) & \text{if } a \in N \\ 0 & \text{otherwise} \end{cases}$$

This approach was later extended to adaptive k, i.e., considering more or less neighbours depending on the requested number of best premises [75].

4.2   Naive Bayes

Given a fact a and a goal to prove g, we try to estimate the probability it is useful based on the features $f_1, ..., f_n$ of the goal g:

$$P(a \text{ is relevant for proving } g)$$
$$= \qquad P(a \text{ is relevant} \mid g' \text{ s features})$$
$$= \qquad P(a \text{ is relevant} \mid f_1, \dots, f_n)$$
$$\propto \qquad P(a \text{ is relevant}) \Pi_{i=1}^{n} P(f_i \mid a \text{ is relevant})$$
$$\propto \#a \text{ is a proof dependency} \cdot \Pi_{i=1}^{n} \frac{\#f_i \text{ appears when } a \text{ is a proof dependency}}{\#a \text{ is a proof dependency}}$$

This approach can be improved by considering features not present in the goal when a was used. An issue with considering the negative case is that there are too many possible features to take into account. So, instead of looking at all features not present when a was used, we only consider the so called *extended features* of a fact, namely the features that do not appear and are related to those that do appear [86]. This allows considering the probabilities of features not appearing in the goal while a dependency was used. These probabilities can be computed/estimated efficiently. Early versions of MaLARea used the implementation in the SNoW [15] toolkit. Kaliszyk later [71,72] implemented a custom naive Bayes for Flyspeck and other experiments.

4.3 Decision Trees

A *decision tree* is a binary tree with nodes labeled by conditions on the values of the feature vectors. Such trees and their *ensembles* are today among the strongest ML methods, which can also be very efficient. Initially, Färber [36] improved on k-NN approaches [71] with *random forests* (ensembles of decision trees) that used k-NN as a secondary classifier. With further modifications, random forests can also be applied directly as, e.g., in [112] where they are used for premise selection in Lean. *Gradient-boosted trees*, as implemented by XGBoost [17] and Light-GBM [81] are useful both for premise selection and for efficient ATP guidance as discussed in the next section. Since they work in a binary (positive/negative) setting, they require (pseudo-)negative examples for training. Piotrowski's *ATP-Boost* [113] defined an infinite MaLARea-style loop that interleaves their training with proving, producing increasingly better positive and negative data for the training. Quite surprisingly, *hashing* the large number (over millions) of sparse symbolic features (such as term walks) into much smaller space (e.g. 32000) allowed efficient training of these toolkits over very large libraries such as full MML, with practically no performance penalties [20,65].

4.4 Neural Methods

The first analysis of the performance of deep neural networks for premise selection was done by the *DeepMath* [2] project. Despite requiring significantly more resources, there was only limited improvement over the simpler methods. However, the work also employed the above mentioned *binary setting*, where conjectures and the potential premises were evaluated together. This allows to meaningfully evaluate premises that have not been seen yet. Methods such as k-NN and naive Bayes typically do not allow that without further data-augmentation tricks.[3] As logical formulas naturally have a tree structure, graph neural networks were soon proposed for premise selection [160] (see also Sect. 3.4). Other alternatives such as graph sequence models [56], directed graph networks [122],

---

[3] Such as adding for each premise as a training example its provability by itself, which has indeed been used from the beginnings of ML-based premise selection [150,151]. Data augmentation in general is another very interesting AI/TP topic, see e.g. [74].

recurrent neural networks [115] and transformers [155] have been experimented with, allowing to take into account also dependencies among the premises [115].

The signature independent GNN described in [106] seems to be the strongest method today, based on a recent large evaluation of many methods over the Mizar dataset [59]. The study also demonstrates the role of *ensembles and portfolios* of different premise selection methods. An interesting phenomenon observed in [155] is that some of the neural methods provide a smooth transition between premise selection and *conjecturing*, i.e., proposing so far unproved "premises" (intermediate conjectures) that split the problem into two [42]. Such neural conjecturing methods are further explored e.g. in [45, 68, 110, 114, 120].

5   Guidance of Saturation-Based ATPs

Several decisions that ATPs need to regularly perform are very similar to premise selection. The selection of the next clause in saturation-based provers or of the next extension step in a connection tableaux calculus both correspond to the selection of relevant premises for a given conjecture. For this reason, many ATP guidance techniques are motivated by premise selection, with the additional caveat that efficiency needs to be much more taken into account.

Arguably the most important heuristic choice point in saturation-based theorem proving is *clause selection*. It is the procedure for deciding, at each iteration of the main proving loop, which will be the next clause to activate and thus participate in generating inferences. A perfect clause selection—obviously impossible to attain in practice—would mean selecting just the clauses of the yet-to-be-discovered proof and would thus completely eliminate search from the proving process. Although state-of-the-art clause selection heuristics are far from this goal, experiments show that even small improvements in its quality can have a huge impact on prover performance [139]. This makes clause selection the natural main target for machine-learned prover guidance.

The central idea behind improving clause selection by ML, which goes back at least to the early work of Schulz [31, 134, 136], is to learn from successes. One trains a binary classifier for recognizing as positive those clauses that appeared in previously discovered proofs and as negative the remaining selected ones. In subsequent runs, clauses classified positively are prioritized for selection.

Particular systems mainly differ in 1) which base prover they attempt to enhance, 2) how they represent clauses for the just described supervised learning task, and 3) which ML technique they use. When aiming to improve a prover in the most realistic, i.e., time constrained, setting, there are intricate trade-offs to be made between faithfulness of the chosen representation, capacity of the trained model and the speed in which the advice can be learned and retrieved. State-of-the-art provers are tightly optimized programs and a higher-quality advice may not lead to the best results if it takes too long to obtain.[4]

For example, the ENIGMA (*Efficient Learning-Based Inference Guiding Machine*) system, extending the ATP E [138], started out with easy-to-compute syntactic clause features (e.g., term-walks, see Sect. 3.2) and a simple but very *efficient* linear classifier [62,63]. At the same time, Loos et al. [94], first experimenting with integrating state-of-the-art neural networks with E, discovered their models to be too slow to simply replace the traditional clause selection mechanism. A similar phenomenon appears in ML-based guidance of tactical ITPs (Sect. 7), where the relatively simple but fast predictors used in systems like TacticToe [44] (HOL4) and Tactician [11] (Coq) are hard to beat by more complicated but significantly slower neural guidance in a fair comparison [128]. In later versions of ENIGMA, these trade-offs were further explored by experimenting with gradient boosted trees and tree neural networks over term structure [20], and finally with graph neural networks [60]. In the meantime, Deepire [144,145], an extension of the prover Vampire [85] used tree neural networks but over clause derivation structure. In the rest of this section, we compare these and other systems also from other angles as we highlight some notable aspects of this interesting technology.

Integrating the Learned Advice. There are several possible ways in which the trained model M can be used to improve clause selection in a guided prover. Typically, one seeks to first turn the advice into a total order on clauses, e.g., for maintaining a queue data structure for a quick retrieval, and, in a second step, to somehow combine this order with the original clause selection heuristic.

In the most general case of a black-box binary classifier, M only suggests whether a clause is good (1) or bad (0). The first ENIGMA [62] used a linear combination of the M's classification and clause's length to come up with an overall score for sorting clauses. Not fully relying on M makes the advice more robust, in particular, it mitigates the damage from having large erroneously positively classified clauses. Another sensible strategy, adopted, for example, by ENIGMA-NG [20], is to have all positively classified clauses precede the negatively classified ones and to use clause generation time-stamps (sometimes referred to as *clause age*) as a tiebreaker on clauses in each of the two classes. This amounts to saying "prefer clauses suggested by M and among those apply the FIFO rule: old clauses first before younger ones". Finally, even a binary classifier often internally uses a continuous domain of assigned values (often referred to as *logits*) which are only turned into a binary decision by, e.g., getting compared

---

[4] This is not only because search and/or backtracking is an indispensable part of any reasoning which is not purely memorized, but also because self-learning systems that use faster guidance will produce more data to learn from. In today's ML, a slightly weaker learner trained on much more data will often be better than a slightly stronger learner trained on much less data. If a self-learning system uses a slightly weaker but much faster learner/predictor to do many iterations of proving and learning, the self-learning system will in a fixed amount of time produce much more data and its ultimate performance will thus be higher.

against a fixed threshold. This is for instance true for models based on neural networks. ENIGMA anonymous [60], as well as the system by Loos et al. [94], used a simple comparison of clause logits for ordering their neural queue.

A clause queue ordered with the help of M (in one of the ways just described) can of course be used to simply replace any original clause selection heuristic in the prover. However, it is often better to combine the original heuristic and the learned one [20,144]. The simplest way of doing this is to alternate, in some pre-selected ratio, between selections from the various clause queues (one of them being the one based on M). A *layered clause selection* mechanism, which lead to the best performance of Deepire [144], applies the full original heuristic to clauses classified positively by M and alternates this with the original heuristic applied to all clauses. An advantage of this approach is that it allows for a *lazy evaluation* trick [144], under which not every clause available for selection has to immediately get evaluated by the relative expensive model M. Very recently, in the context of the iProver (instantiation-based) system, guidance using only the trained GNN model M managed to outperform the combination of M with the original heuristics [21]. One of the reasons for that is an improved learning from *dynamic* proof data, taking into account more proof states than the traditional ENIGMA-style training that only learns from the final proof states.

Signature (in-)dependence. First-order clauses, as syntactic objects, are built from predicate and function symbols specified by the input problem's *signature*. A representation of clauses for machine learning may consider this signature fixed, which is often natural and efficient, but ultimately ties the learned guidance to that signature. E.g., if we train our guidance on problems coming from set theory, the model will not be applicable to problems from other domains.

The early pattern-based guidance by Schulz for E was already signature independent, as well as, e.g., the early concept-alignment methods designed for transfer of knowledge between ITP libraries by Gauthier and Kaliszyk [41]. The ENIGMA systems [20,62,63] relied initially on the fixed signature approach. ENIGMA Anonymous [60] started to abstract from specific symbol identities and replace them with arity abstractions of the term walks and property invariant neural embeddings [106], opening doors to knowledge transfer. A detailed comparison of these ENIGMA abstraction mechanisms with the earlier ones by Schulz, Gauthier and Kaliszyk is discussed in [60] (Appendix B).

An interesting approach from this perspective, is taken by Deepire [144,145], which uses recursive neural networks for classifying the generated clauses based solely on their derivation history. Thus Deepire does not attempt to read "what a clause says", but only bases its decisions on "where a clause is coming from". This makes it trivially independent on problem signature, however, it still relies on a fixed initial axiom set over which all problems are formulated.

Building in Context. A clause C useful for proving conjecture G_1 can be completely useless for proving a different conjecture G_2. While a great prover performance boost via learning can already be achieved *without* taking the conjecture context into account [145] (and, indeed, the standard clause selection heuristics are of this kind), many systems supply some representation of the conjecture as a secondary input to their model to improve the guidance [20,63,94].

Another, more subtle, kind of context is the information of how far the prover is in completing a particular proof. Intuitively, selecting a certain clause could only make sense if some previous clauses have already been selected. This is explored by ENIGMAWatch [49], where a proof state is approximated by a vector of *completion ratios* of a fixed set of previously discovered proofs. A system called TRAIL goes further and allows every processed clause to influence the score of any unprocessed clause through a multiplicative attention mechanism [24]. This level of generality, however, is computationally quite costly, and TRAIL does not manage to improve over plain E prover [138] under practical time constraints.

Looping and Reinforcement. Once we train a model that successfully improves the performance of the base prover, we can collect even more proofs to train on and further improve our guidance in a subsequent iteration. Such re-learning from new proofs, as introduced by MaLARea [156], constitutes a powerful technique for tapping the full potential of a particular guidance architecture [65,145].

Iterative improvement is also at the core of the reinforcement learning (RL) approach [146], which builds on a different conceptual framework than the one we used so far (e.g., agent, action, state, reward – see also Sect. 6), but in the context of saturation-based provers gave so far rise to systems of comparable design [4,24]. The main reason for this is that even systems based on RL reward (and reinforce the selection of) clauses that appeared in the discovered proofs. The alternative of only rewarding the final proof-finishing step and letting the agent to distinguish the good from the bad through trial-and-error would be prohibitively expensive.

Beyond Clause Selection. Although clause selection has been the main focus of research on this front, there are many other ways in which machine learning can be used to improve saturation-based provers. For example, it is possible to predict good term orderings for the underlying superposition calculus [9] or symbol weights [10]. In the later case we still aim to influence clause selection via clause weight (one of the standard heuristics), but only indirectly, in an initialization phase, which comes before the proof search starts. Another class of approaches, which we only briefly mention here, use ML-style techniques for synthesis of ATP strategies and suggesting good targeted strategies or strategy schedules based on input problem features [55,64,87,133,153].

ML-based approaches have also been used to prevent interactions between unsuitable clauses [48], and in general to design iterative algorithms (*Split & Merge*) that repeatedly split the problems into separate reasoning components whose results are again merged after some time [19]. Such algorithms can also be seen as soft/learnable alternatives to ATP methods such as *splitting* [159], and to manual design of theory procedures and their combinations in SMT.

6   Guidance of Tableaux and Instantiation-Based ATPs

While the saturation-based ATPs are today the strongest, the proofs produced by such systems are often at odds with human intuition. Tableaux-based provers produce proofs closer in argumentation to human reasoning; this is likely due to the case-based style of the typical tableaux proof system. Essentially, such systems are performing a sort of model elimination rather than a search for contradiction. An advantage of such approaches is that the proof state is compactly represented, and it is easier to control the number of possible actions in comparison to saturation provers, where the number of mutually resolvable clauses can grow quickly. This has resulted in a lot of recent research in adding ML-based guidance in particular to the *connection tableaux calculus* [108].

First, the MaLeCoP (Machine Learning Connection Prover) system [157] used an external and relatively slow evaluation method to select the extension steps in the leanCoP ATP [107]. This showed that with good guidance, one can avoid 90% of the inferences. This was made much faster by integrating an efficient sparse naive Bayes classifier in an ML-guided OCAML re-implementation of leanCoP (FEMaLeCoP) [73]. As an alternative to the direct selection of *extension steps*, Monte Carlo simulations can be used to select the promising branches (MonteCoP) [37]. A major progress was obtained by removing the *iterative deepening* used by default in leanCoP, and instead using an Alpha-Zero-like architecture for guiding connection tableaux in *rlCoP* [76]. Reinforcement learning of *policy* (action, i.e. inference, selection) and *value* (state, i.e., partial tableaux, evaluation) and their use for intelligent subtree exploration yielded after several iterations of proving and learning on a training set a system that solves 40% more test problems than the default leanCoP strategy . The method still produces rather short proofs and policy guidance methods were later investigated for proofs with thousands of inference steps [166]. Extensions of connection tableaux, such as lazy paramodulation, can also be directly guided using similar methods [123]. Further improvements were recently achieved by integrating signature-independent graph neural networks (Sect. 3.4), e.g. [106,167].

Instantiation-based and SMT (Satisfiability Modulo Theories) systems such as iProver [83], CVC (CVC4/cvc5) [8] and Z3 [28] combine the use of SAT solvers for checking ground (un)satisfiability with various methods for producing suitable ground instantiations of the first-order problems. This approach goes back to the early days of ATP (Sect. 2) and methods such as Gilmore's [46], however they have recently become much more relevant thanks to today's powerful CDCL [143] based SAT solving and other calculus improvements.

In SMT, ML has so far been mainly used for tasks such as portfolio and strategy optimization [5,111,140]. More recent work [13,32,66] has also explored fast non-neural ML guiding methods based on decision trees and manual features. Very recently, the first neural methods for guiding the cvc5 SMT system have started to be developed [109]. Due to the large number of possible instantiations in such settings, this is typically more involved than guiding the clause selection as in the saturation based systems. Interestingly, iProver's instantiation-based calculus is also using the given clause loop, and an efficient neural guidance of its Inst-Gen [84] procedure has recently led to doubling of its performance on the Mizar corpus [21].

7 Tactic Based ITP Guidance

In most interactive theorem provers, proofs are written using meta-programs consisting of *tactics*. A tactic can analyze the current state of the proof and generate a sequence of kernel inference steps or a partial proof term to advance the proof. Actions performed by tactics can range from simple inference steps, to decision procedures, domain-specific heuristics, and even a generic proof search. As an alternative to ATP guidance of low-level steps, it is possible to recommend such tactics and explore the proof space using them. Recommendations are made by analyzing existing tactical proofs (either written by users or found automatically) and learning to predict which tactic performs a useful action on a given proof state. This *mid-level* guidance task falls anywhere between premise selection and the ATP guidance, allowing ML methods that may be slower than those used for ATPs.

7.1 Overview

Early systems in this field include ML4PG [82] for Coq, which gives tactic suggestions by clustering together various statistics extracted from interactive proofs, without trying to finish the proofs automatically. SEPIA [52] provides tactic predictions and also proof search for Coq, which is however only based on tactic traces without considering the proof state. The first system that considers proof states is TacticToe [43], which uses k-NN selection to predict the most likely tactics that complete a goal in the HOL4 proof assistant. Its later versions combined the prediction of promising tactics with Monte-Carlo tree search giving a very powerful method (66% of the library proved) for proof automation [44].

Similar systems have been created for Coq (Tactician [11]) and HOL Light (HOList [7]), as well as frameworks for the exploration of the tactical proof space in Lean [164]. The early PaMpeR [103] system for Isabelle also considers the proof states, however, it only recommends one command for each proof state, leaving its execution to the user (i.e., it lacks the search component). Further (sometimes experimental) systems for Coq include GamePad [58] and CoqGym [163], using deep neural networks and slow (600 s) evaluation mode, as well as Proverbot9001 [129], which was shown to perform well on CompCert. Further Coq-oriented systems (TacTok, Diva, Passport) [39,40,130] are based on the dataset provided by CoqGym. For the Lean proof assistant, further systems include LeanDojo [164] and [89].

7.2   Advantages of Tactic-Based ITP Guidance

There are several advantages associated with using tactics for proof search.

Adaptivity. Tactics provide a higher level and more flexible base for proof search. Generating kernel inference steps or partial proof terms directly is a task that needs high precision. On the other hand, tactics are often adaptive and can perform a sensible high-level action, such as search or decision procedures, on a wider variety of proof states.

Specialization. In highly specialized branches of mathematics, tactics are often specialized to the domain by experts. That is, when the default set of tactics provided by a proof assistant is not satisfactory, a new set of tactics may be written by end-users. This is particularly useful when the mathematical domain requires a deep embedding of a custom logic. An example of this is the Iris separation logic framework [70] for Coq. It includes a set of tactics specifically crafted for working with separation logic. Another example is the CakeML project [88] for HOL using custom tactics, which provides a formally verified compiler for the ML language. Some tactic-based proof search methods, such as Tactician's k-NN model for Coq [11] can learn to use new tactics in real time, which has proven rather powerful [128].

7.3   Challenges in Tactic-Based ITP Guidance

Potential Incompleteness. Contrary to performing basic inference steps for a logic (as e.g. in the ATP calculi), tactics might not represent a minimal and complete set of inference rules. A set of tactics may not be guaranteed to be able to prove every theorem.[5]

Overlap in Functionality. The actions taken by different tactics may have a high degree of overlap. Systems like Proverbot9001 [129] and Tactician [11] attempt to mitigate tactic overlap by *decomposing* and *normalizing* tactic expressions, attempting to eliminate duplicate tactics.

In TacticToe, the *orthogonalization* process is introduced to eliminate redundant tactics [44]. In more detail, TacticToe maintains a database of goal-tactic pairs used for training and this database is subject to the orthogonalization process which is intertwined with the learning. Orthogonalization works as follows: First, each time a new tactic-goal pair (t, g) is extracted from a tactic proof and about to be recorded in the database, we consider if there already exists a better tactic for the goal g in the database. To this end, we organize a competition between the k tactic-goal pairs that are closest[6] to the pair (t,g) (including it). The winner (which is ultimately stored in the database and trained on) is the tactic that subsumes the original tactic t on the goal g and that appears in the

---

[5] This is a theoretical concern that usually is not a problem in practice.

[6] He we already use the learned notion of proximity on the database constructed so far.

largest number of tactic-goal pairs in the database. As a result, already successful tactics with a large coverage are preferred, and new tactics are considered
only if they provide a different contribution.

Diverse Tactic Behavior. Due to the diverse range of tactic behaviors it can
be difficult to tune proof search to appropriately exploit each class of tactics.
For example, while simple tactics are executed rather quickly, more sophisticated tactics may require multiple seconds of execution time before their action
completes. Deciding which tactic to execute and the appropriate allocation of
resources to it is a major challenge.

Tactics Are Designed for Humans. Because tactics are designed for use by
humans, they usually come with a complex language of tactic combinators,
higher-order language features, and syntactic sugar. Typically, in machine learning and reinforcement learning we however prefer to have a relatively simple set
of actions (commands in the ITP setting). Therefore one would like to decompose
the proof scripts into sequences (or more generally trees) of simpler commands.
This is a difficult preprocessing step.

Representation. Simpler tactic-based systems make predictions solely based on
surface-level syntax features of proof states. This makes it difficult to extract
deeper knowledge about the previously introduced concepts. Also, straightforward neural encodings are typically insufficient when new concepts and lemmas
are added on the fly (which is very common in ITP), because it is expensive
to always adapt a large neural model after the addition of such new items. The
most recent work on Graph2Tac [128] learns representations of all definitions
in its dataset and can generate representations for unseen definitions on-the-fly.
This provides more accurate representations of proof states and tactic arguments, resulting in a 50% improvement over baselines that do not incorporate
such background information.

8 Related Symbolic Classification Problems

The majority of investigations discussed above concern the classification of symbolic expressions with the goal of informing a symbolic system, which of a variety
of actions is most likely to result in success. A few exceptions were discussed in
Sect. 3.4, for example [18,35]. Both papers present an embedding and a neural
architecture solving a classification problem but with no intention of integration
within a symbolic system. While this positions such investigations quite far from
the core topic of this Survey, note that these works motivated the approach presented in [20] where the authors used a tree NN to guide clause selection. Thus, it
is likely that future developments improving precise selection and guidance will
be motivated by approaches developed for other symbolic classification problems.

Relatively recent investigations have considered the introduction of probabilities into logic programs [38], i.e., some facts are associated with probability. Such

logic programs allow the introduction of predicates whose definition is a neural network. The authors of *DeepProbLog* [96,97], introduce an approach to train so-called *neural predicates* within the context of a probabilistic logic program. In some sense, this can be viewed as a form of *symbolic guidance* of the training procedure of a statistical model. While the majority of this Survey, including the section on neural classification, focuses on guidance, a few investigations have considered *end-to-end* neural theorem proving [127], that perform a soft unification operation over a vector representation of the proof state and atoms to unify. Investigations have also considered the use of *generative adversarial networks* to train a prover and a teacher simultaneously with the goal of teaching the prover to solve the problems presented by the teacher [117].

ILP is a form of symbolic machine learning whose goal is to derive explanatory hypotheses from sets of examples (denoted E^+ and E^-) together with background knowledge (denoted BK) [23]. Essentially, it is a form of inductive synthesis. Early approaches to providing a statistical characterization of ILP include *nFOIL* [90], which models the search procedure and stopping conditions of FOIL [119] using Naive Bayes. Essentially, nFOIL attempts to maximize the probability that a hypothesis covers the examples. The stopping condition is the score function reaching a user-defined threshold. A similar approach was investigated by integrating Kernel methods and FOIL [91].

More recently, Evans et al. [34] introduced δILP which considers ILP as a satisfiability problem where each proposition denotes a pair of clauses constructable using the *background knowledge*. These clause pairs are used as the definition *predicated template*. The hypothesis consists of a user-defined number of predicated templates. Essentially, a propositional model of this SAT problem is a hypothesis. As a next step, the authors introduce a type of soft inferencing used to compute the evaluations of the propositions composing the SAT problem. Investigations based on the δILP have considered a hierarchical structure of templating [47] and massive predicate invention [116]. Recent work by some of the Authors of δILP illustrates that such *neuro-symbolic* system can be interleaved with a symbolic solver, allowing them to provide feedback during training [33].

While δILP requires learning a propositional model for a SAT encoding of an ILP instance through soft inferencing, the framework is not directly viable as a general *neural* SAT solver. NeuroSAT [141] embeds a SAT instance into a GNN that learns which propositions are required for satisfiability. A more recent approach to neural SAT solving is the SATNet [16,147,161] where satisfiability is formulated as a semidefinite program.

9 Conclusion

In this work, we discussed the main contemporary areas of combining automated reasoning and especially automated theorem proving with machine learning. This includes the early history, characterization of mathematical knowledge, premise selection, ATPs that use machine learning, feedback loops between proving and learning, and some related symbolic classification problems.

Recently, (large) language models (LMs/LLMs) have shown the ability to generate (not necessarily correct) informal math texts. Beyond high-school tasks or mathematics olympics tasks (solutions for which are abundant on the web), their ATP performance in fair-resource evaluations has, however, so far been questionable compared to targeted architectures such as signature-independent graph neural networks [128]. Other research questions currently debated in the context of LLMs are, e.g., the (lack of) *emergence* [132] and the *memorization* of all the benchmark problems, proofs and their informal presentations available on the web [68].[7] Perhaps the most promising uses of (not necessarily large) language models today seem to be the conjecturing tasks mentioned in Sect. 4.4 (e.g., [45, 68, 120, 155]), and in autoformalization [67, 79, 80, 162].

On the other hand, systems like *ChatGPT* have recently convinced a lot of lay and expert audience about the potential of AI systems trained over a lot of informal knowledge. It is likely that this will lead to increased efforts in training AI/TP systems, which not only absorb a lot of knowledge, but also learn to use it correctly and are capable of self-improvement and new discoveries without hallucination, thanks to the ground logical layer. The brief history of the AI/TP field so far demonstrates that perhaps the most interesting systems and research directions emerge when the deductive, search and symbolic methods are in nontrivial ways combined with the learning, inductive and statistical methods, leading to complex and novel AI architectures. In this sense, the future of AI/TP research seems to be bright and very open.

Acknowledgments. We thank Herman Geuvers for creating an environment in his group in Nijmegen that allowed starting and continuing several of the discussed research directions. We also thank the people in Nijmegen with whom we have over the years collaborated on these topics and discussed them, including Tom Heskes, Daniel Kühlwein, Twan van Laarhoven, Evgeni Tsivtsivadze, Jelle Piepenbrock, Jan Heemstra, Freek Wiedijk, Robbert Krebbers and Henk Barendregt.

Finally, we thank the whole AITP community involved in developing these topics, and especially our AI/TP collaborators and colleagues (often from Prague and Innsbruck), including Bob Veroff, Stephan Schulz, Konstantin Korovin, Sean Holden, Jasmin Blanchette, Chad Brown, Mikoláš Janota, Karel Chvalovský, Jiří Vyskočil, Petr Pudlák, Petr Štěpánek, Mirek Olšák, Zsolt Zombori, Geoff Sutcliffe, Christian Szegedy, Tom Hales, Larry Paulson, Adam Pease, Moa Johansson, Sarah Winkler, Sara Loos, Ramana Kumar, Michael Douglas, Michael Rawson, Mario Carneiro, Bartosz Piotrowski, Yutaka Nagashima, Zar Goertzel, Michael Faerber, Shawn Wang, Jan Hůla, Filip Bártek and Liao Zhang.

This work has been partially supported by the COST Action CA20111 (CK, DC, MS), ERC PoC grant no. 101156734 *FormalWeb3* (CK, JJ), project RICAIP no. 857306

---

[7] In particular, while an exact formal proof of a solved problem P may or may not be available on the web or GitHub (and thus in the LLM training data), it is still quite likely that an informal proof with the essential proof ideas for P has been seen during the LLM training. In such cases, the LLM is more likely doing translation (autoformalization) rather than the usual proving of previously unseen problems. Similar issues appear with methods such as (pre-)training on "synthetic" problems, which may be generated in ways that make them close to the target "unseen" problems.

under the EU-H2020 programme (MS), project CORESENSE no. 101070254 under the Horizon Europe programme (MS), ERC-CZ grant no. LL1902 POSTMAN (JJ, TG, JU), Amazon Research Awards (LB, TG, JU), the EU ICT-48 2020 project TAILOR no. 952215 (LB, JU), the ELISE EU ICT-48 project's (no. 951847) Mobility Program (LB), and the Czech Science Foundation project 24-12759S (MS).

References

1. Alama, J., Heskes, T., Kühlwein, D., Tsivtsivadze, E., Urban, J.: Premise selection for math by corpus analysis and kernel methods. JAR **52**(2), 191–213 (2014)
2. Alemi, A.A., Chollet, F., Eén, N., Irving, G., Szegedy, C., Urban, J.: DeepMath - deep sequence models for premise selection. In: NIPS 2016, pp. 2235–2243 (2016)
3. Allamanis, M., Chanthirasegaran, P., Kohli, P., Sutton, C.: Learning continuous semantic representations of symbolic expressions. In: ICML 2017, volume 70 of Proceedings of Machine Learning Research, pp. 80–88. PMLR (2017)
4. Aygün, E., et al.: Proving theorems using incremental learning and hindsight experience replay. In: ICML 2022, vol. 162, pp. 1198–1210 (2022)
5. Balunovic, M., Bielik, P., Vechev, M.T.: Learning to solve SMT formulas. In: NeurIPS, pp. 10338–10349 (2018)
6. Bancerek, G., et al.: Mizar: state-of-the-art and Beyond. In: Kerber, M., Carette, J., Kaliszyk, C., Rabe, F., Sorge, V. (eds.) Intelligent Computer Mathematics. Lecture Notes in Computer Science(), vol. 9150, pp. 261–279. Springer, Cham (2015). https://doi.org/10.1007/978-3-319-20615-8_17
7. Bansal, K., Loos, S., Rabe, M., Szegedy, C., Wilcox, S.: Holist: an environment for machine learning of higher order logic theorem proving. In: ICML 2019, vol. 97, pp. 454–463. PMLR (2019)
8. Barbosa, H., et al.: cvc5: a versatile and industrial-strength SMT solver. In: Fisman, D., Rosu, G. (eds.) Tools and Algorithms for the Construction and Analysis of Systems. Lecture Notes in Computer Science, vol. 13243, pp. 415–442. Springer, Cham (2022). https://doi.org/10.1007/978-3-030-99524-9_24
9. Bártek, F., Suda, M.: Neural precedence recommender. In: Platzer, A., Sutcliffe, G. (eds.) Automated Deduction - CADE 28. Lecture Notes in Computer Science(), vol. 12699, pp. 525–542. Springer, Cham (2021). https://doi.org/10.1007/978-3-030-79876-5_30
10. Bártek, F., Suda, M.: How much should this symbol weigh? A GNN-advised clause selection. In: LPAR 2023, vol. 94 of EPiC, pp. 96–111. EasyChair (2023)
11. Blaauwbroek, L., Urban, J., Geuvers, H.: The tactician - a seamless, interactive tactic learner and prover for Coq. In: Benzmuller, C., Miller, B. (eds.) Intelligent Computer Mathematics. Lecture Notes in Computer Science(), vol. 12236, pp. 271–277. Springer, Cham (2020). https://doi.org/10.1007/978-3-030-53518-6_17
12. Blanchette, J.C., Kaliszyk, C., Paulson, L.C., Urban, J.: Hammering towards QED. J. Formalized Reasoning **9**(1), 101–148 (2016)
13. Blanchette, J.C., El Ouraoui, D., Fontaine, P., Kaliszyk, C.: Machine learning for instance selection in SMT solving. In: AITP 2019 - 4th Conference on Artificial Intelligence and Theorem Proving, Obergurgl, Austria (2019)
14. Bledsoe, W.W., Boyer, R.S., Henneman, W.H.: Computer proofs of limit theorems. Artif. Intell. **3**, 27–60 (1972)
15. Carlson, A., Cumby, C., Rosen, J., Roth, D.: The SNoW learning architecture, vol. 5. Technical report. UIUCDCS-R-99-2101, UIUC Computer Science Department (1999)

16. Chang, O., Flokas, L., Lipson, H., Spranger, M.: Assessing SATNet's ability to solve the symbol grounding problem. In: NeurIPS 2020, vol. 33, pp. 1428–1439 (2020)
17. Chen, T., Guestrin, C.: XGBoost: a scalable tree boosting system. In: Knowledge Discovery and Data Mining 2016, pp. 785–794. ACM (2016)
18. Chvalovský, K.: Top-down neural model for formulae. In: ICLR 2019. OpenReview.net (2019)
19. Chvalovský, K., Jakubův, J., Olšák, M., Urban, J.: Learning theorem proving components. In: Das, A., Negri, S. (eds.) Automated Reasoning with Analytic Tableaux and Related Methods. Lecture Notes in Computer Science(), vol. 12842, pp. 266–278. Springer, Cham (2021). https://doi.org/10.1007/978-3-030-86059-2_16
20. Chvalovský, K., Jakubův, J., Suda, M., Urban, J.: ENIGMA-NG: efficient neural and gradient-boosted inference guidance for E. In: Fontaine, P. (ed.) Automated Deduction - CADE 27. Lecture Notes in Computer Science(), vol. 11716, pp. 197–215. Springer, Cham (2019). https://doi.org/10.1007/978-3-030-29436-6_12
21. Chvalovský, K., Korovin, K., Piepenbrock, J., Urban, J.: Guiding an instantiation prover with graph neural networks. In: LPAR 2023, volume 94 of EPiC Series in Computing, pp. 112–123. EasyChair (2023)
22. Colton, S., Bundy, A., Walsh, T.: Automatic concept formation in pure mathematics. In: IJCAI, pp. 786–793. Morgan Kaufmann (1999)
23. Cropper, A., Dumancic, S.: Inductive logic programming at 30: a new introduction. J. Artif. Intell. Res. **74**, 765–850 (2022)
24. Crouse, M., et al.: A deep reinforcement learning approach to first-order logic theorem proving. In: AAAI 2021, pp. 6279–6287 (2021)
25. Dahn, I., Wernhard, C.: First order proof problems extracted from an article in the MIZAR Mathematical Library. In: International Workshop on First-Order Theorem Proving (FTP'97), pp. 58–62 (1997)
26. Davis, M.: The early history of automated deduction. Handb. Autom. Reasoning **1**, 3–15 (2001)
27. Davis, M., Putnam, H.: A computing procedure for quantification theory. J. ACM (JACM) **7**(3), 201–215 (1960)
28. de Moura, L.M., Bjørner, N.: Z3: an efficient SMT solver. In: Ramakrishnan, C.R., Rehof, J. (eds.) Tools and Algorithms for the Construction and Analysis of Systems. Lecture Notes in Computer Science, vol. 4963, pp. 337–340. Springer, Berlin (2008). https://doi.org/10.1007/978-3-540-78800-3_24
29. Deerwester, S.C., Dumais, S.T., Landauer, T.K., Furnas, G.W., Harshman, R.A.: Indexing by latent semantic analysis. JASIS **41**(6), 391–407 (1990)
30. Denzinger, J., Fuchs, M., Goller, C., Schulz, S.: Learning from previous proof experience. Technical Report AR99-4, Institut für Informatik, TUM (1999)
31. Denzinger, J., Schulz, S.: Learning domain knowledge to improve theorem proving. In: CADE 13, number 1104 in LNAI, pp. 62–76 (1996)
32. El Ouraoui, D.: Méthodes pour le raisonnement d'ordre supérieur dans SMT, Chapter 5. PhD thesis, Université de Lorraine (2021)
33. Evans, R., et al.: Making sense of raw input. Artif. Intell. **299**, 103521 (2021)
34. Evans, R., Grefenstette, E.: Learning explanatory rules from noisy data. J. Artif. Intell. Res. **61**, 1–64 (2018)
35. Evans, R., Saxton, D., Amos, D., Kohli, P., Grefenstette, E.: Can neural networks understand logical entailment? In: ICLR 2018. OpenReview.net (2018)

36. Färber, M., Kaliszyk, C.: Random forests for premise selection. In: Lutz, C., Ranise, S. (eds.) Frontiers of Combining Systems. Lecture Notes in Computer Science(), vol. 9322, pp. 325–340. Springer, Cham (2015). https://doi.org/10.1007/978-3-319-24246-0_20

37. Färber, M., Kaliszyk, C., Urban, J.: Machine learning guidance for connection tableaux. J. Autom. Reason. 65(2), 287–320 (2021)

38. Fierens, D., et al.: Inference and learning in probabilistic logic programs using weighted Boolean formulas. Theory Pract. Log. Prog. 15(3), 358–401 (2015)

39. First, E., Brun, Y.: Diversity-driven automated formal verification. In: Proceedings of the 44th International Conference on Software Engineering, ICSE '22, New York, NY, USA, pp. 749-761. Association for Computing Machinery (2022)

40. First, E., Brun, Y., Guha, A.: TacTok: semantics-aware proof synthesis. Proc. ACM Program. Lang. 4(OOPSLA) (2020)

41. Gauthier, T., Kaliszyk, C.: Aligning concepts across proof assistant libraries. J. Symb. Comput. 90, 89–123 (2019)

42. Gauthier, T., Kaliszyk, C., Urban, J.: Initial experiments with statistical conjecturing over large formal corpora. In: WIP@CIKM'16, volume 1785 of CEUR, pp. 219–228 (2016)

43. Gauthier, T., Kaliszyk, C., Urban, J.: TacticToe: learning to reason with HOL4 tactics. In: LPAR-21, pp. 125–143 (2017)

44. Gauthier, T., Kaliszyk, C., Urban, J., Kumar, R., Norrish, M.: TacticToe: learning to prove with tactics. J. Autom. Reason. 65(2), 257–286 (2021)

45. Gauthier, T., Olšák, M., Urban, J.: Alien coding. Int. J. Approx. Reason. 162, 109009 (2023)

46. Gilmore, P.C.: A proof method for quantification theory: its justification and realization. IBM J. Res. Dev. 4(1), 28–35 (1960)

47. Glanois, C., et al.: Neuro-symbolic hierarchical rule induction. In: ICML 2022, pp. 7583–7615 (2022)

48. Goertzel, Z.A., Chvalovský, K., Jakubův, J., Olšák, M., Urban, J.: Fast and slow enigmas and parental guidance. In: Konev, B., Reger, G. (eds.) Frontiers of Combining Systems. Lecture Notes in Computer Science(), vol. 12941, pp. 173–191. Springer, Cham (2021). https://doi.org/10.1007/978-3-030-86205-3_10

49. Goertzel, Z.A., Jakubův, J., Urban, J.: ENIGMAWatch: ProofWatch meets ENIGMA. In: TABLEAUX 2019, volume 11714 of LNCS, pp. 374–388 (2019)

50. Goller, C.: Learning search-control heuristics with folding architecture networks. In: ESANN 1999, pp. 45–50 (1999)

51. Goller, C., Kuchler, A.: Learning task-dependent distributed representations by backpropagation through structure. In: ICNN'96, pp. 347–352. IEEE (1996)

52. Gransden, T., Walkinshaw, N., Raman, R.: SEPIA: search for proofs using inferred automata. In: Felty, A., Middeldorp, A. (eds.) Automated Deduction - CADE-25. Lecture Notes in Computer Science(), vol. 9195, pp. 246–255. Springer, Cham (2015). https://doi.org/10.1007/978-3-319-21401-6_16

53. Hoder, K., Reger, G., Suda, M., Voronkov, A.: Selecting the selection. In: Olivetti, N., Tiwari, A. (eds.) Automated Reasoning. Lecture Notes in Computer Science(), vol. 9706, pp. 313–329. Springer, Cham (2016). https://doi.org/10.1007/978-3-319-40229-1_22

54. Hoder, K., Voronkov, A.: Sine qua non for large theory reasoning. In: Bjorner, N., Sofronie-Stokkermans, V. (eds.) Automated Deduction - CADE-23. Lecture Notes in Computer Science(), vol. 6803, pp. 299–314. Springer, Berlin (2011). https://doi.org/10.1007/978-3-642-22438-6_23

55. Holden, E.K., Korovin, K.: Heterogeneous heuristic optimisation and scheduling for first-order theorem proving. In: Kamareddine, F., Sacerdoti Coen, C. (eds.) Intelligent Computer Mathematics. Lecture Notes in Computer Science(), vol. 12833, pp. 107–123. Springer, Cham (2021). https://doi.org/10.1007/978-3-030-81097-9_8

56. Holden, E.K.: Korovin, K.: Graph sequence learning for premise selection. CoRR, abs/2303.15642 (2023)

57. Holden, S.B.: Machine learning for automated theorem proving: Learning to solve SAT and QSAT. Found. Trends Mach. Learn. 14(6), 807–989 (2021)

58. Huang, D., Dhariwal, P., Song, D., Sutskever, I.: GamePad: A learning environment for theorem proving. In: ICLR (2019)

59. Jakubův, J., et al.: MizAR 60 for Mizar 50. In: ITP 2023, volume 268 of LIPIcs, pp. 1–22 (2023)

60. Jakubův, J., Chvalovský, K., Olšák, M., Piotrowski, B., Suda, M., Urban, J.: ENIGMA anonymous: symbol-independent inference guiding machine (system description). In: Peltier, N., Sofronie-Stokkermans, V. (eds.) Automated Reasoning. Lecture Notes in Computer Science(), vol. 12167, pp. 448–463. Springer, Cham (2020). https://doi.org/10.1007/978-3-030-51054-1_29

61. Jakubův, J., Kaliszyk, C.: Unified ordering for superposition-based automated reasoning. In: Davenport, J., Kauers, M., Labahn, G., Urban, J. (eds.) Mathematical Software - ICMS 2018. Lecture Notes in Computer Science(), vol. 10931, pp. 245–254. Springer, Cham (2018). https://doi.org/10.1007/978-3-319-96418-8_29

62. Jakubův, J., Urban, J.: ENIGMA: efficient learning-based inference guiding machine. In: Geuvers, H., England, M., Hasan, O., Rabe, F., Teschke, O. (eds.) Intelligent Computer Mathematics. Lecture Notes in Computer Science(), vol. 10383, pp. 292–302. Springer, Cham (2017). https://doi.org/10.1007/978-3-319-62075-6_20

63. Jakubův, J., Urban, J.: Enhancing ENIGMA given clause guidance. In: Rabe, F., Farmer, W., Passmore, G., Youssef, A. (eds.) Intelligent Computer Mathematics. Lecture Notes in Computer Science(), vol. 11006, pp. 118–124. Springer, Cham (2018). https://doi.org/10.1007/978-3-319-96812-4_11

64. Jakubův, J., Urban, J.: Hierarchical invention of theorem proving strategies. AI Commun. 31(3), 237–250 (2018)

65. Jakubův, J., Urban, J.: Hammering Mizar by learning clause guidance (short paper). In: ITP 2019, volume 141 of LIPIcs, pp. 1–8 (2019)

66. Janota, M., Piepenbrock, J., Piotrowski, B.: Towards learning quantifier instantiation in SMT. In: Meel, K.S., Strichman, O. (eds.) 25th International Conference on Theory and Applications of Satisfiability Testing, SAT 2022, August 2-5, 2022, Haifa, Israel, volume 236 of LIPIcs, pp. 1–18. Schloss Dagstuhl - Leibniz-Zentrum für Informatik (2022)

67. Jiang, A.Q., et al.: Draft, sketch, and prove: guiding formal theorem provers with informal proofs. In: ICLR (2023)

68. Johansson, M., Smallbone, N.: Exploring mathematical conjecturing with large language models. In: NeSy, volume 3432 of CEUR Workshop Proceedings, pp. 62–77. CEUR-WS.org (2023)

69. Bayardo Jr, R.J., Schrag, R.: Using CSP look-back for real-world SAT instances. In: AAAI 97, pp. 203–208. AAAI Press/The MIT Press (1997)

70. Jung, R., Krebbers, R., Jourdan, J., Bizjak, A., Birkedal, L., Dreyer, D.: Iris from the ground up: a modular foundation for higher-order concurrent separation logic. J. Funct. Program. 28, e20 (2018)

71. Kaliszyk, C., Urban, J.: Stronger automation for flyspeck: feature weighting and strategy evolution. In: PxTP 2013, volume 14 of EPiC Series in Computing, pp. 87–95. EasyChair (2013)
72. Kaliszyk, C., Urban, J.: Learning-assisted automated reasoning with flyspeck. J. Autom. Reason. **53**(2), 173–213 (2014)
73. Kaliszyk, C., Urban, J.: FEMaLeCoP: fairly efficient machine learning connection prover. In: Davis, M., Fehnker, A., McIver, A., Voronkov, A. (eds.) Logic for Programming, Artificial Intelligence, and Reasoning. Lecture Notes in Computer Science(), vol. 9450, pp. 88–96. Springer, Berlin (2015). https://doi.org/10.1007/978-3-662-48899-7_7
74. Kaliszyk, C., Urban, J.: Learning-assisted theorem proving with millions of lemmas. J. Symb. Comput. **69**, 109–128 (2015)
75. Kaliszyk, C., Urban, J.: MizAR 40 for Mizar 40. J. Autom. Reason. **55**(3), 245–256 (2015)
76. Kaliszyk, C., Urban, J., Michalewski, H., Olšák, M.: Reinforcement learning of theorem proving. In: NeurIPS 2018, pp. 8836–8847 (2018)
77. Kaliszyk, C., Urban, J., Vyskočil, J.: Machine learner for automated reasoning 0.4 and 0.5. In: PAAR@IJCAR, volume 31 of EPiC, pp. 60–66 (2014)
78. Kaliszyk, C., Urban, J., Vyskočil, J.: Efficient semantic features for automated reasoning over large theories. In: IJCAI 2015, pp. 3084–3090. AAAI Press (2015)
79. Kaliszyk, C., Urban, J., Vyskočil, J.: Automating formalization by statistical and semantic parsing of mathematics. In: Ayala-Rincon, M., Munoz, C.A. (eds.) Interactive Theorem Proving. Lecture Notes in Computer Science(), vol. 10499, pp. 12–27. Springer, Cham (2017). https://doi.org/10.1007/978-3-319-66107-0_2
80. Kaliszyk, C., Urban, J., Vyskočil, J., Geuvers, H.: Developing corpus-based translation methods between informal and formal mathematics: project description. In: Watt, S.M., Davenport, J.H., Sexton, A.P., Sojka, P., Urban, J. (eds.) Intelligent Computer Mathematics. Lecture Notes in Computer Science(), vol. 8543, pp. 435–439. Springer, Cham (2014). https://doi.org/10.1007/978-3-319-08434-3_34
81. Ke, G., et al.: LightGBM: a highly efficient gradient boosting decision tree. In: NeurIPS 2017, pp. 3146–3154 (2017)
82. Komendantskaya, E., Heras, J., Grov, G.: Machine learning in proof general: interfacing interfaces. In: UITP, volume 118 of EPTCS, pp. 15–41 (2012)
83. Korovin, K.: iProver - an instantiation-based theorem prover for first-order logic (system description). In: Armando, A., Baumgartner, P., Dowek, G. (eds.) Automated Reasoning. Lecture Notes in Computer Science(), vol. 5195, pp. 292–298. Springer, Berlin (2008). https://doi.org/10.1007/978-3-540-71070-7_24
84. Korovin, K.: Inst-Gen - a modular approach to instantiation-based automated reasoning. In: Voronkov, A., Weidenbach, C. (eds.) Programming Logics. Lecture Notes in Computer Science, vol. 7797, pp. 239–270. Springer, Berlin (2013). https://doi.org/10.1007/978-3-642-37651-1_10
85. Kovács, L., Voronkov, A.: First-order theorem proving and Vampire. In: Sharygina, N., Veith, H. (eds.) Computer Aided Verification. Lecture Notes in Computer Science, vol. 8044, pp. 1–35. Springer, Berlin (2013). https://doi.org/10.1007/978-3-642-39799-8_1
86. Kühlwein, D., Blanchette, J.C., Kaliszyk, C., Urban, J.: MaSh: machine learning for Sledgehammer. In: Blazy, S., Paulin-Mohring, C., Pichardie, D. (eds.) Interactive Theorem Proving. Lecture Notes in Computer Science, vol. 7998, pp. 35–50. Springer, Berlin (2013). https://doi.org/10.1007/978-3-642-39634-2_6
87. Kühlwein, D., Urban, J.: MaLeS: a framework for automatic tuning of automated theorem provers. J. Autom. Reason. **55**(2), 91–116 (2015)

88. Kumar, R., Myreen, M.O., Norrish, M., Owens, S.: CakeML: a verified implementation of ML. In: Principles of Programming Languages (POPL), pp. 179–191. ACM Press (2014)
89. Lample, G., et al.: Hypertree proof search for neural theorem proving. In: NeurIPS (2022)
90. Landwehr, N., Kersting, K., Raedt, L.D.: nFOIL: integrating naïve Bayes and FOIL. In: AAAI 2005, pp. 795–800. AAAI Press/The MIT Press (2005)
91. Landwehr, N., Passerini, A., Raedt, L.D., Frasconi, P.: kFOIL: learning simple relational kernels. In: AAAI 2006, pp. 389–394. AAAI Press (2006)
92. Langley, P.: BACON: a production system that discovers empirical laws. In: International Joint Conference on Artificial Intelligence (1977)
93. Lenat, D.: An artificial intelligence approach to discovery in mathematics. PhD thesis, Stanford University, Stanford, USA (1976)
94. Loos, S.M., Irving, G., Szegedy, C., Kaliszyk, C.: Deep network guided proof search. In: LPAR-21, volume 46 of EPiC Series in Computing, pp. 85–105. EasyChair (2017)
95. López-Hernández, J.C., Korovin, K.: An abstraction-refinement framework for reasoning with large theories. In: Galmiche, D., Schulz, S., Sebastiani, R. (eds.) Automated Reasoning. Lecture Notes in Computer Science(), vol. 10900, pp. 663–679. Springer, Cham (2018). https://doi.org/10.1007/978-3-319-94205-6_43
96. Manhaeve, R., Dumancic, S., Kimmig, A., Demeester, T., Raedt, L.D.: DeepProbLog: neural probabilistic logic programming. In: NeurIPS 2018, pp. 3753–3763 (2018)
97. Manhaeve, R., Dumancic, S., Kimmig, A., Demeester, T., Raedt, L.D.: Neural probabilistic logic programming in DeepProbLog. Artif. Intell. **298**, 103504 (2021)
98. McCune, W.: OTTER 2.0. In: Stickel, M.E. (ed.) 10th International Conference on Automated Deduction. Lecture Notes in Computer Science, vol. 449, pp. 663–664. Springer, Berlin (1990). https://doi.org/10.1007/3-540-52885-7_131
99. Meng, J., Paulson, L.C.: Experiments on supporting interactive proof using resolution. In: Basin, D., Rusinowitch, M. (eds.) Automated Reasoning. Lecture Notes in Computer Science(), vol. 3097, pp. 372–384. Springer, Berlin (2004). https://doi.org/10.1007/978-3-540-25984-8_28
100. Meng, J., Paulson, L.C.: Lightweight relevance filtering for machine-generated resolution problems. J. Appl. Logic **7**(1), 41–57 (2009)
101. Mikolov, T., Chen, K., Corrado, G., Dean, J.: Efficient estimation of word representations in vector space. arXiv preprint: arXiv:1301.3781 (2013)
102. Moskewicz, M.W., Madigan, C.F., Zhao, Y., Zhang, L., Malik, S.: Chaff: engineering an efficient SAT solver. In: DAC 2001, pp. 530–535. ACM (2001)
103. Nagashima, Y., He, Y.: PaMpeR: proof method recommendation system for Isabelle/HOL. In: Huchard, M., Kästner, C., Fraser, G. (eds.) Proceedings of the 33rd ACM/IEEE International Conference on Automated Software Engineering, ASE 2018, Montpellier, France, September 3-7, 2018, pp. 362–372. ACM (2018)
104. Newell, A., Simon, H.: The logic theory machine-a complex information processing system. IRE Trans. Inf. Theory **2**(3), 61–79 (1956)
105. Nieuwenhuis, R., Rubio, A.: Paramodulation-based theorem proving. In: Handbook of Automated Reasoning (in 2 volumes), pp. 371–443 (2001)
106. Olšák, M., Kaliszyk, C., Urban, J.: Property invariant embedding for automated reasoning. In: ECAI 2020, volume 325, pp. 1395–1402. IOS Press (2020)
107. Otten, J., Bibel, W.: leanCoP: lean connection-based theorem proving. J. Symb. Comput. **36**(1–2), 139–161 (2003)

108. Otten, J., Bibel, W.: Advances in connection-based automated theorem proving. In: Hinchey, M., Bowen, J., Olderog, E.R. (eds.) Provably Correct Systems. NASA Monographs in Systems and Software Engineering, pp. 211–241. Springer, Cham (2017). https://doi.org/10.1007/978-3-319-48628-4_9

109. Piepenbrock, J., Janota, M., Urban, J., Jakubův, J.: First experiments with neural cvc5 (2024). http://grid01.ciirc.cvut.cz/~mptp/cvc5-gnn.pdf

110. Piepenbrock, J., Urban, J., Korovin, K., Olšák, M., Heskes, T., Janota, M.: Machine learning meets the Herbrand universe. CoRR, abs/2210.03590 (2022)

111. Pimpalkhare, N., Mora, F., Polgreen, E., Seshia, S.A.: MedleySolver: online SMT algorithm selection. In: Li, C.-M., Manyà, F. (eds.) Theory and Applications of Satisfiability Testing - SAT 2021. pp, pp. 453–470. Springer International Publishing, Cham (2021). https://doi.org/10.1007/978-3-030-80223-3_31

112. Piotrowski, B., Mir, R.F., Ayers, E.: Machine-Learned Premise Selection for Lean. In: Ramanayake, R., Urban, J. (eds.) Automated Reasoning with Analytic Tableaux and Related Methods. Lecture Notes in Computer Science(), vol. 14278, pp. 175–186. Springer, Cham (2023). https://doi.org/10.1007/978-3-031-43513-3_10

113. BPiotrowski, B., Urban, J.: ATPboost: learning premise selection in binary setting with ATP feedback. In: Galmiche, D., Schulz, S., Sebastiani, R. (eds.) Automated Reasoning. Lecture Notes in Computer Science(), vol. 10900, pp. 566–574. Springer, Cham (2018). https://doi.org/10.1007/978-3-319-94205-6_37

114. Piotrowski, B., Urban, J.: Guiding inferences in connection tableau by recurrent neural networks. In: Benzmuller, C., Miller, B. (eds.) Intelligent Computer Mathematics. Lecture Notes in Computer Science(), vol. 12236, pp. 309–314. Springer, Cham (2020). https://doi.org/10.1007/978-3-030-53518-6_23

115. Piotrowski, B., Urban, J.: Stateful premise selection by recurrent neural networks. In: LPAR 2020, volume 73 of EPiC, pp. 409–422. EasyChair (2020)

116. Purgal, S.J., Cerna, D.M., Kaliszyk, C.: Differentiable inductive logic programming in high-dimensional space. CoRR, abs/2208.06652 (2022)

117. Purgal, S.J., Kaliszyk, C.: Adversarial learning to reason in an arbitrary logic. In: FLAIRS 2022 (2022)

118. Purgal, S.J., Parsert, J., Kaliszyk, C.: A study of continuous vector representations for theorem proving. J. Log. Comput. **31**(8), 2057–2083 (2021)

119. Quinlan, J.R.: Learning logical definitions from relations. Mach. Learn. **5**, 239–266 (1990)

120. Rabe, M.N., Lee, D., Bansal, K., Szegedy, C.: Mathematical reasoning via self-supervised skip-tree training. In: ICLR. OpenReview.net (2021)

121. Ramakrishnan, I.V., Sekar, R., Voronkov, A.: Term indexing. In: Handbook of Automated Reasoning (in 2 volumes), pp. 1853–1964 (2001)

122. Rawson, M., Reger, G.: Directed graph networks for logical reasoning (extended abstract). In: PAAR 2020, volume 2752 of CEUR Workshop Proceedings, pp. 109–119. CEUR-WS.org (2020)

123. Rawson, M., Reger, G.: lazyCoP: Lazy paramodulation meets neurally guided search. In: Das, A., Negri, S. (eds.) Automated Reasoning with Analytic Tableaux and Related Methods. Lecture Notes in Computer Science(), vol. 12842, pp. 187–199. Springer, Cham (2021). https://doi.org/10.1007/978-3-030-86059-2_11

124. Reger, G., Suda, M., Voronkov, A.: Playing with AVATAR. In: Felty, A., Middeldorp, A. (eds.) Automated Deduction - CADE-25. Lecture Notes in Computer Science(), vol. 9195, pp. 399–415. Springer, Cham (2015). https://doi.org/10.1007/978-3-319-21401-6_28

125. Robinson, J.A.: A machine-oriented logic based on the resolution principle. J. ACM (JACM) **12**(1), 23–41 (1965)
126. Robinson, J.A., Voronkov, A. (eds.): Handbook of Automated Reasoning (in 2 volumes). Elsevier and MIT Press (2001)
127. Rocktäschel, T., Riedel, S.: End-to-end differentiable proving. In: NeurIPS 2017, pp. 3788–3800 (2017)
128. Rute, J., Olšák, M., Blaauwbroek, L., Massolo, F.I.S., Piepenbrock, J., Pestun, V.: Graph2Tac: learning hierarchical representations of math concepts in theorem proving. CoRR, abs/2401.02949 (2024)
129. Sanchez-Stern, A., Alhessi, Y., Saul, L.K., Lerner, S.: Generating correctness proofs with neural networks. CoRR, abs/1907.07794 (2019)
130. Sanchez-Stern, A., First, E., Zhou, T., Kaufman, Z., Brun, Y., Ringer, T.: Passport: improving automated formal verification using identifiers. ACM Trans. Program. Lang. Syst. **45**(2), 1–30 (2023)
131. Scarselli, F., Gori, M., Tsoi, A.C., Hagenbuchner, M., Monfardini, G.: The graph neural network model. IEEE Trans. Neural Netw. **20**(1), 61–80 (2009)
132. Schaeffer, R., Miranda, B., Koyejo, S.: Are emergent abilities of large language models a mirage? In: Oh, A., Neumann, T., Globerson, A., Saenko, K., Hardt, M., Levine, S. (eds.) Advances in Neural Information Processing Systems, vol. 36, pp. 55565–55581. Curran Associates, Inc. (2023)
133. Schäfer, S., Schulz, S.: Breeding theorem proving heuristics with genetic algorithms. In: GCAI, volume 36 of EPiC, pp. 263–274 (2015)
134. Schulz, S.: Explanation based learning for distributed equational deduction. Diplomarbeit in Informatik, Fachbereich Informatik, Univ. Kaiserslautern (1995)
135. Schulz, S.: Learning Search Control Knowledge for Equational Deduction, volume 230 of DISKI. Infix Akademische Verlagsgesellschaft (2000)
136. Schulz, S.: Learning search control knowledge for equational theorem proving. In: Baader, F., Brewka, G., Eiter, T. (eds.) KI 2001: Advances in Artificial Intelligence. Lecture Notes in Computer Science(), vol. 2174, pp. 320–334. Springer, Berlin (2001). https://doi.org/10.1007/3-540-45422-5_23
137. Schulz, S.: E - a Brainiac theorem prover. AI Commun. **15**(2–3), 111–126 (2002)
138. chulz, S., Cruanes, S., Vukmirovic, P.: Faster, Higher, Stronger: E 2.3. In: Fontaine, P. (ed.) Automated Deduction – CADE 27. Lecture Notes in Computer Science(), vol. 11716, pp. 495–507. Springer, Cham (2019). https://doi.org/10.1007/978-3-030-29436-6_29
139. Schulz, S., Möhrmann, M.: Performance of clause selection heuristics for saturation-based theorem proving. In: Olivetti, N., Tiwari, A. (eds.) Automated Reasoning. Lecture Notes in Computer Science(), vol. 9706, pp. 330–345. Springer, Cham (2016). https://doi.org/10.1007/978-3-319-40229-1_23
140. Scott, J., Niemetz, A., Preiner, M., Nejati, S., Ganesh, V.: MachSMT: a machine learning-based algorithm selector for SMT solvers. In: Groote, J.F., Larsen, K.G. (eds.) Tools and Algorithms for the Construction and Analysis of Systems. pp, pp. 303–325. Springer International Publishing, Cham (2021). https://doi.org/10.1007/978-3-030-72013-1_16
141. Selsam, D., Lamm, M., Bünz, B., Liang, P., de Moura, L., Dill, D.L.: Learning a SAT solver from single-bit supervision. In: ICLR 2019. OpenReview.net (2019)
142. Shawe-Taylor, J., Cristianini, N.: Kernel Methods for Pattern Analysis. Cambridge University Press, New York (2004)
143. Silva, J.P.M., Sakallah, K.A.: GRASP - a new search algorithm for satisfiability. In: ICCAD 1996, pp. 220–227. IEEE Computer Society/ACM (1996)

144. Suda, M.: Improving ENIGMA-style clause selection while learning from history. In: Platzer, A., Sutcliffe, G. (eds.) Automated Deduction - CADE 28. Lecture Notes in Computer Science(), vol. 12699, pp. 543–561. Springer, Cham (2021). https://doi.org/10.1007/978-3-030-79876-5_31

145. Suda, M.: Vampire with a brain is a good ITP hammer. In: Konev, B., Reger, G. (eds.) Frontiers of Combining Systems. Lecture Notes in Computer Science(), vol. 12941, pp. 192–209. Springer, Cham (2021). https://doi.org/10.1007/978-3-030-86205-3_11

146. Sutton, R.S., Barto, A.G.: Reinforcement Learning - An Introduction. Adaptive Computation and Machine Learning. MIT Press, Cambridge (1998)

147. Topan, S., Rolnick, D., Si, X.: Techniques for symbol grounding with SATNet. In: NeurIPS 2021, vol. 34, pp. 20733–20744. Curran Associates, Inc. (2021)

148. Urban, J.: Experimenting with machine learning in automatic theorem proving. Master's thesis, Charles University, Prague (1998). English summary at https://www.ciirc.cvut.cz/~urbanjo3/MScThesisPaper.pdf

149. Urban, J.: Translating Mizar for first order theorem provers. In: Asperti, A., Buchberger, B., Davenport, J.H. (eds.) Mathematical Knowledge Management. Lecture Notes in Computer Science, vol. 2594, pp. 203–215. Springer, Berlin (2003). https://doi.org/10.1007/3-540-36469-2_16

150. Urban, J.: MPTP - motivation, implementation, first experiments. J. Autom. Reasoning 33(3–4), 319–339 (2004)

151. Urban, J.: MPTP 0.2: design, implementation, and initial experiments. J. Autom. Reasoning 37(1–2), 21–43 (2006)

152. Urban, J.: MaLARea: a metasystem for automated reasoning in large theories. In: ESARLT, volume 257 of CEUR. CEUR-WS.org (2007)

153. Urban, J.: BliStr: the blind Strategymaker. In: GCAI 2015, volume 36 of EPiC, pp. 312–319 (2015)

154. Urban, J.: No one shall drive us from the semantic AI paradise of computer-understandable math and science! https://slideslive.com/38909911/no-one-shall-drive-us-from-the-semantic-ai-paradise-of-computerunderstandable-math-and-science (2018). Keynote at the Artificial General Intelligence Conference (AGI'18)

155. Urban, J., Jakubův, J.: First neural conjecturing datasets and experiments. In: Benzmuller, C., Miller, B. (eds.) Intelligent Computer Mathematics. Lecture Notes in Computer Science(), vol. 12236, pp. 315–323. Springer, Cham (2020). https://doi.org/10.1007/978-3-030-53518-6_24

156. Urban, J., Sutcliffe, G., Pudlák, P., Vyskočil, J.: MaLARea SG1 - machine learner for automated reasoning with semantic guidance. In: Armando, A., Baumgartner, P., Dowek, G. (eds.) Automated Reasoning. Lecture Notes in Computer Science(), vol. 5195, pp. 441–456. Springer, Berlin (2008). https://doi.org/10.1007/978-3-540-71070-7_37

157. Urban, J., Vyskočil, J., Štepánek, P.: MaLeCoP machine learning connection prover. In: Brunnler, K., Metcalfe, G. (eds.) Automated Reasoning with Analytic Tableaux and Related Methods. Lecture Notes in Computer Science(), vol. 6793, pp. 263–277. Springer, Berlin (2011). https://doi.org/10.1007/978-3-642-22119-4_21

158. Veroff, R.: Using hints to increase the effectiveness of an automated reasoning program: case studies. J. Autom. Reasoning 16(3), 223–239 (1996)

159. Voronkov, A.: AVATAR: architecture for first-order theorem provers. In: Biere, A., Bloem, R. (eds.) Computer Aided Verification. Lecture Notes in Computer

Science, vol. 8559, pp. 696–710. Springer, Cham (2014). https://doi.org/10.1007/978-3-319-08867-9_46

160. Wang, M., Tang, Y., Wang, J., Deng, J.: Premise selection for theorem proving by deep graph embedding. In: NIPS'17, pp. 2783-2793, Red Hook, NY, USA. Curran Associates Inc. (2017)

161. Wang, P., Donti, P.L., Wilder, B., Kolter, J.Z.: SATNet: bridging deep learning and logical reasoning with a differentiable satisfiability solver. In: ICML 2019, volume 97, pp. 6545–6554. PMLR (2019)

162. Wang, Q., Kaliszyk, C., Urban, J.: First experiments with neural translation of informal to formal mathematics. In: Rabe, F., Farmer, W., Passmore, G., Youssef, A. (eds.) Intelligent Computer Mathematics. Lecture Notes in Computer Science(), vol. 11006, pp. 255–270. Springer, Cham (2018). https://doi.org/10.1007/978-3-319-96812-4_22

163. Yang, K., Deng, J.: Learning to prove theorems via interacting with proof assistants. In: ICML-36, volume 97 of PMLR, pp. 6984–6994 (2019)

164. Yang, K., et al.: LeanDojo: theorem proving with retrieval-augmented language models. arXiv preprint: arXiv:2306.15626 (2023)

165. Zhang, L., Blaauwbroek, L., Kaliszyk, C., Urban, J.: Learning proof transformations and its applications in interactive theorem proving. In: Sattler, U., Suda, M. (eds.) Frontiers of Combining Systems. Lecture Notes in Computer Science(), vol. 14279, pp. 236–254. Springer, Cham (2023). https://doi.org/10.1007/978-3-031-43369-6_13

166. Zombori, Z., Csiszárik, A., Michalewski, H., Kaliszyk, C., Urban, J.: Towards finding longer proofs. In: Das, A., Negri, S. (eds.) Automated Reasoning with Analytic Tableaux and Related Methods. Lecture Notes in Computer Science(), vol. 12842, pp. 167–186. Springer, Cham (2021). https://doi.org/10.1007/978-3-030-86059-2_10

167. Zombori, Z., Urban, J., Olšák, M.: The role of entropy in guiding a connection prover. In: Das, A., Negri, S. (eds.) Automated Reasoning with Analytic Tableaux and Related Methods. Lecture Notes in Computer Science(), vol. 12842, pp. 218–235. Springer, Cham (2021). https://doi.org/10.1007/978-3-030-86059-2_13

Approximation Fixpoint Theory in Coq
With an Application to Logic Programming

Bart Bogaerts[1] and Luís Cruz-Filipe[2(✉)]

[1] Department Computer Science, Vrije Universiteit Brussel (VUB), Pleinlaan 2,
1050 Brussels, Belgium
bart.bogaerts@vub.be
[2] Department Mathematics and Computer Science, University of Southern Denmark,
Campusvej 55, 5230 Odense, Denmark
lcfilipe@gmail.com

Abstract. Approximation Fixpoint Theory (AFT) is an abstract framework based on lattice theory that unifies semantics of different non-monotonic logic. AFT has revealed itself to be applicable in a variety of new domains within knowledge representation. In this work, we present a formalisation of the key constructions and results of AFT in the Coq theorem prover, together with a case study illustrating its application to propositional logic programming.

1 Introduction

Approximation Fixpoint Theory (AFT) is an abstract lattice-theoretic framework originally designed to unify semantics of non-monotonic logics [12]. Its first applications were on unifying all major semantics of logic programming [34], autoepistemic logic (AEL) [27], and default logic (DL) [29], thereby resolving a long-standing issue about the relationship between AEL and DL [13,14,22]. AFT builds on Tarski's fixpoint theory of monotone operators on a complete lattice [32], but crucially moves from the original lattice $\langle L, \leq \rangle$ to the bilattice[1] $\langle L^2, \leq, \leq_p \rangle$, where $\leq$ is just the pointwise extension of the order on L and $\leq_p$ is the precision order defined by $(x, y) \leq_p (u, v)$ if $x \leq u$ and $v \leq y$. Intuitively, a pair $(x, y) \in L^2$ approximates elements in the interval $[x, y] = \{z \in L \mid x \leq z \leq y\}$.

AFT generalizes Tarski's study of monotone operators to non-monotone operators using the notion of *approximator*: a monotone operator $A: L^2 \to L^2$ *approximates* a (possibly non-monotone) $O : L \to L$ if $A(x, x) = (O(x), O(x))$ for all $x \in L$. This simple notion allows us to define a variety of different types of fixpoints of interest; in particular:

---

[1] While a bilattice is usually defined as an arbitrary set with two compatible orders, AFT is only concerned with bilattices that are in fact *square* lattices; i.e., the underlying set is of the form L^2.

This work was partially supported by Fonds Wetenschappelijk Onderzoek – Vlaanderen (project G0B2221N).

© The Author(s), under exclusive license to Springer Nature Switzerland AG 2024
V. Capretta et al. (Eds.): *Logics and Type Systems in Theory and Practice*, LNCS 14560, pp. 84–99, 2024.
https://doi.org/10.1007/978-3-031-61716-4_5

- A *partial supported fixpoint* of A is a fixpoint of A.
- The *Kripke-Kleene fixpoint* of A is the least fixpoint of A, denoted $\mathrm{lfp}(A)$.
- A *partial stable fixpoint* of A is a pair (x, y) such that $x = \mathrm{lfp}(A(\cdot, y)_1)$ and $y = \mathrm{lfp}(A(x, \cdot)_2)$, where $A(\cdot, y)_1$ denotes the function $L \to L \colon z \mapsto A(z, y)_1$, and analogously for $A(x, \cdot)_2$.
- The *well-founded fixpoint* of A is the least precise partial stable fixpoint of A.

When applying this to logic programming, Denecker and his coauthors [12] found that Fitting's four-valued immediate consequence operator Ψ_P [16] is an approximator of Van Emden & Kowalski's [34] immediate consequence operator T_P and that all major semantics of logic programming correspond to the fixpoints defined above. The same kind of situation is observed in other fields, such as AEL and DL. Crucially, all that is required to apply AFT to a formalism and obtain several semantics is to define an appropriate approximating operator $L^2 \to L^2$ on this bilattice; often there is an obvious choice for such an approximating operator. Once this is done, the algebraic theory of AFT directly defines different types of fixpoints that correspond to different types of semantics of the application domain. This immediately give insight into how this domain is positioned with respect to other fields (semantically), but also internally within the domain shines light on how different semantics are related.

The last decade has added new application domains to AFT, such as abstract argumentation [31], extensions of logic programming [1,11,24,28], extensions of autoepistemic logic [36], active integrity constraints [5], and constraint languages for the semantic web [8]. The original theory of AFT has also been extended significantly with new types of fixpoints [9,10], and results on *stratification*, [6,37], *predicate introduction* [38], *strong equivalence* [33], and non-deterministic operators [20]. All of these were developed in the highly general setting of lattice theory, making them directly applicable to all application domains, and such ensuring that researchers do not "reinvent the wheel".

Given the success and wide range of applicability of AFT, it sounded natural to formalise this theory using a suitable theorem prover. We chose Coq [3] for this purpose, as this is the theorem prover we are most familiar with. This option poses some challenges: presentations of AFT routinely use classical reasoning principles and most proofs are by transfinite induction. Our option was to develop constructive alternatives to standard proofs and explicitly include some hypotheses that are classically trivially true where needed. We believe this effort to be meaningful, since an important motivations for studying fixpoints in computer science is their computability.

A natural question that arises is whether there are constructive models of our axioms, and this motivated including some examples in the formalisation. We also point out that most practical applications of AFT (in particular, our larger example described in Sect. 4) use the subset lattice of a finite base set with decidable equality, where most of the working assumptions hold.

2 Ordinals and Lattices

AFT relies heavily on definitions of ordinals and lattices. We formalised these as the first step in our development, choosing to use only results from Coq's standard library on lists and natural numbers. The primary reason for this choice was that we did not initially find any existing formalisations that were easy to reuse for our purposes. It was only later in the process that we discovered other libraries close to our development – see Sect. 5 for a discussion.

Ordinals. We define a type `Ordinal` of (unbounded) ordinals as a record type containing a support `Type`, an equivalence relation `eq` (defined equality), a distinguished element `zero`, a successor function `succ` and a strict total order `lt` that is well-founded and compatible with `eq`.[2]

```
Record Ordinal : Type := { T :> Type;
                           eq : T → T → Prop;
                           zero : T;
                           succ : T → T;
                           lt : relation T;
                           ... }
```

We require `succ x` to be the least element strictly greater than `x`, and that we can decide for each element whether it is of the form `succ x` for some `x`, or not; the elements for which this does not hold are called limits. This case analysis, together with well-foundedness of `lt`, allows us to do proofs about ordinals by transfinite induction, as long as the property being proved is stable under equality (this is due to working with a defined equality).

```
Lemma transfinite_induction (P:O → Prop) :
  (∀ x y, eq x y → P x → P y) →
  (∀ x, P x → P (succ x)) →
  (∀ x, limit x → (∀ y, lt y x → P y) → P x)
  → ∀ x, P x.
```

Intuitively, the elements of `O:Ordinal` are the ordinals smaller than O. As a sanity check, we construct some typical examples, including the first infinite ordinal ω (with support `nat`) and the type of all polynomials in ω (with support `list nat`). We do not deal with arithmetic on ordinals, since this is immaterial for our development.

---

[2] Throughout this presentation we write ... for additional arguments that are left out for conciseness; we leave some arguments implicit when they can be inferred from the context by a human (even if not by Coq); we omit universally quantified variables at the top of lemmas; and we ignore namespace clashes that force us to include module names in the Coq source.

Lattices. (Complete) lattices are similarly defined as a record type consisting of a carrier type C with a defined equality (an equivalence relation compatible with the order), a partial order, and an operator lub computing least upper bounds. Requiring least upper bounds to be computable restricts the kind of lattices that we can define; still we show that we capture e.g. powerset lattices (which appear in all applications of AFT so far).

Record Lattice : Type := { C :> Type;
           eq : relation C;
           leq : relation C;
           lub : (C → Prop) → C;
           ... }

Bottom and joins are defined using lub, while the strict order less is defined as fun x y ⇒ leq x y ∧ ∼eq x y . We prove the usual properties of all these functions, including uniqueness of lub modulo eq. Reversing the order in a lattice yields a dual lattice, and we use a reflection technique to port results about leq and lub to geq and glb.

A standard example of a complete lattice, ubiquous in AFT, is the powerset lattice. We formalise this construction as an operator PowerSet: Type → Lattice, such that PowerSet T has carrier T → Prop.

Given a lattice $\langle L, \leq \rangle$, the corresponding precision lattice is the lattice $\langle L^2, \leq_p \rangle$ where $(x, y) \leq_p (x', y')$ iff $x \leq x'$ and $y' \leq y$. We formalise this construction as an operator BiLattice: Lattice → Lattice.

Definition L2 : Type := L*L.
Definition L2eq := fun x y ⇒ eq (fst x) (fst y) ∧ eq (snd x) (snd y).
Definition leqp := fun x y ⇒ leq (fst x) (fst y) ∧ leq (snd y) (snd x).
Definition L2lub := fun (S:L2 → Prop) ⇒ (lub (fun x ⇒ ∃ y, eq x (fst y) ∧ S y),
                 glb (fun x ⇒ ∃ y, eq x (snd y) ∧ S y)).
Definition BiLattice := Build_Lattice L2 L2eq leqp L2lub ... .

We now prove our first set of results about fixpoints, starting with the Knaster–Tarski theorem. Given a monotonic function f on a lattice L, we define lfp f as the glb of all its prefixpoints, and show that this is the smallest fixpoint of f.

Definition prefp : (f:L → L) → L → Prop := fun x ⇒ leq (f x) x.
Definition lfp (f:L → L) := glb (prefp f).

Lemma fp_lfp : monotonic f → fixpoint f (lfp f).
Lemma lfp_lfp : fixpoint f x → leq (lfp f) x.

Next, we define chain f as a predicate over L that holds for those elements of L that are reachable from bot L by repeated application of f and lubs. We prove that, if f is monotonic, lub (chain f) also satisfies chain, and that it coincides with lfp f. All these proofs mostly follow the standard pen-and-paper argumentation.

Inductive chain (f:L → L) : L → Prop :=
| base x : eq x (bot L) → chain f x

```
| succ x y : chain y → eq x (f y) → chain f x
| lim x (S:L → Prop) : (∀ y, S y → chain f y) → eq x (lub S) → chain f x.
```

Lemma lub_chain_lfp : monotonic f → eq (lfp f) (lub (chain f)).

Finally, given a function f that respects equality, we define iteration of a function f an ordinal number of times using well-founded recursion. This definition is not very informative, and we show that the resulting definition coincides with what is expected, and only generates elements in chain f.

```
Definition iterate_fun (f: L → L) {O:Ordinal} :=
  fun (x:O) (F:(∀ (y:O), lt y x → C L)) ⇒
    match (succ_or_limit x) with
    | inleft (existT _ y H) ⇒ f (F y (succ_lt' _ _ _ H))
    | _ ⇒ (lub (fun z ⇒ ∃ y (H:lt y x), eq z (F y H)))
    end.
```

```
Definition iterate {O:Ordinal} (t:T O) :=
  Fix (lt_wf O) _ iterate_fun t.
```

Lemma iterate_succ : eq (iterate f (succ x)) (f (iterate f x)).
Lemma iterate_limit : limit x →
  eq (iterate f x) (lub (fun z ⇒ ∃ y, lt y x ∧ eq z (iterate f y))).
Lemma iterate_zero : eq (iterate f zero) (bot L).

Lemma iterate_chain : chain f (iterate f t).

O-inductions. In this section, we fix a lattice L, an ordinal o and an operator O:L → L. We write eqL for the equality on L and eqO for the equality on o.

AFT provides an alternative characterisation of lfps using what are called *O*-inductions, which relax the definition of chain by allowing the sequence to grow "slower". In the application of AFT to logic programming this boils down to not necessarily applying all applicable rules at every step.

```
Definition O_refinement (x y:L) := leq x y ∧ leq y (join x (O x)).
Definition O_induction (i:o → L) := (∀ y, O_refinement (i y) (i (succ y)))
  ∧ (∀ y, limit y → eqL (i y) (lub (fun l ⇒ ∃ z, lt z y ∧ eqL l (i z))))
  ∧ ∀ x y, eqO x y → eqL (i x) (i y).
```

An O_induction that cannot be refined further is called terminal. We formalise this notion by explicitly identifying an ordinal that returns its last element. If O is monotonic, then all O_inductions converge to its least fixpoint.

```
Definition terminal (i:o → L) (o':o) := ∀ y, O_refinement (i o') y → eqL y (i o').
Lemma terminal_O_induction_limit : terminal i o' → eqL (i o') (lfp O).
```

One of the classic results in AFT states that every monotonic operator has an O_induction that is terminal. The proof uses the fact that any chain can be injected in some "large enough" ordinal.

```
Definition large_enough := ∀ f, increasing f →
  ∃ (i:O → L), O_induction f i ∧ ∀ y, chain f y → ∃ o, eqL y (i o).
```

Lemma `large_enough_terminal` : `large_enough` → ∀ `f`, `monotonic f` →
∃ `i o`, `O_induction f i` ∧ `terminal f i o`.

We also prove a weaker version of this result, where the size of the ordinal is allowed to depend on `f`.

Lemma `large_enough_terminal'` : `monotonic f` → `O_induction f 0 i` →
(∀ `y`, `chain f y` → ∃ `o`, `eqL y (i o)`) → ∃ `o`, `terminal L f 0 i o`.

We discuss the existence of such ordinals separately. In particular, we prove that the hypotheses of the last lemma hold for any monotonic function on any lattice assuming (i) that the chain is a total order and (ii) a restricted form of Markov's principle.

Hypothesis `Hf` : ∀ `x y:L`, `chain f x` → `chain f y` → `leq x y` ∨ ~`leq x y`.

Hypothesis `forall_exists`: ∀ (`S:L → Prop`),
(~∀ `x`, `S x` → `leq x y`) → ∃ `x`, `S x` ∧ `less y x`.

Lemma `chain_has_large_enough` : `monotonic f` →
∃ `0 i`, `O_induction f 0 i` ∧ (∀ `y`, `chain f y` → ∃ `o`, `eqL y (i o)`).

The construction of the ordinal is a bit involved, and due to space restrictions we omit it here.

3   Approximators and Fixpoints

The bulk of AFT is built on the notion of approximator of an operator O: a (precision-)monotonic function on the billatice that coincides with O on exact values. Throughout this section we assume the lattice L to be fixed, as well as `O:L → L` and `A:BiLattice L → BiLattice L`.

Definition `approximator O A` := `monotonic A` ∧ ∀ `x`, `eqL (A (x,x)) (O x,O x)`.

Two types of fixpoints used in AFT are defined directly from the approximator: supported fixpoints are simply fixpoints of A; the Kripke-Kleene fixpoint is the least fixpoint of A.

Definition `supported_fp (x:BiLattice L)` := `fixpoint A x`.
Definition `Kripke_Kleene_fp` := `lfp A`.

Stable and well-founded fixpoints are defined using the stable revision operator `stable_revision`, which maps (x, y) to $(\text{lfp } A(\cdot, y)_1, \text{lfp } A(x, \cdot)_2)$. The two components of this operator are defined separately using two operators `partial_A_1` and `partial_A_2`. Reasoning about the monotonicity of these allows us to show that `stable_revision` is monotonic, after which we can define the above-mentioned fixpoints and prove their usual relationships.

Definition `stable_fp (x:BiLattice L)` := `fixpoint (stable_revision A) x`.
Definition `wf_fp` := `lfp (stable_revision A)`.

Lemma stable_fp_fp_A : monotonic A → ∀ x, stable_fp A x → supported_fp A x.
Lemma wf_fp_stable : monotonic A → stable_fp A (wf_fp A).
Lemma wf_fp_fp : monotonic A → supported_fp A (wf_fp A).
Lemma Kripke_Kleene_wf_fp : monotonic A → leq (Kripke_Kleene_fp A) (wf_fp A).

An element (x,y) of BiLattice L is consistent (L_consistent) if leq x y, i.e., if the set of elements z it approximates is non-empty, and an operator is consistent (A_consistent) if it maps consistent elements to consistent elements. In particular, all approximators are A_consistent. If A_consistent A, then all values in an O_induction over A are L_consistent; furthermore, if A is symmetric (meaning $A(x,y)_1 = A(y,x)_2$ for all x and y), then the stable revision operator is consistent as well: A_consistent (stable_revision A).

To prove additional results on preservation of consistency we needed to assume the existence of a large enough ordinal.

Hypothesis Ho : large_enough o (BiLattice L).
Lemma consistent_lfp : monotonic A → A_consistent A → L_consistent (lfp A).
Lemma consistent_Kripke_Kleene_fp : L_consistent (Kripke_Kleene_fp A).
Lemma wf_consistent : A_symmetric A → L_consistent (wf_fp A).

These lemmas also follow from the fact that every element in chain A is L_consistent; however, we were not able to prove this result without assuming existence of a large-enough ordinal, either.

Lastly, we show that well-founded fixpoints can be computed via *well-founded inductions* (wf_induction), first defined by Denecker and Vennekens [15]. These are transfinite sequences over BiLattice L that generalise O-inductions: a refinement either updates (x,y) by following the approximator (to something at most as precise as $A(x,y)$), or it decreases the second component, intuitively by removing elements that could only ever be derived by means of ungrounded (self-supporting) reasoning.

Inductive A_refinement (x y : BiLattice L) : Prop :=
| A_application : leq x y → leq y (join x (A x)) → A_refinement x y
| A_unfounded : eqL (fst x) (fst y) → leq (snd y) (snd x) →
                    leq (snd (A y)) (snd y) → A_refinement x y.

The definition of wf_induction is similar to the definition of O_induction, using A_refinement instead of O_refinement. We can similarly define a notion of wf_terminal, and use it to relate each wf_induction with fixpoints of A.

Lemma wf_induction_fp : wf_terminal i o' → fixpoint A (i o').
Lemma wf_terminal_stable_fp : wf_terminal i o' → stable_fp A (i o').
Lemma wf_terminal_wf_fp : wf_terminal i o' → eqL (i o') (wf_fp A).

The last two results are proved in two steps. First, we use transfinite induction to show that any wf_induction always stays below those fixpoints. Secondly, we define an auxiliary notion

Definition prudent a := ∀ x, leq (fst (A (x,snd a))) x → leq (fst a) x.

Intuitively, this tells us that fst a is derived for a "good" reason. Indeed, a is *not* prudent in case there is some x smaller than fst a that is a prefixpoint of

$A(\cdot, a_2)$, meaning that the only way to derive a would be starting from something larger than x in the first place: fst a could not be derived from the ground up.

We then show that all elements of a wf_induction are prudent, which follows from the definition of prudent.

This concludes the first part of the formalisation. After fine-tuning the definitions of ordinals and lattices and finishing the respective formalisations, the development of AFT described in this section was then relatively straightforward. The main challenges were (a) identifying the results that depend on the possibility of building a large-enough ordinal and (b) developing direct arguments for several results where the classical proof is by contradiction. The latter turned out sometimes to be a tricky exercise, but always possible.

4   An Example: Propositional Logic Programming

We now formalise a complete example, to show that our theory is applicable. We focus on propositional logic programming with negation, and show that the standard, classical, semantics correspond to fixpoints in AFT.

Throughout this section we assume a fixed non-empty set symb:Set of propositional symbols with decidable equality. The type of literals is defined as an inductive type with two constructors pos,neg:symb $\rightarrow$ literal, and rule as a list literal (the body of the rule) paired with a single symb (the head of the rule). We add the standard notation h :- b for (b,h):rule. Finally, a program is a list of rules. We define predicates pos_L, pos_R and pos_P for positive literals, rules and programs – those where all literals are built using the constructor pos. We also define the list symbs_in_P P of all symbols in P:program; this list is defined straightforwardly by going through the program and appending any symbols found, with no efforts to remove duplicates.

The next step is defining the standard semantics of logic programming. The type interpretation is simply defined as symb $\rightarrow$ Prop; this is also the carrier type of the lattice PowerSet symb, on which we work later on. Interpretations map literals, rules and programs to propositions in the natural way, and I:interpretation is a model of P:program if I maps P to a true proposition. We also define supported_model: an interpretation I where all true propositional symbols are supported by some rule whose body is also true in I – the function intlL generalizes an interpretation to a list of literals.

Definition supported_model (I:interpretation) (P:program) : Prop :=
    model P $\wedge$ $\forall$ s, I s $\rightarrow$ $\exists$ b, In (s :- b) P $\wedge$ int_lL I b.

A least_model is a model that is smaller (wrt the order in PowerSet symb) than all other models, and a minimal_model is one that is not strictly larger than any other model.

The semantics of programs with negation is classically defined using the notion of reduct [17]. Computing the reduct of a program requires being able to decide whether an interpretation satisfies the negative literals in the body of a rule; instead, we define what it means for a program Pr to be a reduct

of another program P, and show that there is at most one program with this
property (up to reordering and duplication of rules). Furthermore, this program
is always positive.

```
Definition reduct_R (r:rule) := (head r :− pos_atoms (body r)).
```

```
Definition reduct (I:interpretation) (P Pr:program) : Prop :=
  ∀ r, In r Pr ↔ ∃ r', In r' P ∧ all_negs_true r' I ∧ r = reduct_R r'.
```

```
Lemma reduct_unique : reduct I P Pr → reduct I P Pr' → ∀ r, In r Pr → In r Pr'.
Lemma reduct_pos : reduct I P P' → pos_P P'.
```

We can now define stable models, van Emden and Kowalski's immediate consequence operator, and Fitting's consequence operator. The latter works on
BiLattice (PowerSet symb), viewing the pair (I,J) as the knowledge that
everything in I holds and nothing outside J holds.

```
Definition stable (I:interpretation) (P:program) :=
  ∃ Pr, reduct I P Pr ∧ minimal_model I Pr.
```

```
Definition consequence (P:program) : PowerSet symb → PowerSet symb :=
  fun I s ⇒ ∃ r, In r P ∧ s = head r ∧ int_1L I (body r).
```

```
Definition comb_int_L (I J:interpretation) (l:literal) :=
  match l with pos s ⇒ I s | neg s ⇒ ∼J s end.
```

```
Definition comb_int_P (P:program) (I J:interpretation) : interpretation :=
  fun s ⇒ ∃ r, In r P ∧ s = head r ∧ ∀ l, In l (body r) → comb_int_L I J l.
```

```
Definition Fitting_cons (P:program) :
  BiLattice (PowerSet symb) → BiLattice (PowerSet symb) :=
  fun X ⇒ match X with (I1,I2) ⇒ (comb_int_P P I1 I2,comb_int_P P I2 I1) end.
```

From these definitions we can show the classical results that all models of
a program are prefixpoints of the associated immediate consequence operator,
and all supported models are fixpoints. The converse implications require some
classical reasoning.

```
Lemma model_prefp : model I P → prefp (immediate_cons P) I.
Lemma supp_model_fp : supported_model I P → fixpoint (immediate_cons P) I.
```

```
Hypothesis I_classical : ∀ I symb, I symb ∨ ∼I symb.
```

```
Lemma prefp_model : prefp (immediate_cons P) I → model I P.
Lemma fp_supp_model : fixpoint (immediate_cons P) I → supported_model I P.
```

With this assumption we can also show that positive programs have a unique
minimal model, which coincides with lfp (immediate_cons P).

```
Lemma pos_P_minimal_lfp :
  pos_P P → minimal_model I P → eqL (lfp (immediate_cons P)) I.
```

We then move to the Fitting consequence operator. We show that it is monotonic and that it approximates the immediate consequence operator. Furthermore, stable models of a program are stable fixpoints of the Fitting consequence operator; the converse holds provided that reducts always exist.

```
Lemma stable_model_stable_fp_Fitting :
   stable I P → stable_fp (Fitting_cons P) (I,I).
```

```
Hypothesis P_has_reduct : ∀ I, ∃ P', reduct I P P'.
```

```
Lemma stable_fp_Fitting_stable_model :
   stable_fp (Fitting_cons P) (I,I) → stable I P.
```

Next, we turn our attention to the well-founded semantics of logic programs. We formalise the construction of well-founded models described in [35], which we now describe.

Well-founded models are built using a notion of partial interpretation – two disjoint[3] lists of propositional symbols, those known to be true, and those known to be false. We define this a record type. Partial interpretations are ordered by knowledge.

```
Record partial_int := { ppos : list symb;
                        pneg  : list symb;
                        pcons : disjoint ppos pneg }.
```

```
Definition p_le (I J: partial_int) : Prop :=
   incl (ppos I)(ppos J) ∧ incl (pneg I) (pneg J).
```

Satisfaction of literals simply reduces to checking whether the underlying propositional symbol is included in the corresponding list. This notion generalises to rules and programs in the natural way. Satisfaction is decidable, since it is based on finite lists over a decidable type. Falsification (stronger than the negation of satisfaction) is defined similarly, but checking whether the symbol underlying a positive (respectively, negative) literal is in the negative (resp. positive) list of symbols in the partial interpretation.

Well-founded models are built as fixpoints of an operator that works distinctly on the positive and negative parts of a partial interpretation. For the positive part, the authors use an operator T that naturally generalises the immediate consequence operator. (We prepend the authors' initials to the name of the Coq counterparts to these operators.)

```
Fixpoint GRS_T (P: program) (I: partial_int) : list symb :=
   match P with
   | nil ⇒ nil
   | (b,h) :: P' ⇒ if satisfy_lL_dec I b then (h :: GRS_T P' I) else (GRS_T P' I)
   end.
```

GRS_T simply iterates through a program and collects the heads of the rules whose bodies are satisfied in I. We show that it behaves as expected.

---

[3] Disjointness of this lists is called *consistency* in the original reference.

Lemma GRS_T_char : In s (GRS_T P I) → ∃ b, In (b,s) P ∧ satisfy_1L I b.
Lemma GRS_T_char' : In (b,s) P → satisfy_1L I b → In s (GRS_T P I).

For the negative part, the authors use the notion of unfounded set wrt a
partial interpretation I – a set of atoms that can never be proven by extending
I – and define the greatest unfounded set wrt I as the union of all these sets.
The latter definition is not directly translatable to Coq; instead, we construct
this set directly and prove that it has the desired property.

Definition unfounded_R (l: list symb) (I: partial_int) (s: symb) (r: rule) :=
  match r with (b,h) ⇒ s = h → falsify_1L I b ∨ ∃ s', In (pos s') b ∧ In s' l end.

Definition unfounded_P (l: list symb) (I: partial_int) (s: symb) :=
  ∀ r, In r P → unfounded_R l I s r.

Definition unfounded (l: list symb) (I: partial_int) : Prop :=
  ∀ s, In s l → In s (symbs_in_P P) ∧ unfounded_P l I s.

To construct the greatest unfounded set wrt I, we first define a function
remove_not_unfounded that removes all elements from a list l:list symb that are
unfounded wrt I. We then define another function remove_n_times that iterates
the previous function, and prove that this reaches a fixpoint when given length
l as an argument. Finally, gus I is defined by computing this fixpoint from the
list of symbols in P. We show that this is indeed the greatest unfounded set wrt
I according to [35].

Lemma gus_unfounded : unfounded (gus I) I.
Lemma gus_greatest: unfounded l I →
  (∀ s, In s l → In s (symbs_in_P P)) → ∀ s, In s l → In s (gus I).

In the second lemma, the extra condition is needed to account for the fact that
our type symb may include symbols not in the Herbrand base of P.
    The operator U maps each interpretation to its greatest unfounded set.

Definition GRS_U : partial_int → list symb := fun I ⇒ gus I.

Combining T and U yields an operator W whose iterations from the empty
partial interpretation are always partial interpretations (i.e., consistent – cf.
Lemma 3.4 in [35]). In general, though, W does not preserve consistency, and
the proof of the lemma cited uses some properties that hold specifically for
iterates of W. We call partial interpretations with this properties *rich partial
interpretations*.

Record rich_pint : Set :=
  { rpint :> partial_int;
    rgen : ∀ s, In s (ppos rpint) → ∃ b, In (b,s) P ∧ satisfy_1L rpint b;
    runf : ∀ s, In s (pneg rpint) →
              ∃ l, (∀ s, In s l → In s (symbs_in_P P))
              ∧ In s l ∧ unfounded l rpint;
    rprop : ∀ J l p, rpint ≤p J → In p (ppos rpint) → unfounded l J →
  ~In p l}.

For I:rich_pint, property rgen imposes that all elements in ppos I are supported by some rule that is true in I; dually, runf ensures that all elements in pneg I must be in some unfounded set wrt I. The last property states that no element of ppos I can ever become unfounded, in the sense that it can not be in any unfounded set wrt a partial interpretation J extending I.

We show that, for I:rich_pint, the lists GRS_T I and GRS_U I are disjoint and satisfy the properties rgen, runf and rprop. This allows us to define GRS_W_rich as an operator over rich_pints, and iterating it from the empty partial interpretation we obtain GRS_W_aux : nat → rich_pint (the partial interpretation W_n in [35]). Finally, we define GRS_W : nat → partial_int by forgetting the extra structure in GRS_W_aux n.

The next step is showing that GRS_W reaches a fixpoint (the well-founded model of the program). We employ a counting trick: we show that the number of decided symbols in GRS_W n (i.e. the number of symbols appearing in either ppos (GRS_W n) or pneg (GRS_W n)) increases unless GRS_W n and GRS_w (S n) are equal as partial interpretations.

Lemma GRS_W_conv_1 : ∀ (I:rich_pint), ∼I =p preW I →
  ∃ s, decided s (preW I) ∧ ∼decided s I.

Lemma GRS_W_conv_2 : ∼GRS_W n =p GRS_W (S n) →
  S n ≤ size_as_set (ppos (GRS_W (S n)) ++ pneg (GRS_W (S n))).

Lemma GRS_W_fp_char :
  GRS_W (length (symbs_in_P P)) =p GRS_W (length (symbs_in_P P) + 1).

Definition wf_model : partial_int := GRS_W (length (symbs_in_P P)).

The last step is showing that wf_model P and wf_fp (Fitting_cons P) coincide. The hard part of the proof is relating the second component of the stable revision operator built from Fitting_cons P with the notion of greatest unfounded set.

Lemma gus_Fitting : In s (symbs_in_P P) → In s (gus I) ↔
  ∼snd (stable_revision (Fitting_cons P) (p_int_to_pair_int I)) s.

This lemma relies heavily on the fact that I is built from two lists, and therefore we can always decide whether I s or ∼I s holds; this implies that the fixpoint constructions inside the computation of stable_revision converge in a (computable) number of iterations, and that the result is also decidable in the same sense. From this lemma, we prove the final result in our formalisation. The function p_int_to_pair_int is the natural map from partial_ints to BiLattice (PowerSet symb).

Lemma wf_model_wf_fp :
  eqL (p_int_to_pair_int wf_model) (wf_fp (Fitting_cons P)).

5   Related Work

To the best of our knowledge, this is the first time that AFT has been formalised using a theorem prover. The most closely related works that we are aware of deal are formalisations of ordinal theory and transfinite induction, or of results in lattice theory.

Several authors have formalised transfinite induction in Coq [2,19,26,30]. Due to the difficulty of developing such a formalisation satisfactorily in a constructive setting, several of them [26,30] are based on classical set theory.

Barras [2] discusses formalising a model of the Calculus of Inductive Constructions in Coq. His work uses an impredicative definition of ordinals as the least collection of transitive sets such that any set of members of the collection also belongs to the collection, and proves the Knaster-Tarski theorem.

Grimm [19] formalises three different types of ordinals in Coq, including arithmetic operations on them and a notion of order. To be comparable, ordinals need to be reduced to a normal form. The formalisation also includes principles for reasoning using transfinite induction, and the author proves a correspondence to van Neumann ordinals.

Due to the authors' different purposes, all these works end up being difficult to use in our setting, which motivated us to include a novel definition of ordinals in our current contribution.

Other formalisations of Tarki's fixed point theorem are included in the work of Grall [18] and in the CoLoR library [4,25]. However, these works do not deal with ordinals, and even though their approach to lattice theory is similar to ours the overlap is very reduced.

Ordinals have also been formalised in Agda using Homotopy Type Theory (HoTT), where extensional equality occurs naturally. In particular, it has been shown [21] that, in HoTT, the constructive definition of ordinals as a hereditarily transitive set coincides with their definition as a type with a transitive, wellfounded and extensional order relation.

One may question whether formalising AFT requires using ordinals and transfinite induction at all, as the latter can often be replaced by well-founded induction [23]. We chose not to investigate this alternative path, as we were trying to follow existing references on AFT as closely as possible. Furthermore, several concepts in AFT are defined directly using ordinals (e.g. well-founded inductions), so bypassing ordinals would require major changes to the theory.

6   Conclusions

We described a formalisation of approximation fixpoint theory in the theorem prover Coq, together with an example of how it can be applied to propositional logic programming. In our work, we tried to work constructively as much as possible, in the sense that we developed alternative, direct, proofs instead of the standard proofs by contradiction or case analysis that are found in the literature. Where such an approach was not possible, we opted for explicitly identifying our classical assumptions (e.g., decidability of equality on some types). We believe there is value in this exercise, as it increases our understanding of how much AFT depends on classical reasoning.

As an example, we formalised the classical theory of propositional logic programming, including the immediate consequence operator, Fitting's consequence operator and the construction of well-founded models as fixpoints of an operator based on unfounded sets, and showed that these can be characterised through corresponding concepts in AFT. Again, we assume some principles of classical reasoning in these proofs, namely decidability of equality on propositional symbols.

We included a few examples throughout our work to show that there we can actually construct objects of the types we define constructively – in particular, we can construct ordinals such as ω or any polynomial on ω. Our example on propositional logic programming also illustrates that, for the kind of finite structures that are often used in practice, our development applies without resorting to full-blown classical reasoning.

Our development consists of 6850 lines of Coq code divided into 6 files, containing 166 definitions and 426 lemmas. The source code was developed using Coq version 8.18, and is available at https://zenodo.org/records/10709614 [7].

References

1. Antic, C., Eiter, T., Fink, M.: Hex semantics via approximation fixpoint theory. In: Cabalar, P., Son, T.C. (eds.) Logic Programming and Nonmonotonic Reasoning. Lecture Notes in Computer Science(), vol. 8148, pp. 102–115. Springer, Berlin (2013). https://doi.org/10.1007/978-3-642-40564-8_11
2. Barras, B.: Sets in Coq, Coq in sets. J. Formaliz. Reason **3**(1), 29–48 (2010)
3. Bertot, Y., Castéran, P.: Interactive Theorem Proving and Program Development - Coq'Art: The Calculus of Inductive Constructions. Texts in Theoretical Computer Science. An EATCS Series, Springer, Cham (2004)
4. Blanqui, F., Koprowski, A.: CoLoR: a Coq library on well-founded rewrite relations and its application to the automated verification of termination certificates. Math. Struct. Comput. Sci. **21**(4), 827–859 (2011)
5. Bogaerts, B., Cruz-Filipe, L.: Fixpoint semantics for active integrity constraints. Artif. Intell. **255**, 43–70 (2018). https://doi.org/10.1016/j.artint.2017.11.003
6. Bogaerts, B., Cruz-Filipe, L.: Stratification in approximation fixpoint theory and its application to active integrity constraints. ACM Trans. Comput. Log. **22**(1), 1–19 (2021). https://doi.org/10.1145/3430750
7. Bogaerts, B., Cruz-Filipe, L.: A formalisation of approximation fixpoint theory in Coq (2024). https://doi.org/10.5281/zenodo.10709613
8. Bogaerts, B., Jakubowski, M.: Fixpoint semantics for recursive SHACL. In: Formisano, A., et al. (eds.) Proceedings 37th International Conference on Logic Programming (Technical Communications), ICLP Technical Communications 2021, Porto (virtual event), 20-27th September 2021. EPTCS, vol. 345, pp. 41–47 (2021). https://doi.org/10.4204/EPTCS.345.14
9. Bogaerts, B., Vennekens, J., Denecker, M.: Grounded fixpoints and their applications in knowledge representation. Artif. Intell. **224**, 51–71 (2015). https://doi.org/10.1016/j.artint.2015.03.006
10. Bogaerts, B., Vennekens, J., Denecker, M.: Safe inductions and their applications in knowledge representation. Artif. Intell. **259**, 167–185 (2018). http://www.sciencedirect.com/science/article/pii/S000437021830122X

11. Charalambidis, A., Rondogiannis, P., Symeonidou, I.: Approximation fixpoint theory and the well-founded semantics of higher-order logic programs. Theory Pract. Log. Program. **18**(3–4), 421–437 (2018). https://doi.org/10.1017/S1471068418000108

12. Denecker, M., Marek, V., Truszczyński, M.: Approximations, stable operators, well-founded fixpoints and applications in nonmonotonic reasoning. In: Minker, J. (ed.) Logic-Based Artificial Intelligence. The Springer International Series in Engineering and Computer Science, vol. 597, pp. 127–144. Springer, Boston (2000). https://doi.org/10.1007/978-1-4615-1567-8_6

13. Denecker, M., Marek, V., Truszczyński, M.: Uniform semantic treatment of default and autoepistemic logics. Artif. Intell. **143**(1), 79–122 (2003). https://doi.org/10.1016/S0004-3702(02)00293-X

14. Denecker, M., Marek, V., Truszczyński, M.: Reiter's default logic is a logic of autoepistemic reasoning and a good one, too. In: Brewka, G., Marek, V., Truszczyński, M. (eds.) Nonmonotonic Reasoning – Essays Celebrating Its 30th Anniversary, pp. 111–144. College Publications (2011). http://arxiv.org/abs/1108.3278

15. Denecker, M., Vennekens, J.: Well-founded semantics and the algebraic theory of non-monotone inductive definitions. In: Baral, C., Brewka, G., Schlipf, J.S. (eds.) LPNMR. Lecture Notes in Computer Science, vol. 4483, pp. 84–96. Springer, Cham (2007). https://doi.org/10.1007/978-3-540-72200-7_9

16. Fitting, M.: Fixpoint semantics for logic programming – a survey. Theor. Comput. Sci. **278**(1–2), 25–51 (2002). https://doi.org/10.1016/S0304-3975(00)00330-3

17. Gelfond, M., Lifschitz, V.: The stable model semantics for logic programming. In: Kowalski, R.A., Bowen, K.A. (eds.) ICLP/SLP, pp. 1070–1080. MIT Press, Cambridge (1988). http://citeseer.ist.psu.edu/viewdoc/summary?nodoi=10.1.1.24.6050

18. Grall, H.: Proving fixed points. Technical Report. HAL-00507775, HAL archives ouvertes (2010)

19. Grimm, J.: Implementation of three types of ordinals in Coq. Technical Report. RR-8407, INRIA (2013)

20. Heyninck, J., Bogaerts, B.: Non-deterministic approximation operators: ultimate operators, semi-equilibrium semantics and aggregates (full version). CoRR **abs/2305.10846** (2023). https://doi.org/10.48550/arXiv.2305.10846

21. de Jong, T., Kraus, N., Forsberg, F.N., Xu, C.: Set-theoretic and type-theoretic ordinals coincide. In: LICS, pp. 1–13 (2023).https://doi.org/10.1109/LICS56636.2023.10175762

22. Konolige, K.: On the relation between default and autoepistemic logic. Artif. Intell. **35**(3), 343–382 (1988). https://doi.org/10.1016/0004-3702(88)90021-5

23. Kuratowski, C.: Une méthode d'élimination des nombres transfinis des raisonnements mathématiques. Fundamenta Mathematicae **3**(1), 76–108 (1922). http://eudml.org/doc/213282

24. Liu, F., Bi, Y., Chowdhury, M.S., You, J., Feng, Z.: Flexible approximators for approximating fixpoint theory. In: Khoury, R., Drummond, C. (eds.) Advances in Artificial Intelligence. Lecture Notes in Computer Science(), vol. 9673, pp. 224–236. Springer, Cham (2016). https://doi.org/10.1007/978-3-319-34111-8_28

25. Blanqui, F.: https://github.com/fblanqui/color/blob/master/Util/Relation/Tarski.v. Accessed 02 June 2021

26. Castéran, P.: https://github.com/coq-community/hydra-battles/tree/master/theories/ordinals. Accessed 02 June 2021

27. Moore, R.C.: Semantical considerations on nonmonotonic logic. Artif. Intell. **25**(1), 75–94 (1985). https://doi.org/10.1016/0004-3702(85)90042-6
28. Pelov, N., Denecker, M., Bruynooghe, M.: Well-founded and stable semantics of logic programs with aggregates. TPLP **7**(3), 301–353 (2007). https://doi.org/10.1017/S1471068406002973
29. Reiter, R.: A logic for default reasoning. Artif. Intell. **13**(1–2), 81–132 (1980). https://doi.org/10.1016/0004-3702(80)90014-4
30. Simpson, C.: Set-theoretical mathematics in Coq. CoRR **abs/math/0402336** (2004)
31. Strass, H.: Approximating operators and semantics for abstract dialectical frameworks. Artif. Intell. **205**, 39–70 (2013). https://doi.org/10.1016/j.artint.2013.09.004
32. Tarski, A.: A lattice-theoretical fixpoint theorem and its applications. Pac. J. Math. (1955)
33. Truszczyński, M.: Strong and uniform equivalence of nonmonotonic theories - an algebraic approach. Ann. Math. Artif. Intell. **48**(3–4), 245–265 (2006). https://doi.org/10.1007/s10472-007-9049-2
34. van Emden, M.H., Kowalski, R.A.: The semantics of predicate logic as a programming language. J. ACM **23**(4), 733–742 (1976). https://doi.org/10.1145/321978.321991
35. Van Gelder, A., Ross, K.A., Schlipf, J.S.: The well-founded semantics for general logic programs. J. ACM **38**(3), 620–650 (1991). https://doi.org/10.1145/116825.116838
36. Van Hertum, P., Cramer, M., Bogaerts, B., Denecker, M.: Distributed autoepistemic logic and its application to access control. In: Kambhampati, S. (ed.) Proceedings of the Twenty-Fifth International Joint Conference on Artificial Intelligence, IJCAI 2016, New York, NY, USA, 9-15 July 2016, pp. 1286–1292. IJCAI/AAAI Press (2016). http://www.ijcai.org/Abstract/16/186
37. Vennekens, J., Gilis, D., Denecker, M.: Splitting an operator: algebraic modularity results for logics with fixpoint semantics. ACM Trans. Comput. Log. **7**(4), 765–797 (2006). https://doi.org/10.1145/1182613.1189735
38. Vennekens, J., Mariën, M., Wittocx, J., Denecker, M.: Predicate introduction for logics with a fixpoint semantics. Parts I and II. Fund. Inform. **79**(1–2), 187–227 (2007)

Constructing Morphisms for Arithmetic Subsequences of Fibonacci

Wieb Bosma[⊠] and Henk Don

Department of Mathematics, Radboud Universiteit, Nijmegen, The Netherlands
{w.bosma,h.don}@math.ru.nl

Abstract. From a general theorem by Dekking it follows that an arithmetic subsequence of any morphic sequence is morphic again. The construction of such a morphism is not directly obvious. In this note we demonstrate the explicit construction of a morphism generating an arbitrary arithmetic subsequence of the infinite fixed point of the Fibonacci morphism.

Keywords: morphic sequences · automata · Fibonacci

1 Introduction

This paper originates in discussions with Hans Zantema about the complexity of morphic and automatic sequences, see also [17,18]. One of the questions that arose was what the complexity is of the sequence of even (or odd) numbered entries of the well-known Fibonacci morphic sequence; the notion of complexity we use here is that of *morphic complexity*, see the next section for details. Because of a theorem of Dekking (see [1] 7.91), arithmetic subsequences of a morphic sequence are morphic again. The construction of a corresponding morphism is part of Dekking's proof. However, Dekking's theorem is more general and goes beyond arithmetic subsequences. This generality has two consequences. Firstly, it is not so easy to obtain the morphism explicitly in a concrete example involving arithmetic subsequences. Secondly, more efficient constructions can be found when restricting to special cases. In particular, arithmetic subsequences might be generated with less complex morphisms.

Our goal in this paper is to address these two issues. The proof of the main result in this paper (Theorem 1) gives an easy, explicit construction generating any arithmetic subsequence of Fibonacci by a morphism. Along the way, we measure and bound the morphic complexity of these arithmetic subsequences by properties of the construction; the paper [4] by Dekking only gives a general bound on the size of the alphabet.

Although some experimentation shows that the same construction can be applied much more generally (but not universally) to morphic sequences, we restrict to the Fibonacci case in this paper.

© The Author(s), under exclusive license to Springer Nature Switzerland AG 2024
V. Capretta et al. (Eds.): *Logics and Type Systems in Theory and Practice*, LNCS 14560, pp. 100–110, 2024.
https://doi.org/10.1007/978-3-031-61716-4_6

2   Preliminaries and Main Result

Denote the Fibonacci numbers by F_n, with $F_0 = 0$, $F_1 = 1$, and $F_n = F_{n-1} + F_{n-2}$ for $n \geq 2$.

A sequence over an alphabet Σ is called a *pure morphic sequence* if it can be obtained as an infinite fixed point of a morphism (or substitution) $g : \Sigma \rightarrow \Sigma^*$. Our key example is the morphic Fibonacci sequence on $\Sigma = \{0, 1\}$, defined by

$$\sigma = \mathcal{F}^\infty(0) \quad \text{where} \quad \begin{cases} \mathcal{F}(0) = 01, \quad \text{and} \\ \mathcal{F}(1) = 0. \end{cases}$$

An initial segment of σ is

$$01001010010010100101001001010010010100101001001010.$$

The segment above shows σ from $\sigma[0]$ up to $\sigma[49]$, where by $\sigma[j]$ we denote the j-th entry of σ, for $j \geq 0$.

If Σ and $\tilde{\Sigma}$ are alphabets, a *coding* is a map $\tau : \Sigma \rightarrow \tilde{\Sigma}$. A coding does not have to be injective. A sequence $\tilde{\rho} \in \tilde{\Sigma}^\infty$ is a *morphic sequence* if there exists a pure morphic sequence ρ on an alphabet Σ and a coding $\tau : \Sigma \rightarrow \tilde{\Sigma}$ such that $\tilde{\rho}[j] = \tau(\rho[j])$ for all $j \geq 0$. By the complexity of the morphism g, we mean the sum of the word lengths of the images of the letters $\sum_{a \in \Sigma} |g(a)|$. The *morphic complexity* of a morphic sequence is the minimal complexity of a morphism producing this sequence (with or without a coding). Note that there are various alternative notions of complexity for infinite sequences. In this paper, complexity of an infinite sequence always means morphic complexity. Every non-trivial morphic sequence has complexity at least 3. The morphic Fibonacci sequence σ is essentially the only sequence with complexity 3.

Our main result in this paper is the following for the subsequences formed by a given residue class k mod m of the indices.

Theorem 1. *Let σ be the morphic Fibonacci sequence. For any modulus $m > 1$ and any k with $0 \leq k < m$, the arithmetic subsequence $(\sigma[n \cdot m + k])_{n \geq 0}$ is morphic with morphic complexity at most $(m + 1)F_{2m+2}$.*

A sharper but less explicit upper bound for the complexity is $(m + 1)F_{z(m)+2}$, where $z(m)$ is the smallest index j for which F_j is divisible by m; see the discussions later on in this paper, where also a good approximation for the complexity of our morphism is given.

3   Fibonacci Numbers

The proof of Theorem 1 is *constructive* and only uses a few elementary properties of the sequence of Fibonacci numbers. The following result is well-known, see for example [15]; we also include a proof here.

Lemma 1. *For every integer $m > 1$ the sequence $(F_n \bmod m)_{n \geq 0}$ is purely periodic; in particular, there exists $j \geq 2$ such that $m|F_n$ if and only if $j|n$.*

Proof. Extend the definition of F_n to negative n. By this definition, for all $n \in \mathbb{Z}$

$$F_{n+1} \equiv F_n + F_{n-1} \bmod m.$$

The sequence $(F_n \bmod m)_{n \in \mathbb{Z}}$ takes on at most m different values. Therefore, there must exist $u < v$ such that

$$F_u \equiv F_v \bmod m, \qquad F_{u+1} \equiv F_{v+1} \bmod m.$$

The recursive definition now implies for all $n \in \mathbb{Z}$,

$$F_{u+n} \equiv F_{v+n} \bmod m.$$

Hence the sequence $(F_n \bmod m)_{n \geq 0}$ is periodic with period $v - u$, and $F_{v-u} \equiv 0 \bmod m$.

Let $j \geq 2$ be the first strictly positive index with $F_j \equiv 0 \bmod m$. Define $a \equiv F_{j+1} \bmod m$. Then

$$F_j \equiv 0 \equiv aF_0 \bmod m, \qquad F_{j+1} \equiv a \equiv aF_1 \bmod m,$$

and the recursion gives $F_{j+n} \equiv aF_n \bmod m$ for all n. It follows that $F_n \equiv 0 \bmod m$ if and only if $j|n$, since a is coprime to m: any common factor would divide F_{j+1} and F_j, and by the recursive relation also F_{j-1}, and eventually $F_1 = 1$.

The (minimal) period $\pi(m)$ of $(F_n \bmod m)_{n \geq 0}$ is called the Pisano period. Since a pair of consecutive values in $(F_n \bmod m)_{n \geq 0}$ determines the sequence, it is easy to see that $\pi(m)$ is at most the maximum of the distance between equal pairs. This gives $\pi(m) \leq m^2$. Sharper results are available in the literature: it is known that $\pi(m) \leq 6m$, which is tight; see [6, 15].

Clearly, $F_{\pi(m)} \equiv 0 \bmod m$, but there could be a smaller Fibonacci number divisible by m. Let $z(m)$ be the smallest index j for which $F_j \equiv 0 \bmod m$. Then $z(m)$ divides $\pi(m)$. The Fibonacci numbers are known to show quite remarkable behaviour (see for instance [7]) and the divisibility properties are no exception; see Fig. 1 for a visualization of $\pi(m)$ and $z(m)$.

The following result will be useful for bounding the complexity of our morphisms.

Lemma 2 (Sallé, [13]). *Let $m \geq 2$. Then $z(m) \leq 2m$.*

This is the sharpest possible linear upper bound that works for all m. It is sharp if and only if $m = 6 \cdot 5^k$ for some $k \geq 0$, see [9]. However, the bound is quite weak for most values of m, as can already be seen from Fig. 1. Sharper bounds or exact values for different types of composite numbers m can be found in [9] and [10]. Tables with values of $z(m)$ up to $m = 10^5$ were published by the Fibonacci Association in 1965 ([16] and [8]), but they can of course easily be reproduced with the help of modern computers.

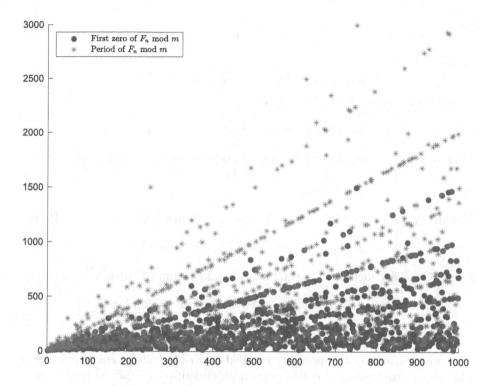

Fig. 1. The index of the smallest n such that $F_n \equiv 0 \bmod m$ and the period of $(F_n \bmod m)_{n \geq 0}$; m is on the horizontal axis.

4   Fibonacci Morphism

We now turn to properties of the morphic Fibonacci sequence. Let s be any word in $\{0,1\}^+$; by $\text{len}(s)$ we denote its length, and by $\text{wt}(s)$ its weight, that is, the number of letters equal to 1 in s. As usual we can apply the morphism $\mathcal{F}$ recursively to any word.

Lemma 3. *For every* $s \in \{0,1\}^+$ *of length* m *and weight* w, *and every* $j \geq 0$:

$$len(\mathcal{F}^j(s)) = F_{j+2} \cdot m - F_j \cdot w.$$

Proof. Note that by construction the lengths of the iterates of $\mathcal{F}$ on the letters are Fibonacci numbers:

$$\text{len}(\mathcal{F}^j(0)) = F_{j+2} \quad \text{and} \quad \text{len}(\mathcal{F}^j(1)) = F_{j+1},$$

justifying the name Fibonacci for this morphism. Also $\text{len}(\mathcal{F}(s)) = 2(m-w) + w = 2m - w$ and $\text{wt}(\mathcal{F}(s)) = m - w$. For $j \geq 2$ we find likewise that

$$\text{len}(\mathcal{F}^j(s)) = 2 \cdot \text{len}(\mathcal{F}^{j-1}(s)) - \text{wt}(\mathcal{F}^{j-1}(s)).$$

Since
$$\text{wt}(\mathcal{F}^{j-1}(s)) = \text{len}(\mathcal{F}^{j-1}(s)) - \text{len}(\mathcal{F}^{j-2}(s)),$$

we get
$$\text{len}(\mathcal{F}^{j}(s)) = \text{len}(\mathcal{F}^{j-1}(s)) + \text{len}(\mathcal{F}^{j-2}(s)).$$

So $\text{wt}(\mathcal{F}^{j}(s)) = \text{len}(\mathcal{F}^{j-2}(s))$, and also
$$\text{wt}(\mathcal{F}^{j}(s)) = \text{wt}(\mathcal{F}^{j-1}(s)) + \text{wt}(\mathcal{F}^{j-2}(s)).$$

That is, both the lengths and the weights satisfy the Fibonacci recurrence.

As $\text{len}(\mathcal{F}^{0}(s)) = m = F_2 \cdot m - F_0 \cdot w$ and $\text{len}(\mathcal{F}^{1}(s)) = 2m - w = F_3 \cdot m - F_1 \cdot w$, the result follows by induction.

Corollary 1. *For every $m \geq 2$ there exists $j \geq 2$ such that for every $s \in \{0,1\}^{+}$:*

$$len(s) = m \quad \Rightarrow \quad len(\mathcal{F}^{j}(s)) \equiv 0 \bmod m.$$

Proof. Choose j such that $F_j \equiv 0 \bmod m$, as in Lemma 1; then
$$len(\mathcal{F}^{j}(s)) = F_{j+2} \cdot m - F_j \cdot \text{wt}(s) \equiv 0 \bmod m$$

by Lemma 3.

A morphism with this property is called an *m-block stable substitution* in [5]. This paper discusses fixed points of m-block substitutions and $\mathcal{F}^3$ is given as an example of a 2-block stable substitution. See also [12].

We also use the following lemma, which says that every subword of σ occurs in each residue class of indices for every modulus. A subword of an infinite word here means a contiguous finite subsequence.

Lemma 4. *Let w be a subword of σ of length $m \geq 2$. For every k with $0 \leq k < m$, there exists $n \equiv k \bmod m$ such that $w = \sigma[n] \cdots \sigma[n + m - 1]$.*

Proof. Suppose that w starts at index n_0, so $w = \sigma[n_0] \cdots \sigma[n_0 + m - 1]$. Note that $\sigma = \mathcal{F}^{\infty}(0) = \mathcal{F}^{\infty}(1)$, so there exists j_0 such that for all $j \geq j_0$,

$$w = \mathcal{F}^{j}(1)[n_0] \cdots \mathcal{F}^{j}(1)[n_0 + m - 1].$$

Since $(F_n \bmod m)_{n \geq 0}$ is purely periodic (Lemma 1), and since $F_1 \equiv 1 \bmod m$, we can find arbitrary large j such that $F_{j+2} \equiv 1 \bmod m$. Take such $j \geq j_0$. Then write
$$\sigma = \mathcal{F}^{j}(\sigma) = \mathcal{F}^{j}(0)\mathcal{F}^{j}(1) \cdots$$

Using that $\text{len}(\mathcal{F}^{j}(0)) = F_{j+2}$, we find that

$$w = \sigma[F_{j+2} + n_0] \cdots \sigma[F_{j+2} + n_0 + m - 1].$$

Let $n_1 := F_{j+2} + n_0$. Clearly $n_1 \equiv F_{j+2} + n_0 \equiv n_0 + 1 \bmod m$. By iterating, we obtain for all k such that $0 \leq k \leq m - 1$ existence of $n \equiv k \bmod m$ such that $w = \sigma[n] \cdots \sigma[n + m - 1]$, which is the statement of the lemma.

By V_m we will denote the set of words of length m over $\{0,1\}$ that occur when we 'chop up' σ in consecutive words of length m:

$$V_m = \{\sigma[0]\cdots\sigma[m-1], \sigma[m]\cdots\sigma[2m-1], \sigma[2m]\cdots\sigma[3m-1], \ldots\}.$$

Note that many of the 2^m possible words over $\{0,1\}$ do not occur in σ. In fact, the Fibonacci sequence is known to be a Sturmian sequence. This means that only $m+1$ different words of length m occur ([1] Ch. 9, 10.5.8). Using Lemma 4, we conclude that $|V_m| = m+1$.

5   Proof of the Theorem

We are now ready to give our constructive proof of Theorem 1. For any $m \geq 2$ and $0 \leq k < m$, we will find an alphabet together with a morphism and coding which generate the arithmetic subsequence $(\sigma[k+n\cdot m])_{n\geq 0}$.

Proof. Let k,m be given. Choose j, as in Lemma 1, such that m divides F_j. Denote the words in V_m by $w_0, \ldots, w_m$. Define a new alphabet $\Sigma = \{a_0, \ldots, a_m\}$ and a map g by setting $g(w_i) = a_i$ for $w_i \in V_m$ and inductively extending g to concatenations of words in V_m: if $g(w)$ is defined, then $g(ww_i) = g(w)a_i$. On Σ, we define a morphism h by

$$h(a_i) = g(\mathcal{F}^j(g^{-1}(a_i))), \qquad 0 \leq i \leq m.$$

Note that this is well-defined, since the argument of g is of the form $\mathcal{F}^j(w_i)$, which is a word in $\{0,1\}^+$ of length a multiple of m by Lemma 3.

Assume without loss of generality that $w_0 = \sigma[0]\cdots\sigma[m-1]$. Define the infinite pure morphic sequence $h^\infty(a_0)$, then

$$
\begin{aligned}
h^\infty(a_0) &= (g \circ \mathcal{F}^j \circ g^{-1})^\infty(a_0) \\
&= (g \circ \mathcal{F}^\infty \circ g^{-1})(a_0) \\
&= (g \circ \mathcal{F}^\infty)(w_0) = g(\sigma).
\end{aligned}
$$

This means that there is a one-to-one correspondence between $h^\infty(a_0)$ and σ. Indeed, the n-th letter of $h^\infty(a_0)$ can be used to recover the n-th block of length m in σ as follows:

$$g^{-1}(h^\infty(a_0)[n]) = \sigma[n\cdot m]\cdots\sigma[(n+1)\cdot m-1].$$

This allows to create a coding $\tau : \Sigma \to \{0,1\}$ defined by

$$\tau(a_i) = (g^{-1}(a_i))[k] = w_i[k],$$

mapping the nth letter in $h^\infty(a_0)$ to $\sigma[n\cdot m+k]$, so $\tau(h^\infty(a_0))$ equals $(\sigma[n\cdot m+k])_{n\geq 0}$. This completes the construction of the morphism generating $(\sigma[n\cdot m+k])_{n\geq 0}$.

Note that $|\Sigma| = |V_m| = m + 1$. For $a_i \in \Sigma$, using that $\text{len}(w_i) = m$ for all i, we obtain by Lemma 3

$$\text{len}(h(a_i)) = \text{len}(g(\mathcal{F}^j(g^{-1}(a_i)))) = \frac{1}{m}\text{len}(\mathcal{F}^j(w_i)) \le F_{j+2}. \qquad (1)$$

The smallest j that works is $z(m)$, which is at most $2m$ by Lemma 2. It follows that the complexity of the morphism is at most $(m+1)F_{2m+2}$.

6   Examples

Example 1. By way of example we will construct the subsequence $(\sigma[4n+3])_{n \ge 0}$. First note that Theorem 1 gives the upper bound $5 \cdot F_{2\cdot4+2} = 5\cdot 55 = 275$ for the complexity of this subsequence. This bound is based on the general Lemma 2. If the first zero $z(m)$ of $(F_n \bmod m)_{n \ge 0}$ is known, we could possibly do better, since we can replace the bound in Theorem 1 by $(m+1)F_{z(m)+2}$. In our current example, the first nonzero Fibonacci number divisible by 4 is $F_6 = 8$. This implies that the upper bound for the complexity can already be improved to $5 \cdot F_{6+2} = 105$.

There are 5 subwords w_0, w_1, w_2, w_3, w_4, of length 4 in V_4:

$$w_0 = 0100, \qquad w_1 = 1010, \qquad w_2 = 0101, \qquad w_3 = 0010, \qquad w_4 = 1001.$$

In fact, they are all present already in the initial segment of σ shown above. They correspond to the letters a_0, a_1, a_2, a_3, a_4 of our new alphabet Σ by the map $g(w_i) = a_i$. Since F_6 is the first positive Fibonacci number divisible by 4, the length of $\mathcal{F}^6(w)$ is a multiple of 4 for all $w \in V_4$ (by Corollary 1). Computing the images $\mathcal{F}^6(w_i)$ and defining the morphism h by $h(a_i) = g(\mathcal{F}^6(w_i))$ gives

$$a_0 \mapsto a_0a_1a_0a_1a_2a_3a_2a_3a_2a_3a_4a_3a_4a_0a_4a_0a_4a_0a_1,$$
$$a_1 \mapsto a_0a_1a_0a_1a_2a_3a_2a_3a_4a_3a_4a_3a_4a_0a_4a_0a_1,$$
$$a_2 \mapsto a_0a_1a_0a_1a_2a_3a_2a_3a_2a_3a_4a_3a_4a_0a_4a_0a_4,$$
$$a_3 \mapsto a_0a_1a_0a_1a_2a_3a_2a_3a_2a_3a_4a_3a_4a_3a_4a_0a_4a_0a_1,$$
$$a_4 \mapsto a_0a_1a_0a_1a_2a_3a_2a_3a_4a_3a_4a_3a_4a_0a_4a_0a_4,$$

so images of this substitution are words of length 19 and 17. Finally, the coding τ sends each letter a_i to the final 'bit' of w_i:

$$\tau(a_i) = (g^{-1}(a_i))[3] = \begin{cases} 0 & \text{for} \quad i = 0, 1, 3, \\ 1 & \text{for} \quad i = 2, 4. \end{cases}$$

Now we can read off the initial segment easily from these:

$$(\sigma[4n+3])_{n \ge 0} = \tau(h^\infty(a_0)) = 0000101010101010100\ 00001010101010100\ 00 \cdots.$$

The complexity of the morphism h is $3 \times 17 + 2 \times 19 = 89$, considerably smaller than the upper bound of 105.

Table 1 gives some statistics for $m = 2, \ldots, 30$:

- the index of the first zero $z(m)$,
- an upper bound for the complexity of $\sigma([n \cdot m + k])_{n \geq 0}$, not using $z(m)$,
- an improved upper bound using $z(m)$,
- an approximation for the complexity of h (see Remarks in the next section),
- the actual complexity of h (with the exception of the case $m = 30$, which is too big to handle).

Table 1. Upper bounds, approximation and true complexity of h for $m = 2, \ldots, 30$

m	$z(m)$	$(m+1)F_{2m+2}$	$(m+1)F_{z(m)+2}$	$[(1-\phi^4)(m+1)F_{z(m)+2}]$	compl(h)
2	3	24	15	13	13
3	4	84	32	27	27
4	6	275	105	90	89
5	5	864	78	67	67
6	12	2639	2639	2254	2255
7	8	7896	440	376	377
8	6	23256	189	161	161
9	12	67650	3770	3220	3210
10	15	194821	17567	15004	15005
11	10	556416	1728	1476	1473
12	12	1578109	4901	4186	4181
13	7	4449354	476	407	407
14	24	12480600	1820895	1555230	1555935
15	20	34852944	283376	242032	242335
16	12	96949079	6409	5474	5473
17	9	268746336	1602	1368	1368
18	12	742675211	7163	6118	6123
19	18	2046683100	135300	115560	115580
20	30	5626200216	45744489	39070458	39088169
21	8	15430992126	1210	1033	1033
22	30	42235173769	50101107	42791454	42764027
23	24	115380647424	2913432	2488368	2488056
24	12	314656725625	9425	8050	8045
25	25	856733282574	5106868	4361786	4359619
26	21	2329224424344	773739	660852	660911
27	36	6323840144076	1094468732	934787896	934658668
28	24	17147315166491	3520397	3006778	3007037
29	14	46440262677600	29610	25290	25281
30	60	125634925674311	125634925674311	107305037048062	?

Theorem 1 gives an upper bound on the complexity of the arithmetic subsequences of Fibonacci; also note that this upper bound is the same for all residue classes with the same modulus. Strong conjectures for the true complexity are obtained by exhaustively searching all possible morphisms of a given complexity. However, strictly speaking such experiments only give lower bounds. Indeed, even for the smallest cases, the true value of the morphic complexity is not rigorously proved. These true values might be different for different residue classes. Our computer experiments give rise to the conjecture that this indeed is the case for modulus $m = 2$.

Example 2. A brute force search (checking morphisms with words up to 3 letters on an alphabet of size 5 and up to 2 letters on an alphabet of size 6) strongly suggested complexity 8 for $(\sigma[2n])_{n\geq 0}$, and complexity 10 for $(\sigma[2n + 1])_{n\geq 0}$. Morphisms of these complexities can be defined as follows:
$\sigma[2n] = f_0^\infty(0)$ where

$$f_0(0) = 01, \quad f_0(1) = 2, \quad f_0(2) = 31, \quad f_0(3) = 04, \quad f_0(4) = 0,$$

with coding τ_0 given by

$$\tau_0(0) = 0, \quad \tau_0(1) = 0, \quad \tau_0(2) = 1, \quad \tau_0(3) = 1, \quad \tau_0(4) = 1$$

and $\sigma[2n + 1] = f_1^\infty(0)$ where

$$f_1(0) = 01, \quad f_1(1) = 21, \quad f_1(2) = 3, \quad f_1(3) = 24, \quad f_1(4) = 51, \quad f_1(5) = 1$$

with coding τ_1 given by

$$\tau_1(0) = 1, \quad \tau_1(1) = 0, \quad \tau_1(2) = 0, \quad \tau_1(3) = 1, \quad \tau_1(4) = 1, \quad \tau_1(5) = 1.$$

Morphisms of the same complexity were independently found by Zantema [17].

   We had no proof for $(\sigma[2n])_{n\geq 0} = \tau_0(f_0^\infty(0))$ and $(\sigma[2n+1])_{n\geq 0} = \tau_1(f_1^\infty(0))$, but merely an agreement on (more than a million) initial terms. Jeffrey Shallit informed us that equality can be proved with the automated theorem prover Walnut [11, 14].

7   Further Remarks

As can be seen from Table 1, the bound $(m+1)F_{z(m)+2}$ seems reasonably sharp, and the approximation is pretty good.

   A few remarks can be made here.

(a) In our proof of Theorem 1, we take the smallest j such that F_j is a multiple of m. This guarantees (Lemma 3) that $\text{len}(\mathcal{F}^j(w)) = F_{j+2}\cdot m - F_j\cdot\text{wt}(w)$ is a multiple of m for all $w \in V_m$. One could wonder about divisibility of $\text{wt}(w)$ by m. Note that if $w_1 = \sigma[n]\cdots\sigma[n+m-1]$ and $w_2 = \sigma[n+1]\cdots\sigma[n+m]$, then $|\text{wt}(w_1)-\text{wt}(w_2)| \leq 1$. The difference cannot be always zero, since that would imply periodicity of σ. All subwords of σ of length m are in V_m (Lemma 4). It follows that there exist $w_1, w_2 \in V_m$ with $|\text{wt}(w_1) - \text{wt}(w_2)| = 1$. In particular, not all weights can be divisible by m. Therefore the condition that F_j is a multiple of m is necessary, and we cannot take a smaller j.

(b) In (1), we bound $\text{len}(\mathcal{F}^j(w))$ by $F_{j+2} \cdot m$, ignoring that Lemma 3 gives a smaller bound. If j and m are not very small, then $F_j \approx \phi \cdot F_{j+1}$ and $\text{wt}(w) \approx \phi^2 \cdot \text{len}(w)$, where $\phi = (\sqrt{5} - 1)/2$ is the positive solution of $\phi^2 = 1 - \phi$. So

$$\text{len}(\mathcal{F}^j(w)) = F_{j+2} \cdot m - F_j \cdot \text{wt}(w) \approx (1 - \phi^4) \cdot F_{j+2} \cdot m.$$

This means that the error in (1) is relatively small ($(1 - \phi^4) \approx 0.854$). This observation can be used to approximate the complexity of the morphism h by $(1 - \phi^4)(m + 1)F_{z(m)+2}$ without explicitly determining h. If m itself is a Fibonacci number, $\text{wt}(w)$ often is a Fibonacci number as well, and in this case the approximation is particularly good. The table above gives these approximations rounded to the nearest integer.

(c) As noted before, our morphism gives upper bounds for the complexity of arithmetic subsequences of Fibonacci. Example 2 indicates that morphisms of lower complexity sometimes do exist, but we do not know if less complex morphisms exist for all m, let alone how to systematically construct them.

(d) To find all words in V_m, one could just examine the subwords of σ starting from the beginning until $m + 1$ different words of length m have been found (cf. Lemma 4). In the cases for m up to 30 we considered for Table 1 this did not take long; compare for example [2], Sect. 3.1.

Alternatively, one could start with $w_0 = \sigma[0] \cdots \sigma[m - 1]$ and $V = \{w_0\}$. Then compute the image $\mathcal{F}^{z(m)}(w_0)$ and add all its subwords to V. Iterate by computing the images of all newly added words and adding their subwords to V as well. Stop if V contains $m + 1$ words. In every iteration, at least one new word will be added to V, so at most m images have to be computed before this procedure terminates.

(e) Jeffrey Shallit reports that the upper bound 2^{4m^2+1} can be deduced from [3] and from the theory behind Walnut [14]. This is much weaker than our upper bound $(m + 1)F_{2m+2}$. On the other hand, Walnut gives examples showing that our bound probably is still far from the truth.

Acknowledgements. The authors thank Hans Zantema for useful discussions about these and related matters. We further thank Jeffrey Shallit for his interest in our work and fruitful remarks. Also, the anonymous referees suggested various improvements.

References

1. Allouche, J.-P., Shallit, J.: Automatic Sequences, Theory Applications Generalizations. Cambridge University Press, Cambridge (2003)
2. Cassaigne, J.: Recurrence in infinite words. In: Ferreira, A., Reichel, H. (eds.) Proceedings of the 18th Annual Symposium on Theoretical Aspects of Computer Science (STACS 2001), Lecture Notes in Computer Science 2010, pp. 1–11 (2001)
3. Charlier, É., Rampersad, N., Rigo, M., Waxweiler, L.: The minimal automaton recognizing $m\mathbb{N}$ in a linear numeration system. Integers 11B (2011)
4. Dekking, F.M.: Iteration of maps by an automaton. Discret. Math. **126**, 181–186 (1994)

5. Dekking, M., Keane, M.: Two-block substitutions and morphic words. Adv. Appl. Math. **148**, 102536 (2023)
6. Freyd, P., Brown, K.S.: Problems and solutions: solutions: E3410. Am. Math. Mon. **99**(3), 278–279 (2013). https://doi.org/10.2307/2325076. JSTOR 2325076
7. Koshy, T.: Fibonacci and Lucas Numbers with Applications. Wiley, New York (2001)
8. Lind, D., Morris, R.A., Shapiro, L.D.: Tables of Fibonacci Entry Points, Part II, Fibonacci Association (1965)
9. Marques, D.: Fixed points of the order of appearance in the Fibonacci sequence. Fibonacci Quart. **50**(4), 346–352 (2012)
10. Marques, D.: Sharper upper bounds for the order of appearance in the Fibonacci sequence. Fibonacci Quart. **51** (2013)
11. Mousavi, H.: Automatic theorem proving in Walnut (2016). arXiv:1603.06017
12. OEIS Foundation Inc. Entry A143667 in the On-Line Encyclopedia of Integer Sequences (2024). https://oeis.org/A143667
13. Sallé, H.J.A.: Maximum value for the rank of apparition of integers in recursive sequences. Fibonacci Quart. **13**(2), 159–161 (1975)
14. Shallit, J.: The Logical Approach To Automatic Sequences: Exploring Combinatorics on Words with Walnut, Vol. 482 of London Math. Society Lecture Note Series. Cambridge University Press, Cambridge (2023)
15. Wall, D.D.: Fibonacci series modulo m. Am. Math. Mon. **67**, 525–532 (1960)
16. Wunderlich, M.: Tables of Fibonacci Entry Points, Part I, Fibonacci Association (1965)
17. Zantema, H.: Characterizing morphic sequences (2023). arXiv:2309.10562
18. Zantema, H., Bosma, W.: Complexity of automatic sequences. Inf. Comput. **288**, 104710 (2022)

A Variation of Reynolds-Hurkens Paradox

Thierry Coquand[(✉)] [ID]

University of Göteborg, Gothenburg, Sweden
coquand@chalmers.se
http://www.cse.chalmers.se/~coquand

Abstract. We present a variation of Reynolds-Hurkens paradox formulated as the non existence of a powerful triple in the proof system HOL.

Keywords: Dependent Type Theory · Paradox

1 Introduction

We present a variation of Hurkens paradox [8], itself being a variation of Reynolds "paradox" [10], as used in [4]. We first explain a related paradox in higher order logic, which can be seen as a variation of Russell's paradox. We then show how this paradox can be formulated in system λU^-. We finally argue that an analysis of the computational behavior of this paradox requires to extend existing type systems with a first class notion of definitions and head linear reductions, as advocated by N.G. de Bruijn [6].

2 Some Paradoxes in Minimal Higher-Order Logic

We first present some paradoxes in some extensions of the system λHOL, minimal Higher-Order logic, described in [7]. This system can be seen as a minimal logic version of higher-order logic introduced by A. Church [1]. With the notation of [7], it has sorts $*, \square, \Delta$ with $* : \square$ and $\square : \Delta$ and the rules

$$(*, *), \ (\square, \square), \ (\square, *)$$

We denote by $X, Y, \dots$ types of this system.

We can define $\mathsf{Pow}\ X : \square$ for $X : \square$ by $\mathsf{Pow}\ X = X \to *$ and $T\ X : \square$ for $X : \square$ by $T\ X = \mathsf{Pow}\ (\mathsf{Pow}\ X)$.

Note that Pow defines a *contravariant judgemental* functor: if $f : X \to Y$ we can define $\mathsf{Pow}\ f : \mathsf{Pow}\ Y \to \mathsf{Pow}\ X$ by $\mathsf{Pow}\ f\ q = \lambda_{x:X} q\ (f\ x)$ and we also have if furthermore $g : Y \to Z$ the judgemental equality (here β-conversion [7]) $\mathsf{Pow}\ (g \circ f) = (\mathsf{Pow}\ f) \circ (\mathsf{Pow}\ f)$ defining $g \circ f$ as $\lambda_{x:X} g\ (f\ x)$.

It follows that T defines a *judgmental* functor: if $f : X \to Y$ we can define $T\ f : T\ X \to T\ Y$ by $T\ f = \mathsf{Pow}\ (\mathsf{Pow}\ f)$, that is

$$T\ f\ F\ q = F\ (\lambda_{x:X} q\ (f\ x))$$

V. Capretta et al. (Eds.): *Logics and Type Systems in Theory and Practice*, LNCS 14560, pp. 111–117, 2024.
https://doi.org/10.1007/978-3-031-61716-4_7

and we also have if furthermore $g : Y \to Z$ the judgemental equality $T\ (g \circ f) = (T\ g) \circ (T\ f)$.

We assume in this section to have a type $A : \Box$ together with two maps intro : $T\ A \to A$ and match : $A \to T\ A$.

We explain now how to derive simple paradoxes assuming some convertibility properties of these maps.

2.1   A Variation of Russell's Paradox

The first version is obtained by assuming that we have match (intro u) convertible to u, i.e. $T\ A$ is a judgemental retract of A.

Intuitively, we expect Pow A to be a retract of $T\ A$, and this would imply that Pow A is a retract of A and we should be able to deduce a contradiction by Russell's paradox. One issue with this argument is that it holds only using some form of *extensional* equalities, and we work in an intensional setting. One way to solve this issue is to work with Partial Equivalence Relations; this is what was done in [4]. The work [8], suggests that there should be a more direct way to express this idea, and this is what we present here.

The contradiction is obtained as follows. We first define a relation C : Pow $A \to$ Pow A

$$C\ p\ x\ =\ p\ x \to \neg(\mathsf{match}\ x\ p)$$

where, as usual, we define $\bot : *$ by $\bot = \forall_{p:*} p$ and $\neg : * \to *$ by $\neg\ p = p \to \bot$. We can then define p_0 : Pow A

$$p_0\ x\ =\ \forall_{p:\mathsf{Pow}\ A} C\ p\ x$$

We can also define $X_0 : T\ A$

$$X_0\ p\ =\ \forall_{x:A} C\ p\ x$$

and $x_0 : A$ as $x_0 = $ intro X_0. We can then build $l_1 : X_0\ p_0 = $ match $x_0\ p_0$

$$l_1\ x\ h\ =\ h\ p_0\ h$$

and $l_2 : p_0\ x_0$ by

$$l_2\ p\ h\ h_1\ =\ h_1\ x_0\ h\ h_1$$

But this is a contradiction since match $x_0 = $ match (intro X_0) $= X_0$ by hypothesis, and hence $l_2\ p_0\ l_2\ l_1$ is of type $\bot$.

We can summarize this discussion as follows.

Theorem 1. *In* λHOL, *we cannot have a type* A *such that* Pow (Pow A) *is a judgemental retract of* A.

This can be seen as a variation of Russell/Cantor's paradox, which states that Pow A cannot be a retract of A. Here we state that $T\ A$ cannot be a retract of A.

2.2   A Refinement

We define $\delta : A \to A$ by $\delta = \mathsf{intro} \circ \mathsf{match}$ and assume the judgemental equality

$$\mathsf{match} \circ \mathsf{intro} = T \; \delta \tag{1}$$

which implies $\mathsf{match}\ (\delta\ x)\ p = \mathsf{match}\ x\ (p \circ \delta)$.

We now (re)define $p_0 : \mathsf{Pow}\ A$

$$p_0 \; x \;=\; \forall_{p:\mathsf{Pow}\ A} \; p \; (\delta \; x) \to \neg(\mathsf{match} \; x \; p)$$

and $X_0 : T \; A$ as before

$$X_0 \; p \;=\; \forall_{x:A} \; p \; x \to \neg(\mathsf{match} \; x \; p)$$

and $x_0 : A$ as $x_0 = \mathsf{intro}\ X_0$. Using the judgemental equality (1), it is possible to build

$$s_1 : \forall_x \; p_0 \; x \to p_0 \; (\delta \; x) \qquad\qquad s_2 : \forall_p \; X_0 \; p \to X_0 \; (p \circ \delta)$$

by $s_1 \; x \; h \; p \;=\; h \; (p \circ \delta)$ and $s_2 \; p \; h \; x \;=\; h \; (\delta \; x)$.

We can now define $l_0 : \forall_{p:\mathsf{Pow}\ A} \; p \; x_0 \to \neg(X_0 \; p)$ by

$$l_0 \; p \; h \; h_0 \;=\; h_0 \; x_0 \; h \; (s_2 \; p \; h_0)$$

using (1) and $l_1 : X_0 \; p_0$ by

$$l_1 \; x \; h \;=\; h \; p_0 \; (s_1 \; x \; h)$$

and $l_2 : p_0 \; x_0$ by $l_2 \; p = l_0 \; (p \circ \delta)$.

For this, we use the judgemental equality $\mathsf{match}\ (\delta\ x)\ p = \mathsf{match}\ x\ (p \circ \delta)$, consequence of (1).

We can then form the term $l_0 \; p_0 \; l_2 \; l_1$ which is of type $\bot$.

We thus get the following result, using $T \; X = \mathsf{Pow}\ (\mathsf{Pow}\ X)$.

Theorem 2. *In $\lambda\mathsf{HOL}$, we cannot have a type A with two maps* $\mathsf{intro} : T \; A \to A$ *and* $\mathsf{match} : A \to T \; A$ *with* $\mathsf{match} \circ \mathsf{intro}$ *convertible to* $T \; (\mathsf{intro} \circ \mathsf{match})$.

3   An Encoding in $\lambda\mathsf{U}^-$

3.1   Weak Representation of Data Type

Using the notations of [7] the system $\lambda\mathsf{U}^-$ has also sorts $*, \square, \Delta$ with $* : \square$ and $\square : \Delta$ and the rules

$$(*, *), \; (\square, \square), \; (\square, *), \; (\Delta, \square)$$

We explain in this section why the refined paradox has a direct encoding in the system $\lambda\mathsf{U}^-$.

As before, T defines a judgemental functor: if $f : X \to Y$ we can define $T\ f : T\ X \to T\ Y$ by
$$T\ f\ F\ q = F\ (\lambda_{x:X} q\ (f\ x))$$
and we also have if furthermore $g : Y \to Z$ the judgemental equality $T\ (g \circ f) = (T\ g) \circ (T\ f)$ defining $g \circ f$ as $\lambda_{x:X} g\ (f\ x)$.

A T-algebra is a type $X : \square$ together with a map $f : T\ X \to X$.

Following Reynolds [10,11], we represent $A : \square$ by
$$A = \Pi_{X:\square}(T\ X \to X) \to X$$

It can be seen as a weak representation of a data type. If we have $X : \square$ and $f : T\ X \to X$ we can define $\iota\ f : A \to X$ by $\iota\ f\ a = a\ X\ f$. We can then define $\mathsf{intro} : T\ A \to A$ by $\mathsf{intro}\ u\ X\ f = f\ (T\ (\iota\ f)\ u)$, and we have the conversion
$$(\iota\ f) \circ \mathsf{intro} = f \circ (T\ (\iota\ f)) \tag{2}$$

This expresses that the following diagram commutes strictly

So A, intro represents a *weak* initial T-algebra.

We define next $\mathsf{match} : A \to T\ A$ by $\mathsf{match} = \iota\ (T\ \mathsf{intro})$. Using the conversion (2), we have
$$\mathsf{match} \circ \mathsf{intro} = (T\ \mathsf{intro}) \circ (T\ \mathsf{match}) = T\ (\mathsf{intro} \circ \mathsf{match})$$

This is the required conversion (1) and we get in this way an encoding of Theorem 2.

3.2 Some Variations

In [8], Hurkens uses instead
$$B = \Pi_{X:\square}(T\ X \to X) \to T\ X \tag{3}$$

He then develops a short paradox using this type B, but with a different intuition, which comes from Burali-Forti paradox. The variation we present in this note starts instead from the remark that $T\ A$ cannot be a retract of A. In [4], we also use this idea, but with a more complex use of partial equivalence relations, in order to build a strong initial T-algebra from a weak initial T-algebra. This was following Reynolds' informal argument in [10,11].

The same argument from Theorem 2 can use the encoding (3) instead. We define then
$$\iota : \Pi_{X:\square}(T\ X \to X) \to B \to X$$

by

$$\iota\ X\ f\ b = f\ (b\ X\ f)$$

and intro $: T\ B \to B$ by

$$\text{intro } v\ X\ f = T\ (\iota\ f)\ v$$

We then have the choice for defining match $: B \to T\ B$. We can use

$$\text{match} = \iota\ (T\ B)\ \text{intro}$$

as before. Maybe surprisingly, we also can use

$$\text{match } b = b\ B\ \text{intro}$$

In both cases, we get the judgemental equality match $\circ$ intro $= T$ (intro $\circ$ match) required for the use of Theorem 2.

Hurkens calls a triple $(A,$ intro, match$)$ satisfying the judgemental equality (1) a *powerful triple*. He notices that if $(A,$ intro, match$)$ is powerful, then so is (Pow A, Pow match, Pow intro). Theorem 2 states that there no triple can be powerful in λHOL.

4   Computational Behavior

For the paradox corresponding to Theorem 1, we have the following looping behavior with a term reducing to itself (in two steps) by *head linear reduction*

$$l_2\ p_0\ l_2\ l_1 \to l_1\ x_0\ l_2\ l_1$$
$$\to l_2\ p_0\ l_2\ l_1$$
$$\to \dots$$

4.1   Family of Looping Combinators

The paradox corresponding to Theorem 2 does not produce a term that reduces to itself

$$l_0\ p_0\ l_2\ l_1 \to l_1\ x_0\ l_2\ (s_2\ p_0\ l_1)$$
$$\to l_2\ p_0\ (s_1\ x_0\ l_2)\ (s_2\ p_0\ l_1)$$
$$\to l_0\ (p_0 \circ \delta)\ (s_1\ x_0\ l_2)\ (s_2\ p_0\ l_1)$$
$$\to s_2\ p_0\ l_1\ x_0\ (s_1\ x_0\ l_2)\ (s_2\ (p_0 \circ \delta)\ (s_2\ p_0\ l_1))$$
$$\to l_1\ (\delta\ x_0)\ (s_1\ x_0\ l_2)\ (s_2\ (p_0 \circ \delta)\ (s_2\ p_0\ l_1))$$
$$\to \dots$$

Like for Hurkens' paradox however, we obtain a term that reduces to itself if we forget types in abstraction [8].

In [2], I analysed another paradox, closer to Girard's original formulation (as was found out later by H. Herbelin and A. Miquel, a slight variation of this paradox can be expressed in System λU^-). At about the same time, A. Meyer and M. Reinholdt [9], suggested a clever use of Girard's paradox for expressing a fixed-point combinator. While implementing this paradox [2], it was possible to check that, contrary to what [9] was hinting, the term representing this paradox was not reducing to itself[1]. A. Meyer found out then that it was however possible to use this paradox and produce a family of looping combinators instead, i.e. a term which has the same Böhm tree as one of a fixed-point combinator. A corollary, following [9], is that type-checking is undecidable for type : type.

4.2   Definitions and Head Linear Reduction

As discussed in [8], using the notion of *definition* is essential, even for "small" terms, for representing these paradoxes in an understandable way. As was discovered in Automath [6], in a type system with *dependent* types, one cannot reduce definitions to abstractions and applications like in simply typed lambda calculus. Indeed, the representation of

$$\text{let } x : A = e_0 \text{ in } e_1$$

by $(\lambda_{x:A} e_1) \, e_0$ can be incorrect, since the definition $x : A = e_0$ can be used in the type-checking of e_1.

Furthermore, in order to understand the computational behavior of the paradox, the use of *head linear reduction*, which plays an important role in [6], is convenient. This is what was done when presenting above the computational behavior of various paradoxes, with a periodic behavior for the first example and a non periodic behavior for the paradox in λU^-. This use may also be relevant for understanding large proofs.

5   Conclusion

In this note, we presented a variation of Hurkens' paradox [8] and a paradox inspired by Reynolds [4]. This paradox can be seen as a refinement of the simple paradox presented in Theorem 1. The problem is that in the encoding in λU^-, we don't get that $T \, A$ is a *judgmental* retract of A^2. It is possible however to still use a weaker judgemental equality and derive a relatively simple paradox[3].

References

1. Church, A.: A formulation of the simple theory of types. J. Symb. Log. **5**, 56–68 (1940)
2. Coquand, T.: An analysis of Girard's paradox. In: Proceedings of the Symposium on Logic in Computer Science (LICS 1986), Cambridge, Massachusetts, USA, 16–18 June 1986, pp. 227–236. IEEE Computer Society (1986)

---

[1] It would be interesting to go back to this paradox and check if it reduces to itself when removing types in abstractions.

[2] This problem was presented in [5] as one main motivation for the primitive introduction of inductive definitions.

[3] We were not able however to refine in a similar way the paradox of trees [3], to obtain a new paradox in λU^-.

3. Coquand, T.: The paradox of trees in type theory. BIT **32**(1), 10–14 (1992)
4. Coquand, T.: A new paradox in type theory. In: Logic, Methodology and Philosophy of Science IX. Proceedings of the Ninth International Congress of Logic, Methodology and Philosophy of Science, Uppsala, Sweden, 7–14 August 1991, pp. 555–570. North-Holland, Amsterdam (1994)
5. Coquand, T., Paulin, C.: Inductively defined types. In: Martin-Löf, P., Mints, G. (eds.) COLOG 1988. LNCS, vol. 417, pp. 50–66. Springer, Heidelberg (1988). https://doi.org/10.1007/3-540-52335-9_47
6. de Bruijn, N.G.: Generalizing Automath by means of a lambda-typed lambda calculus. In: Mathematical Logic and Theoretical Computer Science. Lecture Notes in Pure and Applied Mathematics, vol. 106, pp. 71–92 (1987)
7. Geuvers, H.: (In)consistency of extensions of higher order logic and type theory. In: Altenkirch, T., McBride, C. (eds.) TYPES 2006. LNCS, vol. 4502, pp. 140–159. Springer, Berlin (2007). https://doi.org/10.1007/978-3-540-74464-1_10
8. Hurkens, A.J.C.: A simplification of Girard's paradox. In: Dezani-Ciancaglini, M., Plotkin, G. (eds.) TLCA 1995. LNCS, vol. 902, pp. 266–278. Springer, Heidelberg (1995). https://doi.org/10.1007/BFb0014058
9. Meyer, A.R., Reinhold, M.B.: "Type" is not a type. In: Conference Record of the Thirteenth Annual ACM Symposium on Principles of Programming Languages, St. Petersburg Beach, Florida, USA, January 1986, pp. 287–295. ACM Press (1986)
10. Reynolds, J.C.: Polymorphism is not set-theoretic. In: Kahn, G., MacQueen, D.B., Plotkin, G.D. (eds.) SDT 1984. LNCS, vol. 173, pp. 145–156. Springer, Heidelberg (1984). https://doi.org/10.1007/3-540-13346-1_7
11. Reynolds, J.C., Plotkin, G.D.: On functors expressible in the polymorphic typed lambda calculus. Inf. Comput. **105**(1), 1–29 (1993)

Between Brackets

Tonny Hurkens$^{(\boxtimes)}$

Haps, The Netherlands
hurkens@science.ru.nl

Abstract. Brackets are symbols showing the structure of an expression in a linear notation. A *term* can be represented as a finite ordered tree of which each vertex is labeled with a member of a given set of non-logical constant or function symbols. This labeled tree can be embedded in the plane in such a way that the vertices correspond to labeled points on a circle and the edges to non-crossing line segments.

In this article, we show that in the standard notation (with each subterm between brackets) the matching left and right brackets can be interpreted as edges of the (circular) *dual* of this plane tree. We also present a dual notation in which the matching brackets correspond to the edges of the plane tree itself. Finally, we present a mixed notation in which constants are separated by *exactly* two brackets.

We define these notations for terms in which arguments are split into left and right arguments. This corresponds to outerplanar trees in which child nodes are split into left and right children.

In Polish notation, brackets are avoided by using each constant as a *prefix* operator with a fixed number of arguments. We generalize this to *infix* operators with fixed numbers of left and right arguments.

Keywords: Term syntax · Trees · Planar graphs

1 Introduction

In this article, we study the syntax of terms: notations of structured terms as linear strings of symbols, using some auxiliary symbols. We will treat both strings and terms as finite sequences. A string is a 'flat' finite sequence of *symbols* and a term is a nested finite sequence built from *constants* in a specific way. The structure of a term can be represented as a *finite ordered tree* in several ways.

We present some notations in which *brackets* encode the edges of a *circular embedded tree*: a special case of a graph embedded in the plane in such a way that the edges are represented by plane curves that intersect only at their endpoints.

The nesting of brackets in a string can be visualized by arranging the symbols in a circle and drawing a straight line segment connecting matching brackets. These line segments, referred to as chords, do not cross each other. In an anticlockwise walk around the circle, starting and ending at the top, the 'left' endpoint of a chord is visited before the 'right' endpoint.

T. Hurkens—In celebration of the 60th birthday of my co-thinker Herman Geuvers.

V. Capretta et al. (Eds.): *Logics and Type Systems in Theory and Practice*, LNCS 14560, pp. 118–133, 2024.
https://doi.org/10.1007/978-3-031-61716-4_8

In a *chord diagram* each end of a chord corresponds to an individual bracket. In our visualizations, each end of a chord corresponds to a *block* of adjacent symbols: a number of right brackets, followed by at most one constant and a number of left brackets. Each such block corresponds to a vertex of a graph, optionally labeled with a constant, and each chord corresponds to an edge of that graph. We will only consider directed graphs in which each vertex has at most one outgoing edge. We will use different brackets to distinguish:

- whether a bracket is part of a block with or without a constant
- whether an edge is directed from the left to the right bracket or vice versa

1.1   Constants Between Matching Brackets

In a *standard notation* like $h(e, r(m, a(n)))$, arguments are surrounded by round brackets. We treat each comma between two arguments as syntactic sugar for the combination $)($ of two brackets. So the main constant h precedes the arguments (e) and $(r(m)(a(n)))$. Note that in such prefix notation, the combinations $((\ $ and $()$ do not occur. However, long blocks of right brackets occur in the standard notation of deeply nested terms.

We will also consider notations like $[[h]e(r)(m)]a(n)$ in which the main constant a occurs between a left argument $[[h]e(r)(m)]$ and a right argument (n). We surround left arguments with *square brackets*. Note that in such infix notation, the combinations $](, \ ]]$ and $[]$ do not occur. For a deeply nested term, left square brackets may clutter.

Figure 1 visualizes these examples of standard notation. The double line connects the main constant to the point that corresponds to (the brackets at) both ends of the string. Each edge (chord) is directed towards the endpoint that corresponds to the *right round* bracket or the *left square* bracket. The result is a circular embedded tree in which each edge is directed to the root (at the top). Cluttering of brackets corresponds to a vertex with many children.

If one follows a directed edge, then the symbols of the surrounded argument are on the *right* side of the edge in case of round brackets, and on the *left* side in case of square brackets. In both cases, the main constant is on the other side: it is the unique constant in the string that is not surrounded by brackets.

Note that the edge corresponding to the round brackets surrounding $a(n)$ in the first notation is the same as the edge corresponding to the square brackets surrounding $[h]e(r)(m)$ in the second notation.

The five chords divide the circle disk in six regions, each labeled with a constant. For a notation with only round brackets, the main constant is the label of the first region anticlockwise from the root.

1.2   Matching Brackets Between Constants

We also present a *dual notation* corresponding to the circular dual of this circular embedding. In this dual notation, brackets link the main constant of a term to the main constant of an argument (instead of linking the left end of an argument to

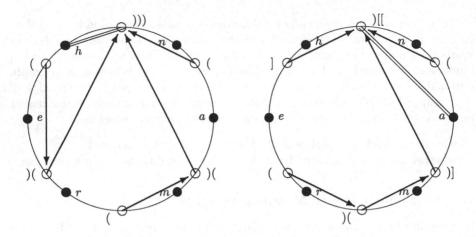

Fig. 1. Circular visualizations of $h(e)(r(m)(a(n)))$ and $[[h]e(r)(m)]a(n)$.

its right end). Now constants are labels of the vertices and cluttering of brackets corresponds to terms with many arguments. Each bracket is part of a block with a constant. We will use *angle brackets* for right arguments and *braces* for left arguments.

We show that this notation is related to *Polish notation*: a notation without brackets for terms in which each constant has a fixed arity. The idea is to treat each block as a single symbol.

Figure 2 shows the dual notations and the circular duals of the embeddings shown in Fig. 1.

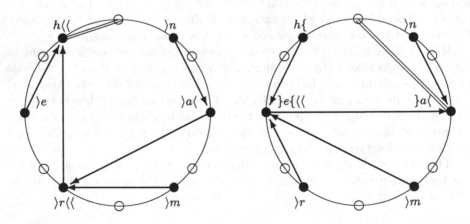

Fig. 2. Circular visualizations of $h\langle\langle\rangle e\rangle r\langle\langle\rangle m\rangle a\langle\rangle n$ and $h\{\}e\{\langle\langle\rangle r\rangle m\}a\langle\rangle n$.

1.3   Two Brackets Between Constants

Finally we present a mix of these notations in which constants are separated by exactly two brackets. For each argument either the brackets of the standard notation are shown, or the brackets of the dual notation. It corresponds to a combination of edges and dual edges of a circular embedded tree such that each vertex has degree at most 2. Figure 3 shows the mixed notations of those shown in the previous figures.

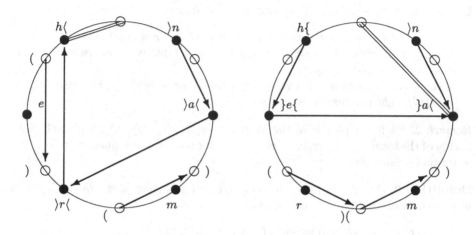

Fig. 3. Circular visualizations of $h\langle(e)\rangle r\langle(m)\rangle a\langle\rangle n$ and $h\{\}e\{(r)(m)\}a\langle\rangle n$.

2   Strings and Terms

Definition 1. *A string is a finite sequence of elements of a set Σ. A language is a set of strings. For each set Σ, the set of all finite sequences $\sigma = \langle s_1, \ldots, s_n \rangle$ of elements of Σ is denoted by $\Sigma^\star$.*

Remark 1. We allow any set Σ as set of *symbols*: a symbol could be a finite sequence. The language $\Sigma^\star$ is closed under concatenation. The *concatenation* of two finite sequences $\langle r_1, \ldots, r_m \rangle$ and $\langle s_1, \ldots, s_n \rangle$ of m and n elements of Σ is the finite sequence $\langle r_1, \ldots, r_m, s_1, \ldots, s_n \rangle$ of length $m + n$. Since concatenation is associative and has the empty sequence as neutral element, concatenation can be defined as operation on finite sequences of finite sequences. The concatenation of $\langle \sigma_1, \ldots, \sigma_n \rangle \in \Sigma^{\star\star}$ is a member of $\Sigma^\star$ that is simply denoted by $\sigma_1 \ldots \sigma_n$ if $n > 1$. The concatenation of $\langle \sigma \rangle$ is σ and the concatenation of $\langle \rangle$ is $\langle \rangle$.

A *term* τ consists of a main constant c and a finite number of arguments. Each argument of τ is the *occurrence* of some (smaller) term τ' at some position relative to c: the same τ' may occur at different positions. Positions have a linear order (from left to right), but we allow arguments at both sides of c.

The constants that occur in a term τ form a string $\gamma(\tau)$, but notations (strings resulting from $\gamma(\tau)$ by inserting or adding extra symbols) are needed to distinguish terms in which the same constants occur in the same linear order, for example $a(h(a))$, $a(h)(a)$, $a([h]a)$, $[a]h(a)$, $[a(h)]a$, $[a][h]a$, and $[[a]h]a$. In the recursive definition of the mixed notation, the *outer arguments* (the first left and last right argument) play a special role. The following definitions facilitate the recursive definitions of notations. Note that these definitions do not define or mention symbols at all.

Definition 2. *We define, by simultaneous recursion:*

- *a* term *is a triple* $\langle l, c, r \rangle$ *where l is a left part and r a right part*
- *a* left part *is either the empty sequence* $\langle \rangle$ *or a tuple* $\langle \tau, l \rangle$ *where τ is a term and l is a (nested) left part*
- *a* right part *is either the empty sequence* $\langle \rangle$ *or a tuple* $\langle r, \tau \rangle$ *where r is a (nested) right part and τ is a term*

Remark 2. Each left part is of the form $\langle \tau_1, \langle \tau_2, \ldots \langle \tau_n, \langle \rangle \rangle \ldots \rangle \rangle$, and each right part is of the form $\langle \langle \ldots \langle \langle \rangle, \tau_1 \rangle \ldots, \tau_{n-1} \rangle, \tau_n \rangle$, for some finite sequence $\langle \tau_1, \ldots, \tau_n \rangle$ of terms (*arguments*).

Definition 3. *We define for each term, left part and right part p the string $\gamma(p)$ of constants occurring in p, as follows:*

- $\gamma(\langle l, c, r \rangle)$ *is the concatenation of $\gamma(l)$, $\langle c \rangle$, and $\gamma(r)$*
- $\gamma(\langle \rangle) = \langle \rangle$
- $\gamma(\langle p, q \rangle)$ *is the concatenation of $\gamma(p)$ and $\gamma(q)$*

For each set C, the set of all terms τ for which $\gamma(\tau) \in C^\star$, is denoted by T_C.

Remark 3. The 'central' part c of a term τ, just like any symbol, could be a finite sequence, but even then it is not 'flattened' in $\gamma(\tau)$.

Each term $\tau \in T_C$ can be constructed using the following steps:

- for each $c \in C$, we have a term $\langle \langle \rangle, c, \langle \rangle \rangle$ (without arguments)
- terms τ and $\langle l, c, r \rangle$ can be combined by adding τ to the other (or same) term as (outer) left argument, resulting in the term $\langle \langle \tau, l \rangle, c, r \rangle$
- terms $\langle l, c, r \rangle$ and τ can be combined by adding τ as (outer) right argument, resulting in the term $\langle l, c, \langle r, \tau \rangle \rangle$

In this way, arguments are added inside-out, but left and right arguments can be added independently, so a term that has both left and right arguments can be constructed in different ways.

Definition 4. *A notation is an injective function $f : T_C \to \Sigma^\star$ for some sets C and Σ with $C \subseteq \Sigma$ such that for each $\tau \in T_C$, the string $\gamma(\tau)$ is the subsequence of $f(\tau)$ obtained by removing the symbols that are not in C.*

A Polish notation is an injective function $f : P \to C^\star$ for some sets P and C with $P \subseteq T_C$ such that $f(\tau) = \gamma(\tau)$ for each $\tau \in P$.

3   Context-Free Languages Representing Terms

The set of ground terms of a first-order language depends on an algebraic signature: a description of the set C of non-logical constant and function symbols (no relation symbols) and the arity (number of arguments) of each $c \in C$.

Example 1. Let H be the language with the following signature:

- a is a function symbol (arity 1)
- h and r are binary function symbols (arity 2)
- e, m, and n are constant symbols (arity 0)

In this example, each constant (or *operator*) has a fixed, known arity, so each term has a unique *Polish notation*: the operator, followed by the arguments in Polish notation. No other symbols are needed. A string represents a term of H in Polish notation if and only if it can be generated by the following context-free grammar:

$$P \longrightarrow aP \mid hPP \mid rPP \mid e \mid m \mid n$$

For example, the strings *herman* and *arnhem* can be derived from the nonterminal P by repeatedly applying an alternative of the production rule for P.

Even if arities are fixed and known, a standard notation like $h(e, r(m, a(n)))$ or $a(r(n, h(e, m)))$ is easier to read: the auxiliary symbols show the nesting of the arguments.

We will only use *brackets* as auxiliary symbols: pairs of matching left (or opening) and right (or closing) brackets. We consider the use of a *comma* to separate consecutive arguments as syntactic sugar for the combination)(of two brackets. In this way, *standard notation* can be defined as *writing each argument between brackets*.

It turns out that in standard *prefix* notation (writing the main constant of a term before its arguments), the brackets surrounding the *last argument* of a term are superfluous. So the notation of the example terms can be simplified to $h(e)r(m)an$ and $ar(n)h(e)m$. In this simplified notation, each term is represented by a string consisting of one or more constants, each separated from the next constant by a single bracket, a comma, or no auxiliary symbol.

In *infix* and *postfix* notations, arguments can be written at the left side of the operator. This is either syntactic sugar or semantically relevant (just like the number and order of arguments). We generalize the standard, simplified, and Polish notation to terms with left and right arguments.

We will use different bracket pairs to distinguish the roles played by brackets. Left arguments will be surrounded with *square brackets*. We introduce *angle brackets* and *braces* that do not *surround* a (right or left) argument τ' of a term τ, but *link* the operator of τ' to the operator of τ. The sharp ends of these brackets 'point' outwards to the linked operator.

3.1   Languages for Terms

We now define some context-free languages that are subsets of $\Sigma^\star$, where Σ is the union of some set C and a set of eight symbols not in C. In the production rules of the grammars, uppercase letters denote non-terminals, ϵ denotes the empty string, and both lowercase letters and the brackets ()[]⟨⟩{} denote terminals. We treat C as non-terminal for a language consisting of strings of length 1. In the examples, C is a finite set of six symbols:

$$C \longrightarrow a \mid h \mid r \mid e \mid m \mid n$$

Definition 5. *The* standard language *for terms (with left and right arguments) is generated by the following grammar with start symbol T:*

$$T \longrightarrow SCR$$

$$S \longrightarrow \epsilon \mid [T]S$$
$$R \longrightarrow \epsilon \mid R(T)$$

Definition 6. *The* dual language *for terms (with left and right arguments) is generated by the following grammar with start symbol D:*

$$D \longrightarrow BCA$$

$$B \longrightarrow \epsilon \mid BC\{AB\}$$
$$A \longrightarrow \epsilon \mid \langle AB \rangle CA$$

Note that a non-terminal like A represents a finite sequence of right arguments. In the second alternative of its production rule, $B\rangle CA$ is the dual notation of the last argument with a right angle bracket inserted before its main constant.

Definition 7. *The* mixed language *for terms (with left and right arguments) is generated by the following grammar with start symbol M:*

$$M \longrightarrow BCA$$

$$B \longrightarrow \epsilon \mid BC\{RS\}$$
$$A \longrightarrow \epsilon \mid \langle RS \rangle CA$$
$$S \longrightarrow \epsilon \mid [M]S$$
$$R \longrightarrow \epsilon \mid R(M)$$

For standard notation, the distinction between square and round brackets is superfluous. In the mixed notation, this distinction is needed to find the split in substrings of the form RS produced by the rules of the non-terminals A and B: such a string is of the form $(M) \ldots (M)[M] \ldots [M]$.

The distinction between braces and angle brackets is needed to find the split in strings of the form BCA produced by the rule of the non-terminal M: such a string is of the form $C\{RS\}C \ldots C\{RS\}C\langle RS\rangle C \ldots C\langle RS\rangle C$.

Note that in the mixed notation, linking brackets are nested in surrounding brackets and vice versa, so neither the distinction between braces and square brackets, nor the distinction between angle and round brackets is needed.

Remark 4. The standard, dual and mixed languages are subsets of the language that is generated by the following grammar with start symbol E:

$$E \longrightarrow FCG$$

$$F \longrightarrow \epsilon \mid [FCG]F \mid FC\{GF\}$$

$$G \longrightarrow \epsilon \mid G(FCG) \mid \langle GF\rangle CG$$

3.2 Notations

Let f_T, f_D and f_M be the functions from T_C to Σ^* that follow the recursive definitions of the standard, dual and mixed language for terms. We claim that each of these functions f_i is a *notation*. The recursive definition of f_i uses an auxiliary function $f_{i,j}$ for each of the non-terminals: for $j = B$ or $j = S$ it is defined on the set of all left parts l for which $\gamma(l) \in C^*$; for $j = A$ or $j = R$ it is defined on the set of all right parts r for which $\gamma(r) \in C^*$. It is clear that the range of such a function is the language generated by the corresponding non-terminal. The non-trivial part is to show that the function is injective.

For each of the three languages the recursive definitions follow a similar pattern:

- $f_i(\langle l, c, r\rangle)$ is the concatenation of $g(l)$, the string $\langle c\rangle$, and $h(r)$ where g and h are auxiliary functions for left and right parts
- the result for an empty left or right part is the empty string
- each non-empty left part is of the form $\langle\langle l', c, r'\rangle, l\rangle$; application of $f_{i,j}$ adds a pair of square brackets (if $j = S$) or braces (if $j = B$), resulting in a string of the form $[\sigma_1 c\sigma_2]\sigma_3$ or $\sigma_1 c\{\sigma_2\sigma_3\}$; here σ_1, σ_2 and σ_3 are the results of recursively applying auxiliary functions to the nested parts l', r', and l
- each non-empty right part is of the form $\langle r, \langle l', c, r'\rangle\rangle$; application of $f_{i,j}$ either adds a pair of round brackets (if $j = R$) or angle brackets (if $j = A$), resulting in a string of the form $\sigma_1(\sigma_2 c\sigma_3)$ or $\langle\sigma_1\sigma_2\rangle c\sigma_3$; here σ_1, σ_2 and σ_3 are the results of recursively applying auxiliary functions to the nested parts r, l', and r'

In order to check that the functions f_i and $f_{i,j}$ are injective, we use the fact that the values assigned to terms, left parts and right parts are strings in the languages generated by the non-terminals E, F and G. Given the value σ of $f_{i,j}$ for some left or right part *and* knowing j, e.g., $j = A$, the string σ must be of the form $\langle\sigma_1\sigma_2\rangle c\sigma_3$. So σ starts with a left angle bracket, so the matching right angle bracket determines σ_3. The border between σ_1 and σ_2 can be found since the languages generated by F and G are very different.

A string like $\langle\rangle c(a)$ in the language generated by non-terminal G as representation for a right part is ambiguous: it could be the result of adding angle brackets with both σ_1 and σ_2 empty and $\sigma_3 = (a)$, or the result of adding round brackets with constant a, i.e., $\sigma_1(\sigma_2 a\sigma_3)$, with $\sigma_1 = \langle\rangle c$ and both σ_2 and σ_3 empty. So it could be a notation of the right part $\langle\langle\rangle, \langle\langle\rangle, c, \langle\langle\rangle, \langle\langle\rangle, a, \langle\rangle\rangle\rangle\rangle\rangle$

(representing a single argument) or of the right part $\langle\langle\langle\rangle, \langle\langle\rangle, c, \langle\rangle\rangle\rangle, \langle\langle\rangle, a, \langle\rangle\rangle\rangle$ (representing two atomic arguments).

In the notations that we presented, these right parts are represented as:

- standard notation: $(c(a))$ and $(c)(a)$
- dual notation: $\langle\rangle c\langle\rangle a$ and $\langle\langle\rangle c\rangle a$
- mixed notation: $\langle\rangle c\langle\rangle a$ and $\langle(c)\rangle a$

3.3   Polish Notations

In order to generalize Polish notation, we assume that each operator has a known number of left arguments (the *left arity*) and right arguments (the *right arity*).

Example 2. Let H' be the language with the following 'two-sided' signature:

- a is a binary function symbol (left arity 1, right arity 1)
- e is a 3-ary function symbols (left arity 1, right arity 2)
- h, m, n, and r are constant symbols (arity 0)

This is not sufficient to represent a term of H' as a string of constants by writing the operator between its arguments. For example, the string *herman* can be interpreted as a term with either e or a as main operator.

In Polish notation, each constant has left arity 0 and is written as prefix operator. In *anti-Polish* or *reversed Polish notation*, each constant has right arity 0 and is written as postfix operator. Even for constants of arity 1, combining a prefix operator like $-$ with a postfix operator like ! is problematic. Should the numerical expression $-0!$ be evaluated to -1 (treating the numerical constant 0 as *left argument* of the operator !) or to 1 (treating 0 as *right argument* of the operator $-$)? We solve this problem by introducing for each constant c a new constant c' (with the same arities). This results for each term with main constant c in a variant with main constant c' that should be used where the term occurs as left argument, e.g. as left argument of the factorial operator: the expression $-0'!$ evaluates to -1 and the expression $-'0!$ evaluates to 1.

So the *generalized Polish language* for terms of H' is generated by the following grammar with start symbol P:

$$P \longrightarrow QaP \mid QePP \mid h \mid m \mid n \mid r$$

$$Q \longrightarrow Qa'P \mid Qe'PP \mid h' \mid m' \mid n' \mid r'$$

Example 3. The strings $h'erm'an$ and $h'e'rman$ are the generalized Polish notations of two terms of H'. The other notations for these terms are:

- standard notation: $[h]e(r)([m]a(n))$ and $[[h]e(r)(m)]a(n)$
- dual notation: $h\{\}e\langle\langle\rangle rm\{\}\rangle a\langle\rangle n$ and $h\{\}e\{\langle\langle\rangle r\rangle m\}a\langle\rangle n$
- mixed notation: $h\{\}e\langle(r)[m]\rangle a\langle\rangle n$ and $h\{\}e\{(r)(m)\}a\langle\rangle n$

In Sect. 1 we visualized the notations of the second term. Figure 2 showed that the string $h\{\}e\{\langle\langle\rangle r\rangle m\}a\langle\rangle n$ is split in six parts: blocks consisting of a constant and some brackets.

Remark 5. Each string in the languages generated by the non-terminals P and Q of the generalized Polish language for H' can be translated to a string in the languages generated by the non-terminals A and B of the dual language: replace each of the constants $a, e, h, m, n, a', e', h', m', n'$ by the corresponding combination of right brackets, constant and left brackets:

$$\}\rangle a\langle, \}\rangle e\langle\langle, \rangle h, \rangle m, \rangle n, \}a\{\langle, \}e\{\langle\langle, h\{, m\{, n\{$$

The number of $\}$ and $\langle$ brackets code the left and right arity. The $\rangle$ before or $\{$ after the constant indicate whether it is occurs as right or left argument.

4   Circular Embedded Trees

An *ordered* (or *plane*) tree combines two relations on the set V of vertices: edges from child to parent and an ordering of children from left to right (resulting in a lexicographical order on V).

In general, the edges of a (finite) *plane* graph are represented by non-crossing plane curves that divide the surface in *regions* (or *faces*), including a single unbounded region: the *outer face*. (Viewed as embedding of a graph on the *sphere*, there is no special face.) Since a tree has no loops or multiple edges, it has a straight-line embedding in the plane (Fáry's theorem [3]). A plane tree 'divides' the plane in a single, unbounded region. A walk around its boundary visits each edge of the tree twice (from parent to child and vice versa) and each vertex at least once.

The embedding of the tree can be chosen in such a way that for each vertex, the ordering of the children is the anticlockwise order of the corresponding edges relative to some other edge. Usually, one choses the edge of the vertex to its parent to turn the cyclic order into a linear one, but this does not work for the root R of the tree. Instead we represent an ordered tree by a *circular embedding*: a graph embedding in which all graph vertices lie on a common circle (see [5]).

Definition 8. *A circular embedded tree is a straight-line circular embedding of a finite ordered tree (V, E) with root R in such a way that, for some other point R' on the circle:*

- *the edges correspond to circle chords that do not cross each other or RR'*
- *the left to right order is the anticlockwise order relative to R'*

The points in V divide the circle in the same number of arcs (from a point P to the next point P' anticlockwise). Let G be the graph that results from a circular embedded tree (V, E) by adding these arcs as edges (even if P is already connected to P'). Just like (V, E) it is an *outerplanar* graph: each point of V is on the boundary of the outer face. But now this boundary is a cycle: a walk around the circle starting and ending at R' visits each point in V once. We do not require that a vertex is visited *before* its children (as in the usual lexicographical order). So the children of any vertex (including the root) are divided into *left* and *right* children.

Proposition 1. *For each inner region there are points P and P' on the circle such that the boundary is formed by the arc from P to P' and the unique path from P' to P via distinct chords.*

Proof. The boundary of an inner region consists of arcs and chords. The chords correspond to the edges of the tree (V, E). Since trees do not have a cycle, each of these boundaries contains at least one arc. Since trees are connected, each of these boundaries contains at most one arc. □

Proposition 2. *Let r and s be inner regions that share a chord PQ as part of their boundaries. Let A and B be circle points on the boundaries of r and s, respectively, that are not in V. Then the chord AB crosses PQ in some point I inside the circle. Except for their endpoints, the line segments AI and IB are completely inside the interiors of r and s, respectively.*

Proof. This follows from the fact that each inner region is convex: the circle disk is divided by straight line segments into inner regions. □

4.1 Dual Structure

Let (V, E) be a circular embedded tree with R, R' and G as before. We now define its *dual*: a straight-line circular embedded graph (V', E') with the same number of vertices. The points in V' are points on the same circle, alternating with the points in V. We assume that $R' \in V'$. The *dual point* $r^\star$ of an inner region r of G is the unique point in V' on the boundary of r. The *dual chord* of a chord PQ of (V, E) is the chord $r^\star s^\star$ where r and s are the two regions that share PQ as part of their boundaries.

Note that, by definition, a chord of (V, E) does not intersect with the interior of any region, a dual chord intersects with the interior of exactly two regions, and RR' only intersects with the region r for which $r^\star = R'$.

Definition 9. *The circular dual (V', E') of (V, E) has an edge between $r^\star$ and $s^\star$ if the following equivalent conditions hold:*

- *the regions r and s share a chord PQ as part of their boundaries*
- *the chord $r^\star s^\star$ crosses exactly one of the chords of (V, E)*

These chords are the diagonals of quadrilateral $Pr^\star Qs^\star$. No other chord or dual chord crosses the interior of this quadrilateral since triangle $PQr^\star$ is part of region r and triangle $PQs^\star$ is part of region s.

Let I be a point in the interior of an inner region r. By definition, I can not be a point on a chord of (V, E). It can only be a point on a chord of (V', E') if $r^\star$ is one of the endpoints of that chord. It can only be a point on RR' if $r^\star = R'$. This implies that the dual chords do not cross each other and do not cross RR'.

Proposition 3. *The circular dual is a tree.*

Proof. The graph (V', E') has no cycle of dual chords, since if it contains the dual $r^\star s^\star$ of chord PQ, then the cycle would have to cross the chord PQ at least twice. It is connected since for each pair of dual points $r^\star$ and $s^\star$, the line segment $r^\star s^\star$ crosses a number of chords of G, i.e. it crosses a number of boundaries between internal regions, so the dual points of these regions are connected by dual edges. □

Let $G^\star$ be the graph that results from (V', E') by adding the circle arcs between the points of V'. The construction of $G^\star$ from G is symmetrical, so $G^{\star\star} = G$.

4.2   Dual and Weak Dual Graphs

The construction of outerplanar graph $G^\star$ from G is similar to the standard construction of the *dual* of a plane graph (see [2]): a plane graph with the same number of edges, in which the roles of vertices and regions are reversed. A plane curve from point P to Q also connects the region r at the right side of the curve to the region s at the left side. The dual of a region is represented by selecting a point in that region and the dual of an edge by a curve connecting the points representing the regions at either side. The difference with the standard construction is that we ignore the unbounded region u (outside the circle) and the circle arcs that form its boundary: we only represent dual points and edges for the *inner* regions and the circle *chords*. So in order to construct the standard dual of G, we should select a point $u^\star$ outside the circle and add non-intersecting curves connecting each dual point $r^\star$ to $u^\star$. In other words: instead of adding the circle arcs as edges to (V', E'), we should add $u^\star$ as dual point and the curves as dual edges of the circle arcs.

Each plane graph G also has a *weak dual*: the subgraph of the dual graph of G that results from removing the dual point $u^\star$ of the unbounded region u of G and removing the dual edges connected to $u^\star$. The removed edges are the duals of the edges forming the boundary of u. As a result, the regions surrounding $u^\star$ in the dual of G (including the unbounded region) are combined to a single region: the unbounded region of the weak dual of G.

So the weak dual of our outerplanar graph G is not $G^\star$ but just the circular embedded tree (V', E').

The (standard) dual of a plane tree like (V, E) is a plane graph with a single vertex (dual to the only region), a number of non-crossing loops (dual to the edges of the tree with the same region at both sides), and a number of regions (dual to the vertices of the tree). The 'special' (unbounded) region corresponds to a 'special' vertex (the root of the tree): the boundary of this region corresponds to the edges between the root and its children. So the *weak* dual of the standard dual of a tree like (V, E) is not (V, E) but the (disconnected) *forest* that results from removing the root from the tree.

However, our construction of the circular dual of a circular embedded tree can be reformulated as the standard construction of the dual of an embedding of a finite ordered tree on the *sphere*:

- each vertex of the tree is mapped to a point on the *equator*
- the edge from a child C to its parent P is represented by a semicircle on the Northern Hemisphere with spherical center on the equator (halfway between C and P)
- the edge from P to C is represented by the semicircle on the Southern Hemisphere with the same spherical center
- the edges correspond to semicircles that do not intersect outside the equator

The left to right order is represented as the anticlockwise order when viewed from above the North Pole, with respect to some other point R' on the equator (that can be connected to the root R without crossing the semicircles).

So each edge is treated as a pair of directed edges and represented by a circle that is the intersection of the sphere with a plane orthogonal to the equator. These circles divide the sphere in symmetrical regions: the southern half of a region is the mirror image of the northern half, but the region is *not* split in two regions by the equator. When viewed from above the North Pole, the equator is a circle and the edges correspond to circle chords. The circle arcs do *not* represent boundaries but are the result of the orthogonal projection of the sphere on the circle disk.

4.3   Squaregraphs

The circular embedded tree (V, E) and its circular dual (V', E') can be combined into the bipartite graph $(V \cup V', D)$ in which the edges in D connect a point $P \in V$ to a dual point $r^\star \in V'$ if and only if P is a point on the boundary of e. Since $r^\star$ is also on the boundary of the convex inner region r, the chords of $(V \cup V', D)$ do not cross each other. Note that RR' is inside the region r with $r^\star = R'$, so RR' is an edge in D.

Let P and Q be different points and r and s different inner regions. Then the following statements are equivalent:

- $r^\star s^\star$ is the dual chord of PQ
- the chords PQ and $r^\star s^\star$ are the crossing diagonals of quadrilateral $Pr^\star Qs^\star$
- the boundary of quadrilateral $Pr^\star Qs^\star$ consists of four chords of D

Since the chords in D do not cross each other, such a quadrilateral is an inner face of $(V \cup V', D)$. It is similar to the *quad-edge data structure* (see [4]) storing both an edge (from origin to destination) and two faces (left, right) of a general subdivision.

A *finite squaregraph* is a finite plane graph in which all inner faces are quadrilaterals and in which all inner vertices have four or more incident edges (see [1]).

Proposition 4. *The outerplanar graph* $(V \cup V', D)$ *is a squaregraph.*

Proof. Since an outerplanar graph has no inner vertices, we only have to prove that each inner face of $(V \cup V', D)$ is a quadrilateral. The boundary of such a face f consists of chords between points and dual points, so there are different points P and Q and a dual point $e^\star$ such that the chords $Pr^\star$ and $r^\star Q$ are part of the boundary of f. So P and Q are on the boundary of region r (of G), just like $r^\star$. There is no point R on the boundary of r between P and Q (relative to $r^\star$) since then $r^\star R$ would be a chord of D dividing face f. So PQ is part of the boundary of region r. Let s be the region (of G) at the other side of PQ. Then face f equals quadrilateral $Pr^\star Qs^\star$.   □

The points of $V \cup V'$ divide the circle in (new) arcs and the chords of D divide the circle disk in (new) regions. Let F be the graph that results from $(V \cup V', D)$ by adding these arcs. Then F is an outerplanar graph.

The circle arc from a point $P \in V$ to the next point $P' \in V$ anticlockwise is part of the boundary of an inner region r of G, so $r^\star$ is on this arc and the chords $Pr^\star$ and $r^\star P'$ are in D. Since the arcs $Pr^\star$ and $r^\star P'$ are also edges of F, this results in faces of F of which the boundary consists of a single arc and a single chord between a point and an adjacent dual point. These arcs are also part of the boundary of the outer face. The other faces of F are the quadrilateral faces of $(V \cup V', D)$.

Note that the weak dual of outerplanar graph F is a tree in which each vertex is either a leaf (dual to a face adjacent to the outer face) or a vertex of degree 4 (dual to a quadrilateral).

Definition 10. *Let C be a set of constants an $\tau \in T_C$. A squaregraph representation of τ is a tuple $(V \cup V', D, R, R', t)$ where:*

- *$(V \cup V', D)$ is the combination of a circular embedded tree (V, E) with its circular dual*
- *$R \in V$ and $R' \in V'$ are the roots*
- *$t : V \to C$ is a labeling of the points with constants*
- *$\tau = \tau_R$ where for each $P \in V$ the term $\tau_P \in T_C$ is defined recursively as the term with main constant $t(P)$ and left and right arguments the terms τ_Q where Q ranges over the left and right children of P (in anticlockwise order relative to R').*

Proposition 5. *Each term τ has a squaregraph representation.*

Proof. Let $\langle c_1, \ldots, c_n \rangle$ be the string $\gamma(\tau)$ of constants occurring in τ. Choose $n+1$ points $R', P_1, \ldots, P_n$ on a circle in anticlockwise order. Define $V = \{P_1, \ldots, P_n\}$ and $t(P_i) = c_i$ for $i = 1, \ldots, n$. Let (V, E) be the tree with an edge (directed to the root) from P_i to P_j if in the dual notation of τ the occurrences c_i and c_j are linked by matching braces (if $i < j$) or angle brackets (if $i > j$).   □

In Sect. 1 we showed visualizations of the standard, dual and mixed notation of two terms. Figure 4 shows the squaregraph representations of these terms.

Each quadrilateral is labeled with the main constant of a nested subterm. The quadrilateral(s) of which the boundary contains the double line are labeled with the main constant of the outer argument(s) of the term. The double line connects the point labeled with the main constant of the term to the unlabeled point at the top: the roots of a tree and its dual.

The *edges* of these dual trees are the (crossing) *diagonals* of the quadrilaterals. In each of the notations, one of the diagonals corresponds to matching brackets.

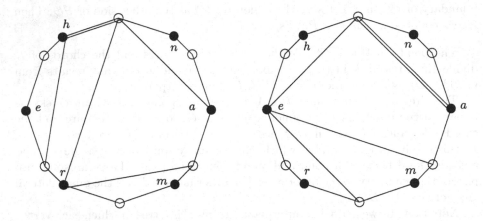

Fig. 4. Squaregraph representations of $h(e)(r(m)(a(n)))$ and $[[h]e(r)(m)]a(n)$.

5   Conclusion

Brackets are auxiliary symbols that can represent a nested structure as a linear string. Matching of brackets is usually defined in terms of the canonical context-free language S of strings with correctly matched brackets: in a string σ of the form $\sigma_1(\sigma_2)\sigma_3$, the shown brackets *match* if and only if the substring σ_2 between the brackets is in S. This occurrence of σ_2 represents a *direct* substructure of σ if both σ_1 and σ_3 are in S. Direct substructures do not overlap and are ordered from left to right. So nested substructures form a finite ordered tree.

A string can contain other symbols than brackets: constants. In this article, we treat $\sigma \in S$ as the *standard notation* of a term if each nested substructure σ' of σ contains a single constant outside its direct substructures, even if it is not the first symbol of σ'. These constants also form a finite ordered tree. There is an edge between two constants if and only if the substring between these constants is of the form $\sigma_1(\sigma_2$ or $\sigma_1)\sigma_2$ for some σ_1 and σ_2 in S.

The main point of this article is a graph-theoretical presentation of this relation between the nesting of brackets in the standard notation of a term and the usual tree structure of that term: matching brackets are treated as *edges* of a tree and this tree is *dual* to the usual tree structure of the term.

It is helpful to visualize a string as a *cyclic* structure and matching brackets as non-crossing circle chords. Matching brackets divide a string in three substrings (one inside and two outside the brackets), but a chord divides a circle in just two arcs. There is a duality between the two ends and the two sides of a chord. In a squaregraph representation of a term, this corresponds to a quadrilateral.

The generalization to terms with left and right arguments results in notations with more symmetry. Dual and mixed notations begin and end with a constant, but this does not mean that in a notation like $c\langle\rangle b\langle\rangle a$ for the nested term $c(b(a))$ with just right arguments, the main constant c and the nested constant a are treated in the same way: adding left arguments results in a notation like $e\{\}d\{\}c\langle\rangle b\langle\rangle a$ for $[[e]d]c(b(a))$ with nested constants at both ends.

The grammar of Definition 5 restricted to terms without left arguments, is:

$$T \longrightarrow CR$$

$$R \longrightarrow \epsilon \mid R(T)$$

The following grammar for the same language looks more natural:

$$T \longrightarrow C \mid T(T)$$

This corresponds to the fact that such terms can be constructed in a *unique* way, step by step, from terms without arguments, by combining two terms. However, as noted in Remark 3, this does not generalize to terms with left and right arguments. Outerplanar squaregraphs can also be constructed step by step (in several ways) from a squaregraph consisting of a single point R connected to a single dual point R', by combining two squaregraphs (via a new quadrilateral). A notation is represented by labeling half of the points with constants and half of the diagonals with brackets.

Disclosure of interests. The author has no competing interests to declare that are relevant to the content of this article.

References

1. Bandelt, H.J., Chepoi, V., Eppstein, D.: Combinatorics and geometry of finite and infinite squaregraphs. SIAM J. Discrete Math. **24**(4), 1399–1440 (2010). https://doi.org/10.1137/090760301
2. Edmonds, J.R.J.: A combinatorial representation for oriented polyhedral surfaces (1960). https://doi.org/10.13016/daw5-mvla
3. Fáry, I.: On straight line representation of planar graphs. Acta Sci. Math. **11**, 229–233 (1948)
4. Guibas, L., Stolfi, J.: Primitives for the manipulation of general subdivisions and the computation of Voronoi. ACM Trans. Graph. **4**(2), 74–123 (1985). https://doi.org/10.1145/282918.282923
5. Weisstein, E.W.: "circular embedding." from mathworld–a wolfram web resource. https://mathworld.wolfram.com/CircularEmbedding.html. Accessed 17 Jan 2024

Some Probabilistic Riddles and Some Logical Solutions

Bart Jacobs[✉]

iHub, Radboud University, Nijmegen, The Netherlands
bart@cs.ru.nl

Abstract. Six riddles in probability theory are solved in a systematic manner. The solutions suggest ingredients for a future symbolic probabilistic logic.

1 Introduction

Herman Geuvers' PhD thesis [6] from 1993 is on logics and type theories. It exploits the idea of seeing formulas as types, especially in the context of the powerful constructive type theory known as 'calculus of constructions'. Actually formalising mathematical arguments via proof assistants is a leading theme in his work, see the overview articles [1,5] and the edited book [17]. Geuvers writes in the introduction of [5]: "We believe that these systems will become the future of mathematics, where definitions, statements, computations and proofs are all available in a computerized form."

This article is about probability theory and its 'logic'. It is an area that is still far away from the level of formalisation that exists in constructive logic and type theory. It is actually somewhat of an embarrassment that there is no formal, symbolic logic for probabilistic reasoning, including updating. Admittedly, the challenges are substantial, since the nature of probabilistic logic makes formal reasoning complicated. Various operations are partial, such as addition of fuzzy, $[0, 1]$-valued, predicates; they carry the structure of an effect module, see *e.g.* [7]. Moreover, updating of probability distributions with a predicate is a partial operation, since it is undefined when the validity of the predicate involved is zero (see Sect. 4 below). What makes it worse, updating, also known as belief revision, makes probabilistic logic non-monotone, since updating may make statements 'less true', see [18]. It does not help that the standard notation and terminology in probability theory are clumsy and inadequate: everything is expressed in terms of probability P and basic constructs are left implicit, like updating of distributions and transformations along channels. The area of probabilistic reasoning is thus in need of its own expressive logic. This matter is urgent since modern AI-techniques are often probabilistic in nature and there are justified requirements that AI-based decisions should be explainable.

This article does not develop such a probabilistic logic. It does illustrate some of the basic constructions that such a logic should have, via examples. It proceeds by formally

B. Jacobs—Dedicated to my dear colleague Herman Geuvers, on the occasion of his 60th birthday.

V. Capretta et al. (Eds.): *Logics and Type Systems in Theory and Practice*, LNCS 14560, pp. 134–149, 2024.
https://doi.org/10.1007/978-3-031-61716-4_9

solving some probabilistic riddles, copied from the BRiddles website[1] which is full
of recreational mathematical puzzles and challenges. These riddles include classical
ones in probability. The outcomes are sometimes unexpected and/or counter-intuitive,
emphasising the need for formal reasoning rules.

This article is based on the author's ongoing research towards a mathematical understanding of the basic reasoning principles of (discrete) probability theory, see *e.g.* [3, 8–
14] and the book-in-preparation [15]. This approach is based on category theory and
is sometimes called categorical probability theory, see *e.g.* [4] for further information.
This paper does not go into the underlying mathematics (or category theory) and has
a more operational focus: it illustrates how to exploit the principles stemming from
this semantics-based approach in order to solve basic riddles. It may thus serve as an
introduction and motivation for this approach. No formal rules are used. However, the
manner in which the semantics is exploited in this paper does suggest rules, notation,
and basic ingredients for a future formal probabilistic logic.

The emphasis is on formalising the (solution of the) riddles. The illustrations below
do involve various discrete calculations. They have been done with our own Python
scripts, based on the EfProb library first described in [2]. How these calculations are
done is not so interesting and is left implicit. Being able to do such calculations should
be part of the evaluation rules in a symbolic probabilistic logic. Such a logic will surely
incorporate parts of probabilistic programming languages, for modeling statistical problems. The calculations in this paper are 'exact' and do not involve approximations via
sampling semantics, as commonly used for probabilistic programs.

The paper starts with a short description of the most basic concepts: multisets, distributions, and predicates. The subsequent Sects. 3–8 are each devoted to a single riddle.
These sections start with a (literal) quote from the website BRiddles, followed by an
analysis. Typically, the analysis involves a bit more theory, and is generalised beyond
the riddle at hand. There are many more riddles on the website BRiddles than the six
ones that we concentrate on here. We have selected these six riddles in order to have a
certain natural build-up and coherence.

2   Preliminaries

In this section we fix some basic notation for multisets, finite discrete probability distributions, and for predicates. We introduce additional constructions later on, at the point
where we need them to solve riddles.

A *multiset* is like a subset except that elements may occur multiple times. Multisets
differ from lists, since the order of elements does not matter. An urn with three red, two
blue and five green balls will be described in 'ket' notation as $3|R\rangle + 2|B\rangle + 5|G\rangle$.
More generally, a multiset over a set X is a finite formal sum of the form $\sum_i n_i|x_i\rangle$,
for elements $x_i \in X$ and natural numbers $n_i \in \mathbb{N}$ describing the multiplicities of these
elements x_i. We shall write $\mathcal{M}(X)$ for the set of such multisets over X.

A multiset $\varphi \in \mathcal{M}(X)$ may equivalently be described in functional form, as func-
tion $\varphi\colon X \to \mathbb{N}$ with finite support: $supp(\varphi) := \{x \in X \mid \varphi(x) \neq 0\}$. Such a function

---

$\varphi\colon X \to \mathbb{N}$ can be written in ket form as $\sum_{x \in X} \varphi(x)|x\rangle$. We switch back and forth between the ket and functional form and use the formulation that best suits a particular situation.

For a multiset $\varphi \in \mathcal{M}(X)$ we write $\|\varphi\| \in \mathbb{N}$ for the *size* of the multiset. It is the total number of elements, including multiplicities:

$$\|\varphi\| := \sum_{x \in X} \varphi(x).$$

For $K \in \mathbb{N}$ we then write $\mathcal{M}[K](X) \subseteq \mathcal{M}(X)$ for the subset of multisets of size K.

Multisets $\varphi, \psi \in \mathcal{M}(X)$ and can be added and compared elementwise, so that $(\varphi + \psi)(x) = \varphi(x) + \psi(x)$ and $\varphi \leq \psi$ means $\varphi(x) \leq \psi(x)$ for all $x \in X$.

A *distribution* is a finite formal sum of the form $\sum_i r_i|x_i\rangle$ with multiplicities $r_i \in [0,1]$ satisfying $\sum_i r_i = 1$. Such a distribution can equivalently be described as a function $\omega\colon X \to [0,1]$ with finite support, satisfying $\sum_x \omega(x) = 1$. We write $\mathcal{D}(X)$ for the set of distributions on X.

Each multiset $\varphi \in \mathcal{M}(X)$ of non-zero size can be turned into a distribution via normalisation. This operation is called frequentist learning, since it involves learning a distribution from a multiset of data, via counting. Explicitly:

$$Flrn(\varphi) := \sum_{x \in X} \frac{\varphi(x)}{\|\varphi\|}\,|x\rangle.$$

For instance, if we learn from the urn mentioned above, we get the probability distribution for drawing a ball of a particular colour from the urn:

$$Flrn\Big(3|R\rangle + 2|B\rangle + 5|G\rangle\Big) = \tfrac{3}{10}|R\rangle + \tfrac{1}{5}|B\rangle + \tfrac{1}{2}|G\rangle.$$

Given two distributions $\omega \in \mathcal{D}(X)$ and $\rho \in \mathcal{D}(Y)$ we can form their parallel product $\omega \otimes \rho \in \mathcal{D}(X \times Y)$, given in functional form as:

$$(\omega \otimes \rho)(x,y) := \omega(x) \cdot \rho(y).$$

Explicitly, for instance,

$$\Big(\tfrac{3}{10}|R\rangle + \tfrac{1}{5}|B\rangle + \tfrac{1}{2}|G\rangle\Big) \otimes \Big(\tfrac{1}{3}|0\rangle + \tfrac{2}{3}|1\rangle\Big)$$
$$= \tfrac{1}{10}|R,0\rangle + \tfrac{1}{15}|B,0\rangle + \tfrac{1}{6}|G,0\rangle + \tfrac{1}{5}|R,1\rangle + \tfrac{2}{15}|B,1\rangle + \tfrac{1}{3}|G,1\rangle.$$

A *predicate* on a set X is a function $p\colon X \to [0,1]$. For a distribution $\omega \in \mathcal{D}(X)$, we write $\omega \models p \in [0,1]$ for the validity (or expected value) of p in ω. It is defined as:

$$\omega \models p := \sum_{x \in X} \omega(x) \cdot p(x). \tag{1}$$

Two special predicates on X are truth and falsity, written as $\mathbf{1}$ and $\mathbf{0}$. They are the constant-1 and constant-0 functions: $\mathbf{1}(x) := 1$ and $\mathbf{0}(x) := 0$. For an element $x \in X$

we write $\mathbf{1}_x \colon X \to [0,1]$ for the point predicate which is 1 only on x, and 0 everywhere else. We have:

$$\omega \models \mathbf{1} = 1 \qquad \omega \models \mathbf{0} = 0 \qquad \omega \models \mathbf{1}_x = \omega(x).$$

For each predicate $p \colon X \to [0,1]$ we define its negation $p^\perp \colon X \to [0,1]$ as $p^\perp(x) = 1 - p(x)$. Then $p^{\perp\perp} = p$ and $\mathbf{1}^\perp = \mathbf{0}, \mathbf{0}^\perp = \mathbf{1}$. Also: $\omega \models p^\perp = 1 - (\omega \models p)$.

For two predicates p on X and q on Y we can form the parallel product $p \otimes q \colon X \times Y \to [0,1]$ as $(p \otimes q)(x,y) := p(x) \cdot q(y)$. Then, for instance, $(\omega \otimes \rho \models p \otimes q) = (\omega \models p) \cdot (\rho \models q)$.

We now turn to the riddles.

3   Hard Conditional Probability Problem

We start with our first riddle[2], quoted in its original form.

> Four friends—Anna, Brian, Christy and Drake are asked to choose any number between 1 and 5. Can you calculate the probability that any of them chose the same number?

We add that the question should be understood as: *at least two* friends choose the same number.

To start, for a natural number $n \geq 1$ we write $[n] := \{1, 2, \ldots, n\}$ for the set with the first n positive natural numbers. It carries a uniform distribution $\upsilon_n \in \mathcal{D}([n])$ of the form:

$$\upsilon_n := \tfrac{1}{n}|1\rangle + \tfrac{1}{n}|2\rangle + \cdots + \tfrac{1}{n}|n\rangle.$$

The above challenge is formulated for $n = 5$, implicitly using the uniform distribution υ_5 for the friend's choices. When we consider the (independent) choices of two friends we use the parallel product $\upsilon_5 \otimes \upsilon_5 = \tfrac{1}{25}|1,1\rangle + \tfrac{1}{25}|1,2\rangle + \cdots + \tfrac{1}{25}|5,4\rangle + \tfrac{1}{25}|5,5\rangle \in \mathcal{D}([5] \times [5])$. Similarly, the distributions for three and four friends are $\upsilon_5 \otimes \upsilon_5 \otimes \upsilon_5$ and $\upsilon_5 \otimes \upsilon_5 \otimes \upsilon_5 \otimes \upsilon_5$. The support of the latter distribution contains $5^4 = 625$ four-tuples, corresponding to the four choices of the four friends.

Suppose first that only two friends play this game. The equality of their choices is expressed by the binary predicate:

$$[5] \times [5] \xrightarrow{Eq_2} [0,1] \quad \text{with} \quad Eq_2(i,j) := \begin{cases} 1 & \text{if } i = j \\ 0 & \text{otherwise.} \end{cases}$$

We claim that the probability that two friends choose the same number can be computed as validity:

$$\upsilon_5 \otimes \upsilon_5 \models Eq_2 \overset{(1)}{=} \sum_{i,j \in [5]} (\upsilon_5 \otimes \upsilon_5)(i,j) \cdot Eq_2(i,j)$$

$$= \sum_{i \in [5]} \upsilon_5(i) \cdot \upsilon_5(i) = \sum_{i \in [5]} \tfrac{1}{5} \cdot \tfrac{1}{5} = \tfrac{1}{5}.$$

---

[2] See www.briddles.com/2014/04/hard-conditional-probability-problem.html. Why this problem has 'conditional probability' in its name is unclear.

When three friends participate, we use the following ternary predicate expressing that at least two choices coincide.

$$[5] \times [5] \times [5] \xrightarrow{Eq_3} [0,1] \quad \text{with} \quad Eq_3(i,j,k) := \begin{cases} 1 & \text{if } i=j \text{ or } i=k \text{ or } j=k \\ 0 & \text{otherwise.} \end{cases}$$

The probability of at least two coinciding choices of three friends is then:

$$v_5 \otimes v_5 \otimes v_5 \models Eq_3 = \tfrac{13}{25}.$$

Generalising this to four friends is now obvious. It yields the desired answer to the riddle:

$$v_5 \otimes v_5 \otimes v_5 \otimes v_5 \models Eq_4 = \tfrac{101}{125}.$$

However, formulating the quaternary predicate Eq_4 becomes a bit awkward since we have to write a disjunction of six equations involving four variables. This does not scale well.

There is an alternative, easier way to formulate the outcome that more easily scales to larger numbers. We notice that for the problem at hand it does not matter in which order the friends choose a number. Thus instead of considering their choices as an (ordered) sequence, in the product set $[5] \times [5] \times [5] \times [5]$, we will treat them as a multiset of size 4, in the set of multisets $\mathcal{M}[4]([5])$. Choosing multiple elements (with replacement) from a distribution is described via the *multinomial* distribution.

In general, for a distribution $\omega \in \mathcal{D}(X)$ and a number $K \in \mathbb{N}$, for the size of the chosen multisets, the multinomial distribution $mn[K](\omega) \in \mathcal{D}(\mathcal{M}[K](X))$ is defined, like in [10, 13], as:

$$mn[K](\omega) := \sum_{\varphi \in \mathcal{M}[K](X)} (\varphi) \cdot \prod_{x \in X} \omega(x)^{\varphi(x)} |\varphi\rangle \quad \text{where} \quad \begin{cases} (\varphi) := \dfrac{\|\varphi\|!}{\varphi\underline{!}} \\ \text{with} \\ \varphi\underline{!} := \prod_x \varphi(x)!. \end{cases}$$

Here is a brief explanation of this formula. Given a multiset $\varphi \in \mathcal{M}[K](X)$ of size K, there are (φ) many sequences in X^K that have occurrences of elements with multiplicities as in φ. These sequences involve the same elements, in different orders. Drawing the K elements of each of these sequences, in order, from ω, yields the same probability, namely $\prod_x \omega(x)^{\varphi(x)}$. This number has to be multiplied by (φ), since there are (φ) many sequences with multiplicities as in φ.

For instance, draws of sizes 2 and 3 from $v_2 = \tfrac{1}{2}|1\rangle + \tfrac{1}{2}|2\rangle$ produce the following distributions over multisets:

$$mn[2](v_2) = \tfrac{1}{4}\big|2|1\rangle\big\rangle + \tfrac{1}{2}\big|1|1\rangle + 1|2\rangle\big\rangle + \tfrac{1}{4}\big|2|2\rangle\big\rangle$$
$$mn[3](v_2) = \tfrac{1}{8}\big|3|1\rangle\big\rangle + \tfrac{3}{8}\big|2|1\rangle + 1|2\rangle\big\rangle + \tfrac{3}{8}\big|1|1\rangle + 2|2\rangle\big\rangle + \tfrac{1}{8}\big|3|2\rangle\big\rangle$$

We use the following generic predicate $Geq_2 \colon \mathcal{M}(X) \to [0,1]$ on multisets, expressing that at least one element occurs twice in a multiset $\varphi \in \mathcal{M}(X)$.

$$Geq_2(\varphi) := \begin{cases} 1 & \text{if } \exists x. \, \varphi(x) \geq 2 \\ 0 & \text{otherwise.} \end{cases}$$

We can now reproduce the above probabilities of having some coincidence of choices, from v_5, for $2, 3, 4$ friends as:

$$mn[2](v_5) \models Geq_2 = \tfrac{1}{5}$$
$$mn[3](v_5) \models Geq_2 = \tfrac{13}{25}$$
$$mn[4](v_5) \models Geq_2 = \tfrac{101}{125}.$$

This formulation in terms of multisets and multinomials scales better, to larger numbers of friends, and thus to larger draw sizes. If a distribution ω has n elements in its support, then the support of the K-fold product distribution $\omega^K = \omega \otimes \cdots \otimes \omega$ has n^K elements, namely all sequences of length K. In contrast, when a set X has n elements, then the number of elements in the set $\mathcal{M}[K](X)$ of K-sized multisets is given by the multichoose coefficient:

$$\left(\binom{n}{K}\right) := \binom{n + K - 1}{K} = \frac{(n + K - 1)!}{K! \cdot (n - 1)!}.$$

For instance,

$$mn[6](v_{10}) \models Geq_2 = \tfrac{1061}{1250}.$$

This is the probability that at least two choices coincide when six people choose from ten equally likely options.

4   Probability of Second Girl Child

We quote the following riddle[3], which provides a standard example in reasoning with conditional probabilities.

Kukki and Fukki are a married couple (don't ask me who he is and who she is). They have two kids, one of them is a girl. Assume safely that the probability of each gender is $1/2$. What is the probability that the other kid is also a girl? Hint: It is not $1/2$ as you would first think.

In order to solve this problem we have to update distributions with predicates. This is sometimes called belief revision. In general, it works as follows.

Let $\omega \in \mathcal{D}(X)$ be a distribution with a predicate $p \colon X \to [0, 1]$, with non-zero validity $\omega \models p \neq 0$. Then we can form the updated distribution $\omega|_p \in \mathcal{D}(X)$ as normalised product:

$$\omega|_p := \sum_{x \in X} \frac{\omega(x) \cdot p(x)}{\omega \models p} |x\rangle. \tag{2}$$

This updated distribution has incorporated the evidence p. One key property is that the validity of p is higher in the updated distribution $\omega|_p$ than in the original ('prior') distribution ω. This means $(\omega|_p \models p) \geq (\omega \models p)$, see e.g. [8, 15]: updating with p makes p more true. This update rule (2) provides the foundation for Bayesian inference, as

---

[3] From: www.briddles.com/2011/07/probability-of-second-girl-child.html.

used for instance in Bayesian networks [16]. This rule is sometimes also called Pearl's update rule, in contrast with an alternative rule of Jeffrey [8,11].

Let's see how this updating works for the challenge at hand. We use the set $\{G, B\}$, where G stands for girl and B for boy. Since both genders are equally likely, we use the uniform distribution $\upsilon = \frac{1}{2}|G\rangle + \frac{1}{2}|B\rangle \in \mathcal{D}(\{G, B\})$. The situation with two children is described by the parallel product $\upsilon \otimes \upsilon = \frac{1}{4}|G, G\rangle + \frac{1}{4}|G, B\rangle + \frac{1}{4}|B, G\rangle + \frac{1}{4}|B, B\rangle$.

We use the following predicates $\{G, B\} \times \{G, B\} \to [0, 1]$ for '2 girls' and 'at least 1 girl', where the latter is the same as 'not 2 boys'.

$$1_G \otimes 1_G \qquad \begin{aligned} \left(1_B \otimes 1_B\right)^{\perp} &= (1_G \otimes 1) + (1_B \otimes 1_G) \\ &= (1_G \otimes 1_G) + (1_G \otimes 1_B) + (1_B \otimes 1_G). \end{aligned}$$

Then, clearly, the probabilities of 'two girls' and 'at least one girl' are:

$$\upsilon \otimes \upsilon \models 1_G \otimes 1_G = \frac{1}{4} \qquad \upsilon \otimes \upsilon \models \left(1_B \otimes 1_B\right)^{\perp} = \frac{3}{4}.$$

The crucial step is that we update the product distribution $\upsilon \otimes \upsilon$ with the predicate $\left(1_B \otimes 1_B\right)^{\perp}$. We mechanically follow the description (2):

$$
\begin{aligned}
\left(\upsilon \otimes \upsilon\right)\big|_{(1_B \otimes 1_B)^{\perp}} &= \sum_{x,y \in \{G,B\}} \frac{(\upsilon \otimes \upsilon)(x, y) \cdot (1_B \otimes 1_B)^{\perp}(x, y)}{\upsilon \otimes \upsilon \models (1_B \otimes 1_B)^{\perp}} |x, y\rangle \\
&= \frac{(\upsilon \otimes \upsilon)(G, G) \cdot (1_B \otimes 1_B)^{\perp}(G, G)}{3/4} |G, G\rangle \\
&\quad + \frac{(\upsilon \otimes \upsilon)(G, B) \cdot (1_B \otimes 1_B)^{\perp}(G, B)}{3/4} |G, B\rangle \\
&\quad + \frac{(\upsilon \otimes \upsilon)(B, G) \cdot (1_B \otimes 1_B)^{\perp}(B, G)}{3/4} |B, G\rangle \\
&\quad + \frac{(\upsilon \otimes \upsilon)(B, B) \cdot (1_B \otimes 1_B)^{\perp}(B, B)}{3/4} |B, B\rangle \\
&= \frac{1/4}{3/4} |G, G\rangle + \frac{1/4}{3/4} |G, B\rangle + \frac{1/4}{3/4} |B, G\rangle \\
&= \frac{1}{3} |G, G\rangle + \frac{1}{3} |G, B\rangle + \frac{1}{3} |B, G\rangle.
\end{aligned}
$$

This gives, as expected, the (normalised) distribution with at least one girl. The answer to the above riddle, that is, the chance of having two girls if we know that there is at least one girl, is the validity in the updated distribution:

$$\left(\upsilon \otimes \upsilon\right)\big|_{(1_B \otimes 1_B)^{\perp}} \models 1_G \otimes 1_G = \tfrac{1}{3}.$$

This answers the riddle.

Like in the previous section, it makes sense to switch to a description in terms of multisets (of girls and boys). We now use the predicates $G_{=n}, G_{\geq n} : \mathcal{M}(\{G, B\}) \to [0, 1]$ for precisely and at least n girls, given on $\varphi \in \mathcal{M}(\{G, B\})$ as:

$$G_{=n}(\varphi) := \begin{cases} 1 & \text{if } \varphi(G) = n \\ 0 & \text{otherwise} \end{cases} \qquad G_{\geq n}(\varphi) := \begin{cases} 1 & \text{if } \varphi(G) \geq n \\ 0 & \text{otherwise.} \end{cases} \qquad (3)$$

Similarly there are predicates $B_{=n}, B_{\geq n} : \mathcal{M}(\{G, B\}) \to [0, 1]$ for boys.

For the riddle at hand we use the following steps to arrive at the same outcome.

$$mn[2](v) = \tfrac{1}{4}\left|2|G\rangle\right\rangle + \tfrac{1}{2}\left|1|G\rangle + 1|B\rangle\right\rangle + \tfrac{1}{4}\left|2|B\rangle\right\rangle$$
$$mn[2](v) \models G_{\geq 1} = \tfrac{1}{4} + \tfrac{1}{2} + 0 = \tfrac{3}{4}.$$

$$mn[2](v)\big|_{G_{\geq 1}} = \frac{1/4}{3/4}\left|2|G\rangle\right\rangle + \frac{1/2}{3/4}\left|1|G\rangle + 1|B\rangle\right\rangle + \frac{0}{3/4}\left|2|B\rangle\right\rangle$$
$$= \tfrac{1}{3}\left|2|G\rangle\right\rangle + \tfrac{2}{3}\left|1|G\rangle + 1|B\rangle\right\rangle$$
$$mn[2](v)\big|_{G_{\geq 1}} \models G_{=2} = \tfrac{1}{3}.$$

This multinomial approach can be generalised easily. For instance, in a family with ten children, if there are at least six girls, the probability that there are at least two boys is obtained by first updating to:

$$mn[10](v)\big|_{G_{\geq 6}} = \tfrac{1}{386}\left|10|G\rangle\right\rangle + \tfrac{5}{193}\left|9|G\rangle + 1|B\rangle\right\rangle + \tfrac{45}{386}\left|8|G\rangle + 2|B\rangle\right\rangle$$
$$+ \tfrac{60}{193}\left|7|G\rangle + 3|B\rangle\right\rangle + \tfrac{105}{193}\left|6|G\rangle + 4|B\rangle\right\rangle.$$

The probability that there are then at least two boys is obtained by summing up the probabilities of the last three terms, giving:

$$mn[10](v)\big|_{G_{\geq 6}} \models B_{\geq 2} = \tfrac{45}{386} + \tfrac{60}{193} + \tfrac{105}{193} = \tfrac{375}{386}.$$

5   Tricky Probability Interview Puzzle

The riddle that we tackle in this section is the following one[4].

> I have two coins.
> - One of the coin is a faulty coin having tail on both side of it.
> - The other coin is a perfect coin (heads on side and tail on other).
>
> I blind fold myself and pick a coin and put the coin on table. The face of coin towards the sky is tail.
>
> What is the probability that other side is also tail?

For the solution of this riddle we use the concept of a *channel*, together with *predicate transformation* along a channel. In general, a channel from a set X to a set Y is a probabilistic computation from X to Y. It is a function of the form $c: X \to \mathcal{D}(Y)$, sending an element $x \in X$ to a distribution $c(x)$ on Y. Traditionally, such a channel is described as a conditional probability distribution $p(y \mid x)$ over Y, indexed by elements $x \in X$. It corresponds to a stochastic matrix and is often called a Markov kernel.

Given a predicate q on the codomain Y of a channel $c: X \to \mathcal{D}(Y)$, we can turn it into a predicate on X, written as $c \lll q$, and given on $x \in X$ by:

$$\left(c \lll q\right)(x) := c(x) \models q = \sum_{y \in Y} c(x)(y) \cdot q(y). \tag{4}$$

---

[4] www.briddles.com/2013/08/tricky-probability-interview-puzzle.html. This is a literal copy, including typos.

We now show how this applies to the above riddle. There are two coins, for which we use the set $\{1,2\}$. For each of the numbers in this set, there is a coin distribution, as described in the dashes in the above quote. This corresponds to a channel $c\colon \{1,2\} \to \mathcal{D}(\{H,T\})$, with:

$$c(1) := 1|T\rangle \qquad\qquad c(2) := \tfrac{1}{2}|H\rangle + \tfrac{1}{2}|T\rangle.$$

Thus, $c(1)$ is the faulty coin that always returns tail T, and $c(2)$ is the fair coin, given by a uniform distribution over heads H and tails T. The two options $1,2$ are equally likely, so we are dealing with a situation $\sigma = \tfrac{1}{2}|1\rangle + \tfrac{1}{2}|2\rangle$.

The tail observation of the coin on the table is captured by the point predicate $1_T\colon \{H,T\} \to [0,1]$. We would like to update the distribution $\sigma \in \mathcal{D}(\{1,2\})$ with this evidence. This σ is a distribution on the domain of the channel c, whereas the observation is a predicate on the channel's codomain. In order to be able to update, we first have to transform this evidence predicate 1_T on $\{H,T\}$ along the channel c into a predicate $c \lll 1_T$ on $\{1,2\}$. This transformed predicate is, according to (4),

$$\left(c \lll 1_T\right)(1) = \sum_{y\in\{H,T\}} c(1)(y) \cdot 1_T(y) = c(1)(T) = 1$$
$$\left(c \lll 1_T\right)(2) = \sum_{y\in\{H,T\}} c(2)(y) \cdot 1_T(y) = c(2)(T) = \tfrac{1}{2}.$$

These numbers describe how compatible the tail observation is with the two numbered coins: coin 1 is 100% compatible, whereas coin 2 is 50% compatible.

We can now compute the validity and update using this transformed predicate:

$$\sigma \models c \lll 1_T = \sigma(1)\cdot(c \lll 1_T)(1) + \sigma(2)\cdot(c \lll 1_T)(2) = \tfrac{1}{2}\cdot 1 + \tfrac{1}{2}\cdot\tfrac{1}{2} = \tfrac{3}{4}$$

$$\sigma|_{c \lll 1_T} \overset{(2)}{=} \frac{\sigma(1)\cdot(c \lll 1_T)(1)}{\sigma \models c \lll 1_T}\,|1\rangle + \frac{\sigma(2)\cdot(c \lll 1_T)(2)}{\sigma \models c \lll 1_T}\,|2\rangle$$
$$= \frac{1/2\cdot 1}{3/4}\,|1\rangle + \frac{1/2\cdot 1/2}{3/4}\,|2\rangle = \tfrac{2}{3}|1\rangle + \tfrac{1}{3}|2\rangle.$$

The riddle asks for the probability that the other side is also tail, that is, the probability of coin 1. The last expression tells that this probability is $\tfrac{2}{3}$.

What just happened may look like magic. Let's cook up another, similar example. Suppose we have different urns containing white and black balls:

- three red urns, each containing 1 white and 4 black balls;
- two blue urns, each containing 3 white and 2 black balls;
- one green urn, containing 5 white balls.

What do we have? A channel $c\colon \{R,B,G\} \to \mathcal{D}(\{W,B\})$ giving for each colour the associated ball distribution, obtained via frequentist learning:

$$c(R) := Flrn\big(1|W\rangle + 4|B\rangle\big) = \tfrac{1}{5}|W\rangle + \tfrac{4}{5}|B\rangle$$
$$c(B) := Flrn\big(3|W\rangle + 2|B\rangle\big) = \tfrac{3}{5}|W\rangle + \tfrac{2}{5}|B\rangle$$
$$c(G) := Flrn\big(5|W\rangle\big) = 1|W\rangle.$$

Suppose we pick one of the urns at random and pull out one ball at random that turns out to be white. Can we compute for each urn colour the likelihood that the white ball was drawn from an urn with that colour? More technically, what is the *posterior* colour distribution after incorporating the white ball evidence?

We first note that the choice of urn colour is determined by the colour distribution $\sigma = Flrn\left(3|R\rangle+2|B\rangle+1|G\rangle\right) = \frac{1}{2}|R\rangle+\frac{1}{3}|B\rangle+\frac{1}{6}|G\rangle$. We proceed via what is called *backward reasoning*, as we have done before: we compute the predicate transformation $c \lll 1_W$ and update the *prior* distribution σ with this predicate. First:

$$\left(c \lll 1_W\right)(R) = \frac{1}{5} \qquad \left(c \lll 1_W\right)(B) = \frac{3}{5} \qquad \left(c \lll 1_W\right)(G) = 1.$$

The posterior colour distribution is: $\sigma|_{c\lll 1_W} = \frac{3}{14}|R\rangle + \frac{3}{7}|B\rangle + \frac{5}{14}|G\rangle$. Now, the blue urn has become most likely, intuitively, since there are two of them and they have relatively many white balls.

In the same way one computes the posterior colour distribution after drawing a black ball. It yields: $\sigma|_{c\lll 1_B} = \frac{3}{4}|R\rangle + \frac{1}{4}|B\rangle$. The green urn has probability zero in this posterior since it has no black balls, so the drawn black ball cannot come from a green urn.

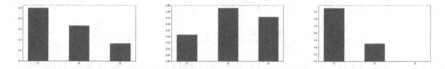

Fig. 1. Plot of the prior colour distribution on the left, with the posterior distribution after observing a white ball in the middle, and the posterior after a black one on the right.

Figure 1 contains plots of the prior distribution $\sigma \in \mathcal{D}(\{R, B, G\})$ together with the two updated posteriors $\sigma|_{c\lll 1_W}$ and $\sigma|_{c\lll 1_B}$. The actual probabilities do not matter that much. The plots illustrate the relative sizes.

6   Life or Death

We now investigate the following riddle[5].

> You are a prisoner sentenced to death. The Emperor offers you a chance to live by playing a simple game. He gives you 50 black marbles, 50 white marbles and 2 empty bowls. He then says, 'Divide these 100 marbles into these 2 bowls. You can divide them any way you like as long as you use all the marbles. Then I will blindfold you and mix the bowls around. You then can choose one bowl and remove ONE marble. If the marble is WHITE you will live, but if the marble is BLACK ... you will die.'
> How do you divide the marbles up so that you have the greatest probability of choosing a WHITE marble?

---

[5] See briddles.com/2011/06/life-or-death-3-june.html.

A moment's thought reveals the best marble division: put one white marble in the first bowl, and all the 99 others in the second bowl. Then, if you (blindly) choose the first bowl, you live, and if you choose the second one, you still have a $\frac{49}{99}$ chance of living. Of course, you can equivalently put one white marble in the second bowl and all the others in the first one. This informal argument shows that your best chance is $\frac{1}{2} \cdot 1 + \frac{1}{2} \cdot \frac{49}{99} = \frac{148}{198} = \frac{74}{99}$.

How to formalise this? First, we take $\{W, B\}$ as set of colours for the marbles. We use a pair of numbers (w, b) for the white and black marbles in the first bowl. The pair of white and black marbles in the second bowl is then $(50 - w, 50 - b)$. Explicitly, we use as space MD of marble divisions the set:

$$MD := \{(w, b) \mid 0 \leq w, b \leq 50 \text{ with } 0 < w + b < 100\}.$$

The requirement $0 < w+b$ ensures that there is at least one marble in the first bowl, and, similarly, $w + b < 100$ guarantees that the second bowl is non-empty. This requirement is implicit in the above formulation, because it must be possible to remove at least one marble from any of the bowls. This set MD has $51^2 - 2 = 2599$ elements.

We define two channels $c_1, c_2 \colon MD \to \mathcal{D}(\{W, B\})$ to assign a white-black distribution to a marble division, where c_i captures a choice of marble from bowl i. Explicitly:

$$c_1(w, b) := \frac{w}{w + b}|W\rangle + \frac{b}{w + b}|B\rangle$$

$$c_2(w, b) := \frac{50 - w}{100 - w - b}|W\rangle + \frac{50 - b}{100 - w - b}|B\rangle.$$

The blind choice of bowls followed by the removal of a marble is then modelled via 50–50 mixture of these channels, that via the (pointwise) convex sum:

$$c := \tfrac{1}{2} \cdot c_1 + \tfrac{1}{2} \cdot c_2, \tag{5}$$

so that:

$$c(w, b) := \frac{50w + (25 - w)(w + b)}{(w + b)(100 - w - b)}|W\rangle + \frac{50b + (25 - b)(w + b)}{(w + b)(100 - w - b)}|B\rangle.$$

We can now find the maximum probability for drawing a white marble by going through all marble divisions. Then indeed:

$$\max\{c(w, b)(W) \mid (w, b) \in MD\} = \tfrac{74}{99}.$$

This maximum is reached for the two marble divisions $(1, 0), (49, 50) \in MD$, corresponding to putting one white marble in the first bowl or in the second bowl.

We can also use the backward reasoning technique introduced in the previous section. We first take the uniform distribution v on MD, of the form:

$$v = \tfrac{1}{2599}|0, 1\rangle + \tfrac{1}{2599}|0, 2\rangle + \cdots + \tfrac{1}{2599}|50, 48\rangle + \tfrac{1}{2599}|50, 49\rangle.$$

We use the white marble point predicate 1_W and apply predicate transformation to get the predicate $c \lll 1_W$ on MD. We can now update the uniform distribution v, giving:

$$v' := v|_{c \lll 1_W} \in \mathcal{D}(MD). \tag{6}$$

It gives the updated probabilities for marble divisions, given that a white marble was drawn. Its plot is on the left in Fig. 2. With a bit of effort one can recognise that the maximal probabilities are reached at the marble divisions $(1,0), (49,50) \in MD$ that we already identified before.

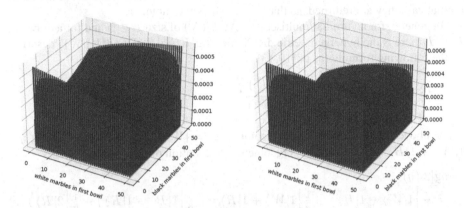

Fig. 2. Two-dimensional plots of the updated uniform distribution on marble divisions $MD \subseteq \{0, 1, \ldots, 50\} \times \{0, 1, \ldots, 50\}$, given that a white marble was drawn. The plot on the left describes the distribution v' from (6), for a $\frac{1}{2}$ chance of choosing the first bowl in (5). The plot on the right describes v'' from (7) with a $\frac{3}{4}$ probability for the first bowl.

In the end we slightly change the set-up: instead of an equal probability of chosing the bowls, we now set the probability of choosing the first bowl at $\frac{3}{4}$, with probability $\frac{1}{4}$ for the second bowl. This means that we change the convex combination of channels in (5) to:

$$d := \frac{3}{4} \cdot c_1 + \frac{1}{4} \cdot c_2 \quad \text{resulting in} \quad v'' := v|_{d \ll 1_W} \in \mathcal{D}(MD). \qquad (7)$$

Now there is only one marble division with highest probability, namely $(1,0)$, with one white marble in the first bowl. The associated survival probability rises from $\frac{74}{99}$ to: $\frac{3}{4} \cdot 1 + \frac{1}{4} \cdot \frac{49}{99} = \frac{173}{198}$. The plot on the right in Fig. 2 illustrates this. It is no longer symmetric, like the distribution on the left.

7 Paradox Probability Puzzle

The riddle is formulated as follows[6].

> This is a famous paradox which has caused a great deal of argument and disbelief from many who cannot accept the correct answer. Four balls are placed in a hat. One is white, one is blue and the other two are red. The bag is shaken and someone draws two balls from the hat. He looks at the two balls and announces that at least one of them is red. What are the chances that the other ball he has drawn out is also red?

---

[6] www.briddles.com/2011/10/paradox-probability-puzzle.html.

Answering this challenge is not so difficult, given what we have already seen. The new aspect is the use of the hypergeometric draw distribution, which involves 'draw-and-delete', whereas the multinomial distribution that we saw in Sect. 3 works via 'draw-and-replace'. More concretely, in hypergeometric draws the drawn elements are removed from the urn (or hat), so that the urn decreases in size, whereas in the multi-nomial case they are returned, so that the urn stays the same.

In general, for an urn/hat/multiset $\psi \in \mathcal{M}[L](X)$ of size $L \in \mathbb{N}$ and for a draw size $K \leq L$ there is the hypergeometric distribution $hg[K](\psi) \in \mathcal{D}(\mathcal{M}[K](X))$ described by the following formula (as in [10,13]).

$$hg[K](\psi) := \sum_{\varphi \in \mathcal{M}[K](X),\, \varphi \leq \psi} \frac{\binom{\psi}{\varphi}}{\binom{L}{K}} |\varphi\rangle \quad \text{where} \quad \binom{\psi}{\varphi} := \prod_{x \in X} \binom{\psi(x)}{\varphi(x)}.$$

In the above riddle we have a set of colours $X = \{W, B, R\}$ and a hat multiset $\eta = 1|W\rangle + 1|B\rangle + 2|R\rangle$. The hypergeometric distribution of draws of size 2 is:

$$hg[2](\eta)$$
$$= \tfrac{1}{6} \Big| 1|W\rangle + 1|B\rangle \Big\rangle + \tfrac{1}{3} \Big| 1|W\rangle + 1|R\rangle \Big\rangle + \tfrac{1}{3} \Big| 1|B\rangle + 1|R\rangle \Big\rangle + \tfrac{1}{6} \Big| 2|R\rangle \Big\rangle.$$

We use predicates on multisets like in (3), so that $R_{\geq n} : \mathcal{M}(\{W, B, R\}) \to [0, 1]$ satisfies $R_{\geq n}(\varphi) = 1$ precisely if $\varphi(R) \geq n$, and $R_{\geq n}(\varphi) = 0$ otherwise. There is a similar predicate $R_{=n}$. We use such predicates $R_{\geq n}, R_{=n}$ also for other colours than red (R).

The riddle involves an update of the draws to those with at least one red, giving:

$$hg[2](\eta)\big|_{R_{\geq 1}} = \tfrac{2}{5} \Big| 1|B\rangle + 1|R\rangle \Big\rangle + \tfrac{2}{5} \Big| 1|W\rangle + 1|R\rangle \Big\rangle + \tfrac{1}{5} \Big| 2|R\rangle \Big\rangle.$$

The desired riddle solution is the probability obtained as validity:

$$hg[2](\eta)|_{R_{\geq 1}} \models R_{=2} = \tfrac{1}{5}.$$

We can now make endless variations. Suppose our hat has five balls of each colour, and we draw ten balls. We would like to know the probability of drawing precisely three blue balls if we know that we have drawn at least four red balls. It is formulated as:

$$hg[10]\Big(5|W\rangle + 5|B\rangle + 5|W\rangle\Big)\Big|_{R_{\geq 4}} \models B_{=3} = \tfrac{100}{217}.$$

This approach is very similar to the one at the end of Sect. 3, except that we use the hypergeometric distribution instead of the multinomial one.

8 Tricky Problem on Probability

The final riddle that we look at is of the following form[7].

> There was a blind man. He had four socks in his drawer either black or white. He opened it and took out two socks. Now the probability that it was a pair of white socks is $1/2$.
>
> Can you find out the probability that he had taken out a pair of black socks?

---
[7] briddles.com/2014/08/tricky-problem-on-probability.html.

We analyse the various options in the table below.

4 socks in the drawer	prob. 2 white draw	prob. 2 black draw
$4\|W\rangle$	1	0
$3\|W\rangle + 1\|B\rangle$	$\frac{3}{4} \cdot \frac{2}{3} = \frac{1}{2}$	0
$2\|W\rangle + 2\|B\rangle$	$\frac{1}{2} \cdot \frac{1}{3} = \frac{1}{6}$	$\frac{1}{2} \cdot \frac{1}{3} = \frac{1}{6}$
$1\|W\rangle + 3\|B\rangle$	0	$\frac{3}{4} \cdot \frac{2}{3} = \frac{1}{2}$
$4\|B\rangle$	0	1

The table tells us that the probability of (hypergeometrically) drawing two white socks is $\frac{1}{2}$ only in the second case, when the drawer contains three white socks and one black—as captured by the multiset $3\|W\rangle+1\|B\rangle$. In that case the probability of drawing two black socks is zero. This is the answer to the riddle.

We zoom out and make the general picture explicit. We have a distribution $\omega \in \mathcal{D}(X)$, where the set X comes with two predicates $p, q\colon X \to [0,1]$, together with a 'probability predicate' $PP\colon [0,1] \to [0,1]$, in a situation:

$$[0,1] \xleftarrow{\;\;p\;\;} X \xrightarrow{\;\;q\;\;} [0,1] \xrightarrow{\;\;PP\;\;} [0,1]$$

In this abstract situation we first update ω with $PP \circ q$ and then take the validity of p, as in: $\omega|_{PP \circ q} \models p$.

In the concrete situation of the above socks riddle we have:

- $X = \mathcal{M}[4](\{W, B\}) = \{4\|W\rangle, 3\|W\rangle + 1\|B\rangle, 2\|W\rangle + 2\|B\rangle, 1\|W\rangle + 3\|B\rangle, 4\|B\rangle\}$
- ω is the uniform distribution on X, that is,

$$\omega = \tfrac{1}{5}\Big|4\|W\rangle\Big\rangle + \tfrac{1}{5}\Big|3\|W\rangle + 1\|B\rangle\Big\rangle + \tfrac{1}{5}\Big|2\|W\rangle + 2\|B\rangle\Big\rangle$$
$$+ \tfrac{1}{5}\Big|1\|W\rangle + 3\|B\rangle\Big\rangle + \tfrac{1}{5}\Big|4\|B\rangle\Big\rangle$$

- $p\colon X \to [0,1]$ gives the probability of drawing two black socks from a drawer $\psi \in X$, so
$$p(\psi) := hg[2](\psi)(2\|B\rangle).$$

- $q\colon X \to [0,1]$ gives the probability of drawing two white socks, so
$$q(\psi) := hg[2](\psi)(2\|W\rangle).$$

- $PP\colon [0,1] \to [0,1]$ is the $\frac{1}{2}$ point predicate, so for $r \in [0,1]$,

$$PP(r) := \mathbf{1}_{\frac{1}{2}}(r) = \begin{cases} 1 & \text{if } r = \frac{1}{2} \\ 0 & \text{otherwise.} \end{cases}$$

Thus, the predicate q corresponds to the middle row in the above table, and p to the right row. The composite $PP \circ q\colon X \to [0,1]$ is the point predicate $\mathbf{1}_{3|W\rangle+1|B\rangle}$, so that:

$$\omega|_{PP\circ q} = 1 \left| 3|W\rangle + 1|B\rangle \right\rangle \qquad \text{and thus}$$

$$\omega|_{PP\circ q} \models p = hg[2]\big(3|W\rangle + 1|B\rangle\big)(2|B\rangle) = 0.$$

Let's finish with our own, adapted riddle. Suppose there are ten socks in the drawer, either white or black. We would like to know the probability of drawing two black socks, given that the probability of drawing four white socks is more than 10%.

We now have $X = \mathcal{M}[10](\{W,B\})$ with uniform distribution ω. The predicate p is as above, for the original riddle, whereas $q\colon X \to [0,1]$ together with $PP\colon [0,1] \to [0,1]$ are now:

$$q(\psi) := hg[4](\psi)(4|W\rangle) \qquad PP(r) := \begin{cases} 1 & \text{if } r \geq \frac{1}{10} \\ 0 & \text{otherwise.} \end{cases}$$

For each drawer $\psi \in X = \mathcal{M}[10](\{W,B\})$ we compute the probability $q(\psi)$ of drawing four white socks, and check if this probability is above $\frac{1}{10}$. This is the case on the left below, but not on the right.

$$hg[4]\big(6|W\rangle + 4|B\rangle\big)(4|W\rangle) = \tfrac{1}{14}$$

$$hg[4]\big(5|W\rangle + 5|B\rangle\big)(4|W\rangle) = \tfrac{1}{42}$$

$$hg[4]\big(10|W\rangle\big)(4|W\rangle) = 1$$
$$hg[4]\big(4|W\rangle + 6|B\rangle\big)(4|W\rangle) = \tfrac{1}{210}$$
$$hg[4]\big(9|W\rangle + 1|B\rangle\big)(4|W\rangle) = \tfrac{3}{5}$$
$$hg[4]\big(3|W\rangle + 7|B\rangle\big)(4|W\rangle) = 0$$
$$hg[4]\big(8|W\rangle + 2|B\rangle\big)(4|W\rangle) = \tfrac{1}{3}$$
$$hg[4]\big(2|W\rangle + 8|B\rangle\big)(4|W\rangle) = 0$$
$$hg[4]\big(7|W\rangle + 3|B\rangle\big)(4|W\rangle) = \tfrac{1}{6}$$
$$hg[4]\big(1|W\rangle + 9|B\rangle\big)(4|W\rangle) = 0$$
$$hg[4]\big(10|B\rangle\big)(4|W\rangle) = 0.$$

Thus, we get as updated drawer distribution:

$$\omega|_{PP\circ q} = \tfrac{1}{4}\left| 10|W\rangle \right\rangle + \tfrac{1}{4}\left| 9|W\rangle + 1|B\rangle \right\rangle + \tfrac{1}{4}\left| 8|W\rangle + 2|B\rangle \right\rangle + \tfrac{1}{4}\left| 7|W\rangle + 3|B\rangle \right\rangle.$$

Only in the last two cases one can draw two black socks, with probabilities:

$$hg[2]\big(8|W\rangle + 2|B\rangle\big)(2|B\rangle) = \tfrac{1}{45} \quad hg[2]\big(7|W\rangle + 3|B\rangle\big)(2|B\rangle) = \tfrac{1}{15}.$$

Thus, we get as answer to our adapted riddle:

$$\omega|_{PP\circ q} \models p = \tfrac{1}{4} \cdot \tfrac{1}{45} + \tfrac{1}{4} \cdot \tfrac{1}{15} = \tfrac{1}{4} \cdot \tfrac{4}{45} = \tfrac{1}{45}.$$

Hopefully these riddle solutions will inspire Herman Geuvers and others to contribute to the development of a symbolic probabilistic logic supported by proof assistants.

References

1. Barendregt, H., Geuvers, H.: Proof-assistants using dependent type systems. In: Robinson, A., Voronkov, A. (eds.) Handbook of Automated Reasoning, pp. 1149–1238. Elsevier Science Publishers (2001). https://doi.org/10.5555/778522.778527
2. Cho, K., Jacobs, B.: The EfProb library for probabilistic calculations. In: Bonchi, F., König, B. (eds.) Conference on Algebra and Coalgebra in Computer Science (CALCO 2017), Volume 72 of LIPIcs. Schloss Dagstuhl (2017). https://doi.org/10.4230/LIPIcs.CALCO.2017.25
3. Cho, K., Jacobs, B.: Disintegration and Bayesian inversion via string diagrams. Math. Struct. Comput. Sci. **29**(7), 938–971 (2019). https://doi.org/10.1017/s0960129518000488
4. Fritz, T.: A synthetic approach to Markov kernels, conditional independence, and theorems on sufficient statistics. Adv. Math. **370**, 107239 (2020). https://doi.org/10.1016/J.AIM.2020.107239
5. Geuvers, H.: Proof assistants: history, ideas and future. Sādhanā **34**(1), 3–25 (2009). https://doi.org/10.1007/s12046-009-0001-5
6. Geuvers, J.H.: Logics and type systems. Ph.D. thesis, Univ. Nijmegen (1993)
7. Jacobs, B.: New directions in categorical logic, for classical, probabilistic and quantum logic. Log. Methods Comput. Sci. **11**(3) (2015). https://doi.org/10.2168/lmcs-11(3:24)2015
8. Jacobs, B.: The mathematics of changing one's mind, via Jeffrey's or via Pearl's update rule. J. Artif. Intell. Res. **65**, 783–806 (2019). https://doi.org/10.1613/jair.1.11349
9. Jacobs, B.: A channel-based perspective on conjugate priors. Math. Struct. Comput. Sci. **30**(1), 44–61 (2020). https://doi.org/10.1017/S0960129519000082
10. Jacobs, B.: From multisets over distributions to distributions over multisets. In: Logic in Computer Science. IEEE, Computer Science Press (2021). https://doi.org/10.1109/lics52264.2021.9470678
11. Jacobs, B.: Learning from what's right and learning from what's wrong. In: Sokolova, A. (ed.) Mathematical Foundations of Programming Semantics, Number 351 in Electronic Proceedings in Theoretical Computer Science, pp. 116–133 (2021). https://doi.org/10.4204/EPTCS.351.8
12. Jacobs, B.: Partitions and Ewens distributions in element-free probability theory. In: Logic in Computer Science. IEEE, Computer Science Press (2022). Article No. 23. https://doi.org/10.1145/3531130.3532419
13. Jacobs, B.: Urns & tubes. Compositionality **4**(4) (2022). https://doi.org/10.32408/compositionality-4-4
14. Jacobs, B.: A principled approach to Expectation Maximisation and Latent Dirichlet Allocation using Jeffrey's update rule. In: Hansen, H., Scedrov, A., de Queiroz, R. (eds.) WoLLIC 2023. LNCS, vol. 13923, pp. 256–273. Springer, Cham (2023). https://doi.org/10.1007/978-3-031-39784-4_16
15. Jacobs, B.: Structured probabilistic reasoning (2023). Forthcoming book. http://www.cs.ru.nl/B.Jacobs/PAPERS/ProbabilisticReasoning.pdf
16. Jacobs, B., Zanasi, F.: The logical essentials of Bayesian reasoning. In: Barthe, G., Katoen, J.-P., Silva, A. (eds.) Foundations of Probabilistic Programming, pp. 295–331. Cambridge University Press (2021). https://doi.org/10.1017/9781108770750.010
17. Nederpelt, R., Geuvers, H. (eds.): Type Theory and Formal Proof. Cambridge University Press, Cambridge (2014). https://doi.org/10.1017/CBO9781139567725
18. Pearl, J.: Probabilistic semantics for nonmonotonic reasoning: a survey. In: Brachman, R., Levesque, H., Reiter, R. (eds.) First International Conference on Principles of Knowledge Representation and Reasoning, pp. 505–516. Morgan Kaufmann (1989)

It's All a Game
Apartness and Bisimilarity

Jeroen J. A. Keiren$^{(\boxtimes)}$ [ID] and Tim A. C. Willemse [ID]

Eindhoven University of Technology, Eindhoven, The Netherlands
{j.j.a.keiren,t.a.c.willemse}@tue.nl

Abstract. We study the connection between apartness and bisimulation games. Strong apartness has been proposed as a relation for distinguishing states in a labelled transition system. Prior work has shown that there is a clear connection between Hennessy-Milner logic, strong bisimilarity and strong apartness. We show that in a bisimulation game, winning strategies for SPOILER can be obtained from apartness proofs, and, *vice versa*, apartness proofs can be produced from winning SPOILER strategies.

1 Introduction

In the study of reactive systems, implementation and specification are commonly described using labelled transition systems (LTSs). Behavioural equivalences are used to determine whether such systems exhibit the same behavior. Sometimes these equivalences allow for abstracting from (internal) implementation details.

Behavioural equivalences and their properties have been well-studied by, *e.g.*, van Glabbeek [6,9]. They each have different characteristics and capture how users can use observations and interactions to distinguish between systems.

The strongest meaningful behavioural equivalence is strong bisimulation [7]. This not only preserves the events that happen, but also the branching structure (and thereby the moments of choice) of the transition systems. Strong bisimulation traditionally has a coinductive definition. Stirling gave an alternative characterisation using Ehrenfeucht-Fraïssé games [8]. Strong bisimulation also has a modal characterisation: if two systems are bisimilar, they satisfy the same set of Hennesy-Milner logic (HML) formulas, and the converse holds provided that the systems are image-finite.

An alternative relation on transition systems is *strong apartness* [4]. Two transition systems are (strongly) apart if and only if they are not strongly bisimilar. As a consequence, when transition systems are apart, there is a distinguishing formula in Hennessy-Milner logic that captures the distinction. A direct proof of the same is given by Geuvers [2].

There is a clear connection between strong bisimilarity, games and HML formulas. However, for apartness only the relation with HML formulas has been studied; the relation with a game-based characterisation was not explored. In this

V. Capretta et al. (Eds.): *Logics and Type Systems in Theory and Practice*, LNCS 14560, pp. 150–167, 2024.
https://doi.org/10.1007/978-3-031-61716-4_10

paper we close this gap. Concretely, we show that SPOILER-winning strategies in bisimulation games correspond to apartness proofs and *vice versa*. We do so using constructive proofs for both directions. The upshot of this is that apartness proofs can be constructed efficiently by computing SPOILER-winning strategies.

Our proofs are general, in the sense that they work for infinite-state LTSs, and do not need assumptions about the branching degree of the transition relation.

Special Thanks

We dedicate this article to Herman Geuvers on the occasion of his 60th birthday. We have known Herman for a long time. For both of us, our first encounter with Herman was as a lecturer at Eindhoven University of Technology (TU/e), albeit for different courses, and at different stages of Herman's professional life. Later, we have had the opportunity to work with Herman as a colleague in the Formal System Analysis (FSA) group at TU/e, where he continues to educate our students in using proof assistants.

It is great to see that Herman (still) has an innate connection with TU/e through the study of formulas as types and their history in the work of De Bruijn [5]. Closer to our own research, Herman has recently shown an increased interest in (branching) bisimulation, apartness, and their connection to distinguishing formulas [2–4].

Herman, thank you for your lessons, and we hope to continue our collaborations and discussions for a long time to come.

2 Preliminaries

We are concerned here with relations on labelled transition systems.

Definition 1. *A Labelled Transition System (LTS) is a structure* $\langle S, A, \rightarrow \rangle$ *where: S is a set of states, A is a set of actions, $\rightarrow \subseteq S \times A \times S$ is the transition relation.*

We typically write $s \xrightarrow{a} t$ instead of $(s, a, t) \in \rightarrow$. For relations $R \subseteq S \times S$ on states, we write the infix $s \, R \, t$ instead of $(s, t) \in R$.

Strong bisimilarity [7] can be used to characterise when two states in an LTS "have the same behaviour". It is the commonly agreed that strong bisimilarity is the finest meaningful behavioural equivalence on states of an LTS.

Definition 2 ([7]). *A symmetric relation $R \subseteq S \times S$ is said to be a* strong bisimulation *whenever for all $s \, R \, t$, if $s \xrightarrow{a} s'$ then there exists a state t' such that $t \xrightarrow{a} t'$ and $s' \, R \, t'$.*

Two states $s, t \in S$ are strongly bisimilar, *denoted by $s \leftrightarrow t$, iff there is a strong bisimulation relation R such that $s \, R \, t$.*

In this paper we only deal with strong bisimulation, so we often refer to it as *bisimulation*.

2.1 Bisimulation Games

An alternative characterisation of bisimilarity can be given using two-player games also known as *bisimulation games* [8]. The bisimulation game is a token-passing game, played on an arena where the configurations derive from positions (pairs of states) (s, t) in an LTS. Player SPOILER attempts to disprove that s and t are bisimilar, whereas DUPLICATOR attempts to prove they are. The game is played as follows.

To establish whether s and t are bisimilar, a token is placed on SPOILER-owned configuration $[(s, t)]$, and SPOILER selects an outgoing transition from either s or t and poses this as a challenge to DUPLICATOR. Suppose SPOILER selected transition $s \xrightarrow{a} s'$; this challenge is recorded in DUPLICATOR-owned configuration $\langle (s, t), (a, s') \rangle$ and from this configuration, DUPLICATOR needs to respond with an a-transition from t. If DUPLICATOR can respond with $t \xrightarrow{a} t'$, the play continues in SPOILER-owned configuration $[(s', t')]$. If SPOILER elected to play from t, the continuation is symmetric.

The player that at some point in the play gets stuck loses that play, in which case the opponent wins the play. In case the play continues indefinitely, DUPLICATOR wins. A configuration is won by a player if that player has a fail-safe strategy, allowing her to win all plays that follow that strategy. In case DUPLICATOR wins a configuration, the two states are bisimilar, but in case SPOILER wins the configuration, the states can be distinguished. The game is formally defined as follows.

Definition 3. *A* bisimulation (sb) game *on an LTS L is played by players* SPOILER *and* DUPLICATOR *on an arena of* SPOILER-*owned configurations* $[(s, t)]$ *and* DUPLICATOR-*owned configurations* $\langle (s, t), ch \rangle$, *where* $(s, t) \in$ Position *and* $ch \in$ Challenge, *and* Position $= S \times S$ *is the set of* positions *and* Challenge $= (A \times S)$ *the set of* pending challenges.

- SPOILER *moves from a configuration* $[(s, t)]$ *by:*
    1. *selecting* $s \xrightarrow{a} s'$ *and moving to* $\langle (s, t), (a, s') \rangle$, *or*
    2. *selecting* $t \xrightarrow{a} t'$ *and moving to* $\langle (t, s), (a, t') \rangle$.
- DUPLICATOR *responds from a configuration* $\langle (u, v), (a, u') \rangle$ *by playing* $v \xrightarrow{a} v'$ *and continuing in configuration* $[(u', v')]$.

A play is a maximal sequence of configurations played according to these rules. Finite plays are won by the opponent of the player who owns the last configuration on the play. Infinite plays are won by DUPLICATOR.

To generalise the concept of winning from plays to configurations in the bisimulation game, we use *strategies*. It is well-known that bisimulation games are memoryless and determined. Therefore, it suffices to consider strategies that assign to every configuration owned by a particular player a unique successor configuration (if it exists) to which the player will move next.

Definition 4 (Strategies/Winning).

- *A strategy of a player is a partial function that maps configurations owned by that player to one of the adjacent configurations. We write $\sigma(c) = \bot$ if strategy σ is undefined for configuration c; we also abuse notation and write $\bot$ for the strategy that is undefined for all positions.*
- *A play is consistent with player-strategy σ if, for every player-owned configuration c on the play, if $\sigma(c)$ is defined, the next configuration on the play is $\sigma(c)$. We refer to a play that is consistent with σ by a σ-play.*
- *Strategy σ for a given player is said to be winning for that player from a given configuration c iff every play starting in c, consistent with σ is won by that player.*
- *We say that DUPLICATOR wins the sb-game for a position (s,t), if she has a strategy σ that is winning for her from the configuration $[(s,t)]$; in this case, we write $s \equiv_s t$. Otherwise, we say that SPOILER wins that game.*

The strategy is a partial function for two reasons. First, SPOILER can get stuck in $[(s,t)]$ if neither s nor t has an outgoing transition, and the configuration has no successor; similarly, DUPLICATOR can get stuck if there is no transition matching SPOILER's challenge. Second, in later proofs we inductively define SPOILER-strategies for a subset of the vertices in the game, where it is useful to leave strategies undefined.

We illustrate the above game using a small example.

Example 1. Consider the four-state transition system depicted below (left). Here c denotes a commute, e and n denote teaching in Eindhoven and Nijmegen, respectively, and r indicates relocating.

Observe that states B and C are not bisimilar. This can be seen by the game, depicted right, in which SPOILER plays her winning strategy, first challenging DUPLICATOR to match a c-transition, and, when she does so by moving to D, switch positions and challenging her to match an r-transition (which DUPLICATOR cannot).

From this play it can also be seen that D and A are not bisimilar.   □

It is well-known that the bisimulation game characterises bisimilarity.

Theorem 1. *Let $s, t \in S$ be states in an LTS. Then we have*

$$s \equiv_s t \iff s \leftrightarrow t$$

2.2 Apartness

Apartness is the dual of bisimilarity. It was first explored by Jacobs in unpublished work in 1995, and studied more recently by Geuvers and others [2,4].

Definition 5 ([2]). *A symmetric relation $Q \subseteq S \times S$ is a strong apartness whenever for all $s \, Q \, t$, if $s \xrightarrow{a} s'$ then for all t' satisfying $t \xrightarrow{a} t'$, also $s' \, Q \, t'$.*

Two states $s, t \in S$ are apart, *which we denote by $s \# t$ iff for all relations Q, if Q is a strong apartness, then $Q(s, t)$.*

In the remainder of this paper we typically write apartness instead of strong apartness.

Since apartness is inductive, it can be derived using the two derivation rules depicted in Fig. 1.

$$(\text{sym}_\#)\, \frac{t \# s}{s \# t} \qquad\qquad (\text{in}_\#)\, \frac{s \xrightarrow{a} s' \quad \forall t' : t \xrightarrow{a} t' \implies s' \# t'}{s \# t}$$

Fig. 1. Proof rules for apartness [2].

The relation between apartness and bisimilarity is as expected: two states in an LTS are apart if and only if they are not bisimilar.

Theorem 2 ([2]). *Let $s, t \in S$ be states in an LTS. Then we have*

$$s \leftrightarrow t \iff \neg(s \# t)$$

So, as a corollary, it immediately follows that SPOILER wins the bisimulation game for configuration $[(s, t)]$ if and only s and t are apart.

Corollary 1. *Let $s, t \in S$ be states in an LTS. Then we have*

$$s \equiv_s t \iff \neg(s \# t)$$

Before we move on to the technical development in the remainder of the paper, it can be insightful to reflect on the derivation rules.

In particular, the $(\text{in}_\#)$ rule contains the premise $\forall t' : t \xrightarrow{a} t' \implies s' \# t'$, and we need to decide how to handle this universal quantification. We interpret this as meta-notation that, effectively, abbreviates a possibly infinite set of premises: for every a-successor t' of t, we get one premise $s' \# t'$. This can give rise to an uncountably infinite set of premises. This is illustrated by the following example.

Example 2. Consider the LTS with set of states $\mathbb{R} \cup \{s\}$ and transitions which are given by: $s \xrightarrow{a} s$ and for all $x \in \mathbb{R}$ $(x \neq 0)$, there are transitions $0 \xrightarrow{a} x$.

State 0 has uncountably many outgoing a transitions. Observe that states s and 0 are apart. This also follows from the following derivation.

$$(\text{in}_\#)\, \frac{s \xrightarrow{a} s \qquad (\text{in}_\#)\, \dfrac{s \xrightarrow{a} s \quad \checkmark}{\forall x : 0 \xrightarrow{a} x \implies s \# x}}{s \# 0}$$

Note that we abuse notation in order to finitely represent the derivation with infinitely many premises. For each of the premises $s \# x$, the result follows immediately from the observation that s can do an a-transition to itself. However, x cannot do an a-transition, so the formula $\forall x' : x \xrightarrow{a} x' \implies s \# x'$ is vacuously satisfied, indicated by $\checkmark$. $\square$

The previous example demonstrates that the derivation can be infinitely wide. Similarly, the height of the derivation can be unbounded. However, in order to obtain a successful apartness proof, each of the branches of the derivation must end in an application of the $(\mathrm{in}_\#)$ rule where the universal quantification is vacuously satisfied. In other words, the branches of successful derivations must be well-founded. We rely on the corresponding induction principle on derivations in our proofs later in the paper. We illustrate the unboundedness of the height of derivations in the following example.

Example 3. Consider the following LTS. It has states s, t, and for all $n \in \mathbb{N}$, states s_n; it has transitions $t \xrightarrow{a} t$, and $s \xrightarrow{a} s_n$ and $t \xrightarrow{a} s_n$ for $n \in \mathbb{N}$, and for $m \in \mathbb{N}$ such that $m > 0$, $s_m \xrightarrow{a} s_{m-1}$. See also (a fragment of) the transition system below:

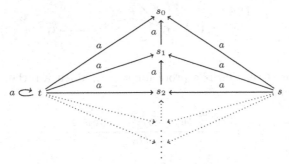

We have $t \# s$. The last step in the apartness proof of $t \# s$ is as follows.

$$(\mathrm{in}_\#) \frac{t \xrightarrow{a} t \qquad \forall s' : s \xrightarrow{a} s' \implies t \# s'}{t \# s}$$

Note that $s \xrightarrow{a} s'$ if and only if $s' = s_i$ for $i \in \mathbb{N}$. Now, for any $i \in \mathbb{N}$ such that $0 < i$, the relevant step in the apartness proof of $t \# s_i$ is as follows. We use that $s_i \xrightarrow{a} s'$ iff $s' = s_{i-1}$ to simplify the presentation.

$$(\mathrm{in}_\#) \frac{t \xrightarrow{a} t \qquad t \# s_{i-1}}{t \# s_i}$$

Since s_0 does not have an outgoing a-transition, for s_0, the apartness proof is:

$$(\mathrm{in}_\#) \frac{t \xrightarrow{a} t \qquad \checkmark}{t \# s_0}$$

By composing these proof steps we obtain the apartness proof of $t \# s$. Clearly, it is of unbounded height, since each derivation for $t \# s_i$ is of height $i + 1$, and the derivation of $t \# s$ requires the derivations of all $t \# s_i$. $\square$

3   From Bisimulation Games to Apartness and Back

In the remainder of this paper we focus on the correspondence between winning SPOILER-strategies in the bisimulation game and apartness proofs. From such a SPOILER-strategy an apartness proof can be constructed and *vice versa*.

3.1   SPOILER-Strategies Constructed from Apartness Proofs

We first show that from any apartness proof we can construct a SPOILER-strategy that is winning for SPOILER in the bisimulation game. Our construction, and the associated proof of correctness, is by induction on the derivation. Inductively obtained SPOILER strategies, however, cannot be combined naively: if we are not careful, the combined SPOILER-strategy could inadvertently lead to the introduction of cycles that are winning for DUPLICATOR. This is illustrated by the following example.

Example 4. Consider the following (disconnected) LTS.

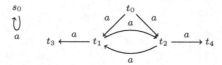

The following derivation is a proof for $s_0 \underline{\#} t_0$, albeit a rather naive one. We omit the rule names.

$$\frac{s_0 \xrightarrow{a} s_0 \quad s_0 \underline{\#} t_1 \quad s_0 \underline{\#} t_2}{s_0 \underline{\#} t_0}$$

Where the derivation for $s_0 \underline{\#} t_1$ is the following:

$$\cfrac{s_0 \xrightarrow{a} s_0 \qquad \cfrac{t_2 \xrightarrow{a} t_4 \quad \cfrac{\cfrac{s_0 \xrightarrow{a} s_0 \quad \checkmark}{s_0 \underline{\#} t_4}}{t_4 \underline{\#} s_0}}{t_2 \underline{\#} s_0}}{s_0 \underline{\#} t_2} \qquad \cfrac{s_0 \xrightarrow{a} s_0 \quad \checkmark}{s_0 \underline{\#} t_3}$$
$$\overline{\qquad\qquad\qquad\qquad s_0 \underline{\#} t_1 \qquad\qquad\qquad\qquad}$$

The derivation for $s_0 \underline{\#} t_2$ is the following:

$$\cfrac{s_0 \xrightarrow{a} s_0 \qquad \cfrac{t_1 \xrightarrow{a} t_3 \quad \cfrac{\cfrac{s_0 \xrightarrow{a} s_0 \quad \checkmark}{s_0 \underline{\#} t_3}}{t_3 \underline{\#} s_0}}{t_1 \underline{\#} s_0}}{s_0 \underline{\#} t_1} \qquad \cfrac{s_0 \xrightarrow{a} s_0 \quad \checkmark}{s_0 \underline{\#} t_4}$$
$$\overline{\qquad\qquad\qquad\qquad s_0 \underline{\#} t_2 \qquad\qquad\qquad\qquad}$$

Both (sub)derivations contain two different derivations for $s_0 \# t_1$ and two different derivations for $s_0 \# t_2$. Based on the subderivation for $s_0 \# t_1$, a natural choice would be for SPOILER to play $[(s_0, t_1)] \to \langle (s_0, t_1), (a, s_0) \rangle$ and $[(s_0, t_2)] \to \langle (t_2, s_0), (a, t_4) \rangle$. Similarly, based on the subderivation for $s_0 \# t_2$, it would be a natural choice for SPOILER to play $[(s_0, t_1)] \to \langle (t_1, s_0), (a, t_3) \rangle$, and $[(s_0, t_2)] \to \langle (s_0, t_2), (a, s_0) \rangle$. We cannot simply choose arbitrary successors for $[(s_0, t_1)]$ and $[(s_0, t_2)]$ based on the inductively constructed strategies when defining the strategy for the entire derivation. Suppose that we would let SPOILER play $[(s_0, t_1)] \to \langle (s_0, t_1), (a, s_0) \rangle$ based on the first subderivation, and $[(s_0, t_2)] \to \langle (s_0, t_2), (a, s_0) \rangle$ based on the second subderivation. In this case, DUPLICATOR can decide to play $\langle (s_0, t_1), (a, s_0) \rangle \to [(s_0, t_2)]$ and $\langle (s_0, t_2), (a, s_0) \rangle \to [(s_0, t_1)]$, the resulting play would be infinite, and hence be won by DUPLICATOR. So, the constructed SPOILER-strategy is not winning for SPOILER.

If, instead, we prioritise the choices made in the derivation of $s_0 \# t_1$ (the first subderivation), we get a strategy for SPOILER that ensures she plays $[(s_0, t_1)] \to \langle (s_0, t_1), (a, s_0) \rangle$ and $[(s_0, t_2)] \to \langle (t_2, s_0), (a, t_4) \rangle$, instead. In fact, in this case, both subderivations lead to strategies that are defined for the same sets of configurations, so the second subderivation can be ignored when defining the SPOILER-strategy. Observe that our choice to prioritise the choices made in the first subderivation over the second subderivation is completely arbitrary. □

We now formalise these intuitions and give an inductive construction of a strategy σ for SPOILER on the bisimulation game, that corresponds closely to the provided apartness proof. To accommodate for all apartness proofs, even ones that contain unnecessary steps such as repeatedly taking a transition in a loop, before ultimately breaking out of the loop and completing the proof, we cannot simply reason about the structure of the labelled transition systems, and we reason directly about derivations instead.

Using the well-ordering theorem, every set can be well-ordered. We here use the well-ordering theorem to obtain a well-ordering $<$ on the (generally infinite) set of apartness proofs for a given LTS. In particular, we use this to derive a combined strategy from the apartness proofs in the premises corresponding to $\forall t' : t \xrightarrow{a} t' \implies s' \# t'$ in the applications of the (in$_\#$) rule. The combined strategy assigns to a SPOILER-owned configuration in the bisimulation game the successor that corresponds to the strategy derived from the smallest (w.r.t. $<$) relevant apartness proof in the premise. We finally use this to extend the strategy using the final step in the derivation.

We first define when a strategy is consistent with a derivation.

Definition 6. *Let H be a derivation.*

- $\mathrm{Conf}_{Sp}(H) = \{[(s, t)] \mid s \# t \text{ is a node in } H\}$ *denotes the set of* SPOILER-*configurations associated with derivation H.*
- $\mathrm{Conf}_{Du}(H) = \{\langle (s, t), (a, s') \rangle \mid s \xrightarrow{a} s' \ \forall t' : t \xrightarrow{a} t' \implies s' \# t' \text{ is a premise that is used in } H\}$ *denotes the set of* DUPLICATOR-*configurations associated with derivation H.*

- $\text{Conf}(H) = \text{Conf}_{Sp}(H) \cup \text{Conf}_{Du}(H)$
- *Strategy σ is H-consistent iff for all $[(s,t)] \in \text{Conf}_{Sp}(H)$, $\sigma([(s,t)]) \in$ $\text{Conf}_{Du}(H)$.*

We first define $\sigma(H)$, the H-consistent SPOILER-strategy constructed from derivation H. The intuitions behind the construction are as follows. When the last step in the derivation of $s\#t$ is an application of the $(\text{in}_\#)$ rule, we first combine the inductively constructed SPOILER-strategies from the set of premises $\mathcal{H}$ in such a way that for a SPOILER-configuration $[(p,q)]$, the combined strategy $\sigma(\mathcal{H})$ is consistent with the *first* premise whose strategy is defined for $[(p,q)]$. Note that, to allow for infinitely many premises that can appear due to the infinite branching in labelled transition systems, we use transfinite recursion when combining the strategies in this case. The strategy $\sigma(H)$ for the entire derivation is then this combined strategy $\sigma(\mathcal{H})$. If $\sigma(\mathcal{H})$ is not defined for $[(s,t)]$, $\sigma(H)([(s,t)])$ is set in accordance with the choices made in the last step in the derivation, that is, SPOILER chooses the a-transition to s', and the configuration becomes $\langle (s,t), (a,s') \rangle$.

If the last step in the derivation is an application of the $(\text{sym}_\#)$ rule, the strategy is inherited from the inductively constructed strategy $\sigma(H')$ of the premise H'. If the goal is $[(s,t)]$ and $\sigma(H')$ is not defined for $[(s,t)]$, we know that it must be defined for $[(t,s)]$, and $\sigma(H)([(s,t)])$ is set to $\sigma(H')([(t,s)])$. This way, the strategy that is chosen "earlier" in the derivation is preferred over choices that are made later.

Definition 7. *We construct $\sigma(H)$ by induction on the derivation. To this end, we consider the last step in the derivation.*

- *The last step in the derivation is*

$$\frac{s \xrightarrow{a} s' \qquad \forall t' : t \xrightarrow{a} t' \implies s'\#t'}{s\#t}.$$

This means for every t' such that $t \xrightarrow{a} t'$ we have an apartness proof that witnesses $s'\#t'$. Let $H(s',t')$ denote the corresponding derivation. Note that by induction $\sigma(H(s',t'))$ is an $H(s',t')$-consistent SPOILER-strategy. Let $\mathcal{H} = \{H(s',t') \mid t \xrightarrow{a} t'\}$, the set of derivations witnessing $\forall t' : t \xrightarrow{a} t' \implies s'\#t'$. Recall that this set is well-ordered by $<$.

We first combine the separate strategies $\sigma(H(s',t'))$ into a single strategy $\sigma(\mathcal{H})$ using transfinite recursion. Note that there is an ordinal, say β, in bijective correspondence with $\mathcal{H}$; we write H_α for the α-element of $\mathcal{H}$. We define ordinal-indexed sequences $(C_\alpha, \sigma_\alpha)$ of pairs of sets of configurations in the bisimulation game, and SPOILER-strategies, with the property that σ_α is a SPOILER-winning strategy on C_α.

The definition is as follows.

$$C_\alpha = C_{<\alpha} \cup C'_\alpha \text{ where } \begin{cases} C_{<\alpha} = \bigcup_{\alpha' < \alpha} \mathrm{Conf}(H_{\alpha'}) \\ C'_\alpha = \mathrm{Conf}(H_\alpha) \setminus C_{<\alpha} \end{cases}$$

$$\sigma_\alpha([\,(p,q)\,]) = \begin{cases} \sigma_{\alpha'}([\,(p,q)\,]) & \text{if } [\,(p,q)\,] \in C_{<\alpha}, \\ & \text{and } \alpha' = \min\{\alpha'' \mid [\,(p,q)\,] \in \mathrm{Conf}(H_{\alpha''})\} \\ \sigma_{H_\alpha}([\,(p,q)\,]) & \text{if } [\,(p,q)\,] \in C'_\alpha \\ \bot & \text{otherwise} \end{cases}$$

From this definition it follows that $C_{<\beta} = \bigcup\{\mathrm{Conf}(H) \mid H \in \mathcal{H}\}$, which we abbreviate to $\mathrm{Conf}(\mathcal{H})$. Also $\sigma_{<\beta}$ is a SPOILER-*strategy for* $\mathrm{Conf}(\mathcal{H})$*, which we denote $\sigma(\mathcal{H})$.*
We use this to define $\sigma(H)$ as follows.

$$\sigma(H)([\,(p,q)\,]) = \begin{cases} \sigma(\mathcal{H})([\,(p,q)\,]) & \text{if } [\,(p,q)\,] \in \mathrm{Conf}(\mathcal{H}) \\ \langle (s,t),(a,s') \rangle & \text{if } p = s, \; q = t, \text{ and } [\,(p,q)\,] \notin \mathrm{Conf}(\mathcal{H}) \\ \bot & \text{otherwise} \end{cases}$$

– *If the last step in the derivation is*

$$\frac{t \# s}{s \# t}.$$

Let H' be the apartness proof of $t \# s$. We define $\sigma(H)$ as follows:

$$\sigma(H) = \begin{cases} \sigma(H') & \text{if } [\,(s,t)\,] \in \mathrm{Conf}(H') \\ \sigma(H')[[\,(s,t)\,] := \sigma(H')([\,(t,s)\,])] & \text{otherwise} \end{cases}$$

Here $\sigma(H)[c := c']$ denotes the strategy that maps configuration c to c', and behaves like $\sigma(H)$ for all other configurations.

We illustrate the definition using the derivation from Example 4.

Example 5. Recall the derivation witnessing $s_0 \# t_0$ in Example 4. For the sake of the example, we use ordering $<$ that ensures that for the applications of the $(\mathrm{in}_\#)$ rule, the first subderivation is ordered before the second subderivation. For example, $H_1(s_0, t_2) < H_1(s_0, t_3)$. We refer to this derivation using H, the subderivation for $s_0 \# t_1$ as H_1, and the subderivation for $s_0 \# t_2$ as H_2. For each of the subderivations, we write $H_i(s,t)$ for the subderivation of H_i concluding that $s \# t$.

We build strategy $\sigma(H_1)$ starting from the top of the derivation.

– For the subderivation of $s_0 \# t_4$ we get the strategy $\sigma(H_1(s_0, t_4))$ that is only defined for $[\,(s_0, t_4)\,]$; it is $\sigma(H_1(s_0, t_4))([\,(s_0, t_4)\,]) = \langle (s_0, t_4),(a, s_0) \rangle$.
– For $t_4 \# s_0$, we get $\sigma(H_1(t_4, s_0))([\,(s_0, t_4)\,]) = \sigma(H_1(t_4, s_0))([\,(t_4, s_0)\,]) = \sigma(H_1(s_0, t_4))([\,(s_0, t_4)\,]) = \langle (s_0, t_4),(a, s_0) \rangle$. So, it inherits the strategy for $[\,(s_0, t_4)\,]$, and it chooses the same strategy for $[\,(t_4, s_0)\,]$.

- For $\sigma(H_1(t_2, s_0))$, again for $[(t_4, s_0)]$ and $[(s_0, t_4)]$, the choice is inherited from $\sigma(H_1(t_4, s_0))$, and $\sigma(H_1(t_2, s_0))([(t_2, s_0)]) = \langle (t_2, s_0), (a, t_4) \rangle$.
- Strategy $\sigma(H_1(s_0, t_2))([(p, q)]) = \sigma(H_1(t_2, s_0))([(q, p)])$ if $(p, q) = (s_0, t_2)$, and $\sigma(H_1(s_0, t_2))([(p, q)]) = \sigma(H_1(t_2, s_0))([(p, q)])$ otherwise.
- For the subderivation of $s_0 \# t_3$ we immediately get $\sigma(H_1(s_0, t_3))([(s_0, t_3)]) = \langle (s_0, t_3), (a, s_0) \rangle$.
- The combination of the strategies for $H_1(s_0, t_2)$ and $H_1(s_0, t_3)$ is the strategy $\sigma(\{H_1(s_0, t_2), H_1(s_0, t_3)\})$, which is defined as $\sigma(\{H_1(s_0, t_2), H_1(s_0, t_3)\})(c) = \sigma(H_1(s_0, t_2))(c)$ if $\sigma(H_1(s_0, t_2))(c)$ is defined, and $\sigma(H_1(s_0, t_3))(c)$ otherwise.
- Finally, we get that $\sigma(H_1(s_0, t_1))(c) = \langle (s_0, t_1), (a, s_0) \rangle$ if $c = [(s_0, t_1)]$, and $\sigma(\{H_1(s_0, t_2), H_1(s_0, t_3)\})(c)$ otherwise.

The definitions for H_2 are similar. When inferring the strategy corresponding to H, priority is given to the definition in H_1, similar to the last two steps in the construction of the strategy for H_1. In Table 1 we list the strategies for H_1 and H_2, the strategy that combines the strategies for H_1 and H_2 in the last application of the ($\mathrm{in}_{\#}$) rule in derivation H, and the strategy for H.

Table 1. Strategies inferred from the proof in Example 4.

c	$\sigma(H_1)(c)$	$\sigma(H_2)(c)$	$\sigma(\{H_1, H_2\})(c)$	$\sigma(H)(c)$
$[(s_0, t_0)]$	$\bot$	$\bot$	$\bot$	$\langle (s_0, t_0), (a, s_0) \rangle$
$[(s_0, t_1)]$	$\langle (s_0, t_1), (a, s_0) \rangle$	$\langle (t_1, s_0), (a, t_3) \rangle$	$\langle (s_0, t_1), (a, s_0) \rangle$	$\langle (s_0, t_1), (a, s_0) \rangle$
$[(s_0, t_2)]$	$\langle (t_2, s_0), (a, t_4) \rangle$	$\langle (s_0, t_2), (a, s_0) \rangle$	$\langle (t_2, s_0), (a, t_4) \rangle$	$\langle (t_2, s_0), (a, t_4) \rangle$
$[(s_0, t_3)]$	$\langle (s_0, t_3), (a, s_0) \rangle$	$\langle (s_0, t_3), (a, s_0) \rangle$	$\langle (s_0, t_3), (a, s_0) \rangle$	$\langle (s_0, t_3), (a, s_0) \rangle$
$[(s_0, t_4)]$	$\langle (s_0, t_4), (a, s_0) \rangle$	$\langle (s_0, t_4), (a, s_0) \rangle$	$\langle (s_0, t_4), (a, s_0) \rangle$	$\langle (s_0, t_4), (a, s_0) \rangle$
$[(t_0, s_0)]$	$\bot$	$\bot$	$\bot$	$\bot$
$[(t_1, s_0)]$	$\bot$	$\langle (t_1, s_0), (a, t_3) \rangle$	$\langle (t_1, s_0), (a, t_3) \rangle$	$\langle (t_1, s_0), (a, t_3) \rangle$
$[(t_2, s_0)]$	$\langle (t_2, s_0), (a, t_4) \rangle$	$\bot$	$\langle (t_2, s_0), (a, t_4) \rangle$	$\langle (t_2, s_0), (a, t_4) \rangle$
$[(t_3, s_0)]$	$\bot$	$\langle (s_0, t_3), (a, s_0) \rangle$	$\langle (s_0, t_3), (a, s_0) \rangle$	$\langle (s_0, t_3), (a, s_0) \rangle$
$[(t_4, s_0)]$	$\langle (s_0, t_4), (a, s_0) \rangle$	$\bot$	$\langle (s_0, t_4), (a, s_0) \rangle$	$\langle (s_0, t_4), (a, s_0) \rangle$

Observe that the strategy that is constructed in this way is indeed winning for SPOILER from $[(s_0, t_0)]$. □

We next show that, indeed, the strategy $\sigma(H)$ that is constructed from derivation H is SPOILER-winning. Note that this requires both induction on the derivation, as well as transfinite induction on the hypotheses in the ($\mathrm{in}_{\#}$) step.

Theorem 3. *Let H be a derivation, then $\sigma(H)$ is a SPOILER-winning SPOILER-strategy from $\mathrm{Conf}(H)$.*

Proof. We proceed by induction on H.

- If the last step of H is of the form

$$\frac{s \xrightarrow{a} s' \quad \forall t' : t \xrightarrow{a} t' \implies s' \# t'}{s \# t}$$

Let $\mathcal{H} = \{H(s',t') \mid t \xrightarrow{a} t'\}$ as before. The proof proceeds in two steps:

1. Establish that $\sigma(\mathcal{H})$ is SPOILER-winning from $\text{Conf}(\mathcal{H})$.
2. Establish that $\sigma(H)$ is SPOILER-winning from $\text{Conf}(H)$.

The proofs of the steps are as follows.

1. $\sigma(\mathcal{H})$ **is Spoiler-winning from** $\text{Conf}(\mathcal{H})$. The proof proceeds by transfinite induction. Pick arbitrary α, and assume that for all $\alpha' < \alpha$, $\sigma_{\alpha'}$ is a winning strategy for SPOILER from $C_{\alpha'}$.

   First, observe that it follows from the induction hypothesis that, for all $[(p,q)] \in C_{<\alpha}$, if $\alpha' = \min\{\alpha'' \mid [(p,q)] \in \text{Conf}(H_{\alpha''})\}$, $\sigma_{\alpha'}$ is SPOILER-winning from $[(p,q)]$. As $\sigma_{\alpha'}$ stays within $C_{\alpha'} \subseteq C_{<\alpha}$, σ_α is SPOILER-winning from $[(p,q)]$.

   We now show that σ_α is SPOILER-winning for $\text{Conf}(H_\alpha)$ by yet another induction, in this case on H_α.

   - The last step of H_α is of the form

$$\frac{p \xrightarrow{b} p' \quad \forall q' : q \xrightarrow{b} q' \implies p'\#q'}{p\#q}$$

   Let $\overline{\mathcal{H}} = \{H(p',q') \mid q \xrightarrow{b} q'\}$ the set of derivations witnessing $\forall q' : q \xrightarrow{b} q' \implies p'\#q'$. According to the inner induction hypothesis, σ_α is SPOILER-winning from each $[(p',q')]$ such that $H(p',q') \in \overline{\mathcal{H}}$. From this and the definition of the bisimulation game, it also follows immediately that σ_α is SPOILER-winning from $\langle (p,q),(b,p')\rangle$. If $[(p,q)] \in C_{<\alpha}$, the result follows immediately from our earlier observations. So, assume $[(p,q)] \notin C_{<\alpha}$. It must be the case that $[(p,q)] \in C'_\alpha$, and $\sigma_\alpha([(p,q)]) = \sigma_{H_\alpha}([(p,q)])$. Now, if $[(p,q)] \in \text{Conf}(\overline{\mathcal{H}})$ then $\sigma_\alpha([(p,q)])$ is winning from $[(p,q)]$ follows immediately from the induction hypothesis. Otherwise, it must be the case that $\sigma_\alpha([(p,q)]) = \sigma(H_\alpha)([(p,q)]) = \langle (p,q),(b,p')\rangle$. From the observation that σ_α is SPOILER-winning from $\langle (p,q),(b,p')\rangle$ it follows immediately that σ_α is SPOILER-winning from $[(p,q)]$.

   - The last step of H_α is of the form

$$\frac{q\#p}{p\#q}$$

   We distinguish cases based on the definition of σ_α.

   * If $[(p,q)] \in C_{<\alpha}$, the result again follows immediately from the (inner) induction hypothesis.
   * Otherwise, $\sigma_\alpha([(p,q)]) = \sigma(H_\alpha)([(p,q)])$. Let H'_α be the derivation of $q\#p$. If $[(p,q)] \in \text{Conf}(H'_\alpha)$, it follows immediately from the induction hypothesis that σ_α is SPOILER-winning from $[(p,q)]$. If $[(p,q)] \notin \text{Conf}(H'_\alpha)$, then $\sigma_\alpha([(p,q)]) = \sigma(H_\alpha)([(p,q)]) = \sigma(H_\alpha)([(q,p)])$. This must correspond to a DUPLICATOR-configuration $\langle (q,p),(b,q')\rangle$ such that $q \xrightarrow{b} q'$ and $\forall p' : p \xrightarrow{b} p' \implies q'\#p'$ is a premise in H'_α, so according to the

induction hypothesis, σ_α is SPOILER-winning from $\langle (q,p),(b,q') \rangle$ according to the induction hypothesis, hence σ_α is SPOILER-winning from $[(p,q)]$.

2. $\sigma(H)$ **is Spoiler-winning from** Conf(H). If $[(s,t)] \in$ Conf$(\mathcal{H})$, the result follows immediately from the observation that $\sigma(\mathcal{H})$ is SPOILER-winning from Conf$(\mathcal{H})$ and the definition of $\sigma(H)$. Otherwise, we have $\sigma(H)([(s,t)]) = \langle (s,t),(a,s') \rangle \in$ Conf$(\mathcal{H})$, and as $\sigma(\mathcal{H})$ is winning from Conf$(\mathcal{H})$, $\sigma(H)$ is winning from Conf(H).

– If the last step of H is of the form

$$\frac{t\#s}{s\#t}$$

Let H' be the derivation of $t\#s$. By our induction hypothesis, $\sigma(H')$ is SPOILER-winning from Conf$(\overline{H'})$. We now distinguish cases based on the definition of $\sigma(H)$. If $[(s,t)] \in$ Conf(H'), then Conf$(H) =$ Conf(H') and $\sigma(H) = \sigma(H')$, and the result follows immediately from the induction hypothesis. If $[(s,t)] \notin$ Conf(H'), if follows that $[(t,s)] \in$ Conf(H'), and $\sigma(H')([(t,s)]) \in$ Conf(H'), so strategy $\sigma(H')$ is SPOILER-winning from $\sigma(H')([(t,s)])$. Note that by definition Conf$(H) = \{[(t,s)]\} \cup$ Conf(H') and $\sigma(H)([(s,t)]) = \sigma(H')([(t,s)])$, and for all other configurations c, $\sigma(H)(c) = \sigma(H')(c)$, so $\sigma(H)$ is SPOILER-winning from Conf(H).   □

The following corollary now follows immediately.

Corollary 2. *If there is a derivation with conclusion $s\#t$, then $[(s,t)]$ is a* SPOILER-*configuration won by* SPOILER.

3.2   Spoiler-Strategy Induced Deductive Proofs of Apartness

We next show that for a configuration $[(s,t)]$ won by SPOILER in a bisimulation game, we can construct a proof of apartness for $s\#t$ using SPOILER's strategy.

Consider a bisimulation game with set of configurations Conf, partitioned in Conf$_{Sp}$ and Conf$_{Du}$ of SPOILER and DUPLICATOR configurations. From hereon, assume that SPOILER wins at least one SPOILER-configuration in the bisimulation game. This implies that SPOILER has a winning strategy $\sigma \neq \bot$. Let D denote the set of DUPLICATOR configurations that are immediately winning for SPOILER.

$$D = \{\langle (s,t),(a,s') \rangle \in \text{Conf}_{Du} \mid s \xrightarrow{a} s' \wedge t \xrightarrow{a}\!\!\!\!\!/\, \}$$

Intuitively, a strategy σ for SPOILER is winning any configuration from which SPOILER can force play to the set D in a finite number of steps, using σ; and these are the only ones won by her using σ. This suggests an inductive way of defining which configurations are SPOILER-winning using a strategy σ: those that are σ-*attracted* to D. We define the σ-attractor $Attr_\sigma$ to set D as the set of all SPOILER-configurations $Attr_\sigma^\gamma$, where γ is the smallest ordinal satisfying $Attr_\sigma^\gamma = Attr_\sigma^{\gamma+1}$ and, for successor ordinal α and limit ordinal β we define:

$$Attr_\sigma^0 \quad = \{c \in \text{Conf}_{Sp} \mid \sigma(c) \in D\}$$
$$Attr_\sigma^{\alpha+1} = Attr_\sigma^\alpha \cup \{c \in \text{Conf}_{Sp} \mid \forall c' \in \text{Conf}_{Sp} : \sigma(c) \to c' \Rightarrow c' \in Attr_\sigma^\alpha\}$$
$$Attr_\sigma^\beta \quad = \bigcup_{\alpha < \beta} Attr_\sigma^\alpha$$

It may not be straightforward to see that there is a need for limit ordinals in σ-attractors. The following example may shed some light on the matter.

Example 6. Reconsider the LTS of Example 3. Recall that we have $t \underline{\#} s$. Note that in this LTS, all states have an outgoing a-transition (and no other outgoing transitions), except for state s_0, which has no outgoing transition at all. Hence, the set D of DUPLICATOR-owned configurations that are immediately winning for SPOILER is as follows:

$$D = \{\langle (t, s_0), (a, t) \rangle, \langle (s_i, s_0), (a, s_{i-1}) \rangle,$$
$$\langle (s_0, t), (a, t) \rangle, \langle (s_0, s_i), (a, s_{i-1}) \rangle \mid i > 0\}$$

Consider the (partial) strategy $\sigma([(t, s)]) = \langle (t, s), (a, t) \rangle$ and $\sigma([(t, s_i)]) = \langle (t, s_i), (a, t) \rangle$. Observe that σ is SPOILER-winning from $[(t, s)]$: any play that emerges is finite, ending in configuration $\langle (t, s_0), (a, t) \rangle \in D$. Note that for any $\alpha \in \mathbb{N}$ and i, we have:

$$[(t, s_i)] \in Attr_\sigma^\alpha \text{ iff } i \le \alpha$$

As a consequence, we cannot conclude that there is some $\alpha \in \mathbb{N}$ such that $[(t, s)] \in Attr_\sigma^\alpha$, since this requires that for this α we would have $[(t, s_i)] \in Attr_\sigma^\alpha$ for all $i \in \mathbb{N}$; *i.e.*, also for all $i > \alpha$. Clearly, this only happens at the first limit ordinal ω. □

The following lemma confirms that all SPOILER-owned configurations won by SPOILER using σ, are part of the σ-attractor; the proof of this is similar to the correctness proof of the standard attractor for two-player games, see *e.g.* [10].

Lemma 1. *Let W_{Sp} be the set of SPOILER-configurations won by her using strategy σ. Then $W_{Sp} \subseteq Attr_\sigma$.*

Using Lemma 1, we can use the σ-attractor to associate a proof with each SPOILER-owned configuration won by her using σ. We formalise this link in the following definition and theorem. Since the σ-attractor is defined using transfinite induction, the definition and theorem necessarily use transfinite induction, too. We do remark that when the transition system is image-finite, *i.e.*, when for all states $s \in S$ and all actions $a \in A$ in the transition system, the set $\{t \in S \mid s \xrightarrow{a} t\}$ is finite, a standard induction suffices.

Definition 8. *We use transfinite induction to associate a set of partial derivations Pr_σ with the set $Attr_\sigma$; we define Pr_σ as the set Pr_σ^γ, where γ is the smallest ordinal satisfying $Pr_\sigma^\gamma = Pr_\sigma^{\gamma+1}$, and:*

– the set Pr_σ^0 associated with $Attr_\sigma^0$ is as follows:

$$\left\{ \frac{s \xrightarrow{a} s' \quad \checkmark}{s \# t} \;\middle|\; \sigma([\,(s,t)\,]) \in Attr_\sigma^0 \wedge \sigma([\,(s,t)\,]) = \langle\,(s,t),(a,s')\,\rangle \right\}$$

$\cup$

$$\left\{ \frac{\dfrac{t \xrightarrow{a} t' \quad \checkmark}{t \# s}}{s \# t} \;\middle|\; \sigma([\,(s,t)\,]) \in Attr_\sigma^0 \wedge \sigma([\,(s,t)\,]) = \langle\,(t,s),(a,t')\,\rangle \right\}$$

– the set $Pr_\sigma^{\alpha+1}$ associated with $Attr_\sigma^{\alpha+1}$ is defined as follows:

Pr_σ^α

$\cup$

$$\left\{ \frac{s \xrightarrow{a} s' \quad \forall t' : t \xrightarrow{a} t' \Rightarrow s' \# t'}{s \# t} \;\middle|\; \begin{array}{l} \sigma([\,(s,t)\,]) = \langle\,(s,t),(a,s')\,\rangle \\ \wedge\,[\,(s,t)\,] \in Attr_\sigma^{\alpha+1} \end{array} \right\}$$

$\cup$

$$\left\{ \frac{\dfrac{t \xrightarrow{a} t' \quad \forall s' : s \xrightarrow{a} s' \Rightarrow t' \# s'}{t \# s}}{s \# t} \;\middle|\; \begin{array}{l} \sigma([\,(s,t)\,]) = \langle\,(t,s),(a,t')\,\rangle \\ \wedge\,[\,(s,t)\,] \in Attr_\sigma^{\alpha+1} \end{array} \right\}$$

– the set Pr_σ^β associated with $Attr_\sigma^\beta$, for limit ordinal β is defined as: $\bigcup_{\alpha<\beta} Pr_\sigma^\alpha$

Observe that by construction, $[\,(s,t)\,] \in Attr_\sigma$ implies that there is some partial derivation in Pr_σ with conclusion $s \# t$ and *vice versa*.

Theorem 4. *For all conclusions $s \# t$ of partial derivations in Pr_σ a proof can be built only using the partial derivations in Pr_σ.*

Proof. We show, by means of a transfinite induction, that all conclusions $s \# t$ of partial derivations in Pr_σ^γ have a proof using the partial derivations in Pr_σ^γ only.

– Base case $\gamma = 0$. We consider two cases, based on the shape of the partial derivations in Pr_σ^0.
  • Suppose the partial derivation is as follows:

$$\frac{\dfrac{t \xrightarrow{a} t' \quad \checkmark}{t \# s}}{s \# t}$$

Then it must be the case that both $\sigma([\,(s,t)\,]) \in Attr_\sigma^0$ and $\sigma([\,(s,t)\,]) = \langle\,(t,s),(a,t')\,\rangle$. Hence, $\langle\,(t,s),(a,t')\,\rangle \in D$ and thus $t \xrightarrow{a} t'$ and $s \xrightarrow{a}\!\!\!\!\!/\;$ hold true. Then we may conclude that the partial derivation is a valid proof with conclusion $s \# t$, with the first rule applied being an instance of rule $(in_\#)$, and the bottom rule applied being rule $(sym_\#)$.

- Suppose the partial derivation is as follows:

$$\frac{s \xrightarrow{a} s' \quad \checkmark}{s \underline{\#} t}$$

This case is similar to the previous case, except that there is no need for an application of rule $(\text{sym}_{\underline{\#}})$.

- Case γ is a successor ordinal $\alpha + 1$. If we have a partial derivation with conclusion $s \underline{\#} t$ in Pr_σ^α, then, by induction, we also have a proof with conclusion $s \underline{\#} t$ built from partial derivations in Pr_σ^α and therefore also in Pr_σ^γ. So, consider a partial derivation with conclusion $s \underline{\#} t$ in Pr_σ^γ for which we do not yet have a proof.

  - Suppose we have the following partial derivation:

$$\frac{s \xrightarrow{a} s' \quad \forall t' : t \xrightarrow{a} t' \Rightarrow s' \underline{\#} t'}{s \underline{\#} t}$$

Then $\sigma([\,(s,t)\,]) \in Attr_\sigma^{\alpha+1}$ and $\sigma([\,(s,t)\,]) = \langle (s,t), (a, s') \rangle$. By definition of $Attr_\sigma^{\alpha+1}$ and the bisimulation game, the set of successors of $\langle (s,t), (a, s') \rangle$ is $\{[\,(s',t')\,] \mid t \xrightarrow{a} t'\}$ and $\{[\,(s',t')\,] \mid t \xrightarrow{a} t'\} \subseteq Attr_\sigma^\alpha$. By our induction hypothesis, there must then, for all t' such that $t \xrightarrow{a} t'$, be some proper proof with conclusion $s' \underline{\#} t'$, constructed from the partial derivations in Pr_σ^α. But then we also obtain a proper proof of $s \underline{\#} t$ using the partial derivation as an instance of rule $(\text{in}_{\underline{\#}})$.

  - Suppose the partial derivation with conclusion $s \underline{\#} t$ is as follows:

$$\frac{\dfrac{t \xrightarrow{a} t' \quad \forall s' : s \xrightarrow{a} s' \Rightarrow t' \underline{\#} s'}{t \underline{\#} s}}{s \underline{\#} t}$$

Similar to the previous case, with an additional appeal to rule $(\text{sym}_{\underline{\#}})$; see also the base case.

- Case γ is a limit ordinal. Then for every $[\,(s,t)\,] \in \text{Pr}_\sigma^\gamma$, there must be some proof with conclusion in Pr_σ^α, for some $\alpha < \gamma$. So, by induction, we have a proper proof with conclusion $s \underline{\#} t$ in Pr_σ^γ. $\qquad\square$

We illustrate the construction of the apartness proof from a SPOILER-winning strategy, used in the proof, through the following example.

Example 7. We set out to construct an apartness proof showing $s_0 \underline{\#} t_0$ for the transition system of Example 4 by first constructing a SPOILER-winning strategy for configuration $[\,(s_0, t_0)\,]$, and subsequently converting that using the construction provided in the proof of Theorem 4. Consider the strategy σ defined as follows:

$$\sigma([\,(s_0, t_0)\,]) = \langle (s_0, t_0), (a, s_0) \rangle \qquad \sigma([\,(s_0, t_1)\,]) = \langle (t_1, s_0), (a, t_3) \rangle$$
$$\sigma([\,(s_0, t_2)\,]) = \langle (t_2, s_0), (a, t_4) \rangle \qquad \sigma([\,(t_3, s_0)\,]) = \langle (s_0, t_3), (a, s_0) \rangle$$
$$\sigma([\,(t_4, s_0)\,]) = \langle (s_0, t_4), (a, s_0) \rangle$$

Note that in all (save one) of the above target DUPLICATOR configurations, DUPLICATOR's moves are forced. In configuration $\langle (s_0, t_0), (a, s_0) \rangle$, DUPLICATOR may play either to $[(s_0, t_1)]$ or $[(s_0, t_2)]$ giving rise to two proof obligations in the apartness proof. The apartness proof that is induced by σ is as follows:

$$
\cfrac{s_0 \xrightarrow{a} s_0 \qquad \cfrac{t_1 \xrightarrow{a} t_3 \qquad \cfrac{\cfrac{s_0 \xrightarrow{a} s_0 \quad \checkmark}{s_0 \# t_3}}{t_3 \# s_0}}{t_1 \# s_0} \qquad \cfrac{t_2 \xrightarrow{a} t_4 \qquad \cfrac{\cfrac{s_0 \xrightarrow{a} s_0 \quad \checkmark}{s_0 \# t_4}}{t_4 \# s_0}}{t_2 \# s_0}}{}
$$

$$
\cfrac{s_0 \# t_1 \qquad\qquad\qquad\qquad s_0 \# t_2}{s_0 \# t_0}
$$

$\square$

4   Conclusions and Future Work

We have studied the connection between two approaches for reasoning about the (in)equivalence of states in a labelled transition system: bisimulation games and strong apartness. While both approaches can easily be related by exploiting their relation to strong bisimilarity, establishing a direct connection is delicate. This is particularly true when moving from a proof of apartness to a winning strategy for SPOILER, which shares issues with the related problem of moving from a winning memory strategy to a winning memoryless strategy for a given player in two-player games. Interestingly, our construction also allows for computing a derivation of apartness, by simply constructing the bisimulation game and computing a winning strategy for SPOILER.

We expect our results can be generalised to games for equivalences (and simulation relations) that deal with abstraction, see [1]. One complication that we foresee is that these games involve Büchi winning conditions. It is not immediately clear to us how these can be incorporated into the proof rules of [3].

References

1. de Frutos Escrig, D., Keiren, J.J.A., Willemse, T.A.C.: Games for bisimulations and abstraction. Log. Methods Comput. Sci. **13**(4) (2017). https://doi.org/10.23638/LMCS-13(4:15)2017
2. Geuvers, H.: Apartness and distinguishing formulas in Hennessy-Milner logic. In: Jansen, N., Stoelinga, M., van den Bos, P. (eds.) A Journey from Process Algebra via Timed Automata to Model Learning. LNCS, vol. 13560, pp. 266–282. Springer, Cham (2022). https://doi.org/10.1007/978-3-031-15629-8_14
3. Geuvers, H., Golov, A.: Directed branching bisimulation via apartness and positive logic (2022). https://doi.org/10.48550/arXiv.2210.07380
4. Geuvers, H., Jacobs, B.: Relating apartness and bisimulation. Log. Methods Comput. Sci. **17**(3) (2021). https://doi.org/10.46298/lmcs-17(3:15)2021

5. Geuvers, H., Nederpelt, R.: Characteristics of de Bruijn's early proof checker Automath. Fundam. Informaticae **185**(4), 313–336 (2022). https://doi.org/10.3233/FI-222112

6. van Glabbeek, R.J.: The linear time–branching time spectrum II. In: Best, E. (ed.) CONCUR 1993. LNCS, vol. 715, pp. 66–81. Springer, Heidelberg (1993). https://doi.org/10.1007/3-540-57208-2_6

7. Park, D.: Concurrency and automata on infinite sequences. In: Deussen, P. (ed.) TCS 1981. LNCS, vol. 104, pp. 167–183. Springer, Heidelberg (1981). https://doi.org/10.1007/BFb0017309

8. Stirling, C.: Bisimulation, modal logic and model checking games. Log. J. IGPL **7**(1), 103–124 (1999). https://doi.org/10.1093/jigpal/7.1.103

9. van Glabbeek, R.J.: The linear time - branching time spectrum. In: Baeten, J.C.M., Klop, J.W. (eds.) CONCUR 1990. LNCS, vol. 458, pp. 278–297. Springer, Berlin (1990). https://doi.org/10.1007/BFb0039066

10. Zielonka, W.: Infinite games on finitely coloured graphs with applications to automata on infinite trees. Theor. Comput. Sci. **200**(1–2), 135–183 (1998)

Multisets and Distributions

Dexter Kozen[(✉)] and Alexandra Silva

Cornell University, Ithaca, NY, USA
kozen@cs.cornell.edu

Abstract. We give a lightweight alternative construction of Jacobs's distributive law for multisets and distributions that does not involve any combinatorics. We first give a distributive law for lists and distributions, then apply a general theorem on 2-categories that allows properties of lists to be transferred automatically to multisets. The theorem states that equations between 2-cells are preserved by epic 2-natural transformations. In our application, the appropriate epic 2-natural transformation is defined in terms of the Parikh map, familiar from formal language theory, that takes a list to its multiset of elements.

1 Introduction

A notoriously difficult challenge in the semantics of probabilistic programming is the combination of probability and nondeterminism. The chief difficulty is the lack of an appropriate distributive law between the powerset and distribution monads [2]. Many researchers have grappled with this issue over the years, proposing various workarounds [1,3,4,6,7,9–16]. In contrast, if instead of the powerset one considers the multiset monad, it has been known for a while that indeed the multiset monad does distribute over the probability distributions monad. This observation appeared in e.g. [4,5,9], but its origin is hard to trace as it seems to have been folklore knowledge for some time. Whereas the aforementioned references justified the existence of the law quite convincingly from an operational point of view, what was missing for a long time was an actual explicit description.

Recently, an important breakthrough was provided by Jacobs [8], who presented multiple explicit descriptions of the distributive law for finite multisets over finite distributions. His treatment is a combination of categorical and combinatorial reasoning, involving the commutativity of multinomial distributions with *hypergeometric distributions*, discrete probability distributions that calculate the likelihood an event occurring k times in n trials when sampling from a small population without replacement.

In this paper, we give a lighter-weight alternative to Jacobs's proof that does not involve any combinatorics. Our approach is based on the premise that

This paper is dedicated to Herman Geuvers on the occasion of his 60th birthday. Both authors spent time in Nijmegen and enjoyed many conversations with Herman, who has always been curious on the many aspects of programming semantics and categorical aspects. We hope he enjoys this simplified presentation of a complicated law, followed by a categorical generalization!

V. Capretta et al. (Eds.): *Logics and Type Systems in Theory and Practice*, LNCS 14560, pp. 168–187, 2024.
https://doi.org/10.1007/978-3-031-61716-4_11

lists are more intuitive and notationally less cumbersome than multisets. There is an intuitively simple distributive law for lists over distributions: given a list of distributions, sample them independently to obtain a list of outcomes; the probability of that list of outcomes is the product of the probabilities of the components. We have thus turned a list of distributions into a distribution over lists. We show that this operational intuition is correct using purely equational reasoning, then apply a general theorem on 2-categories that maps properties of lists down to properties of multisets.

The general theorem, which may be of independent interest, states that *equations between natural transformations are preserved by epic 2-natural transformations*. In our application, the appropriate epic 2-natural transformation is defined in terms of the *Parikh map*, familiar from formal language theory, that takes a list to its multiset of elements. This theorem allows monad and distributive laws involving multisets to be obtained automatically from monad and distributive laws involving lists.

The contributions of this paper are twofold. On the one hand, we present a concrete construction of an important distributive law between multisets and distributions. Jacobs [8] presents multiple constructions, all of which required more involved machinery. Here our argument is that a simpler presentation, without combinatorics, of the Beck distributive law is possible, though the alternative ones in [8] are valuable in different ways (as the author also argues). In particular, we do not claim that our approach is useful for calculating probabilities of events; indeed, such calculations require combinatorics, which we have eliminated in our approach. On the other hand, our general result provides a useful tool that may be applicable in other contexts.

We start with some brief preliminaries in Sect. 2 introducing the multiset and distribution monads as submonads of a common monad. In Sect. 3 we present the ingredients needed to define the desired distributive law between multisets and distributions and show how it can be obtained from a distributive law between lists and distributions via the Parikh map. In Sect. 4 we present a general theorem on 2-categories and show that many of the concrete results of the previous section are instances of the general theorem.

2   The Distribution and Multiset Monads

The distribution and multiset monads are both submonads of a common generalization. Consider the endofunctor

$$F : \mathsf{Set} \to \mathsf{Set} \qquad FX = \{f \in \mathbb{R}_+^X \mid |\, \mathsf{supp}\, f| < \infty\},$$

the set of nonnegative real-valued functions $f : X \to \mathbb{R}$ with finite support, and for $h : X \to Y$,

$$Fh : FX \to FY \qquad Fh(f) = \lambda b \in Y . \sum_{h(a)=b} f(a).$$

Note that F is covariant, unlike the usual hom-functor $\text{Hom}(-, \mathbb{R}_+)$. For $f \in FX$, let $|f| = \sum_{a \in X} f(a)$.

The functor F carries a monad structure (F, μ, η) with $\eta : I \to F$ and $\mu : F^2 \to F$, where for $a \in X$ and $f \in F^2X$,

$$\eta_X(a) = \lambda b \in X . [b = a]$$

$$\mu_X(f) = \lambda a \in X . \sum_{g \in FX} f(g) \cdot g(a).$$

The notation [a=b] makes use of the so-called Iverson bracket, defined, for a predicate φ as $[\varphi] = 1$ if φ is true, 0 if false.

Restricting the range of functions to $\mathbb{N}$, we get a submonad (M, μ, η), where

$$MX = \{m \in FX \mid \forall a \in X \; m(a) \in \mathbb{N}\}.$$

This is the *multiset monad*. The value $m(a)$ is the multiplicity of $a \in X$ in the multiset m. On the other hand, restricting to convex functions, we get a submonad (D, μ, η), where

$$DX = \{d \in FX \mid \sum_{a \in X} d(a) = 1\}.$$

This is the *distribution monad*. The value $d(a)$ is the probability of $a \in X$ under the distribution d.

Clearly $\mu^M : M^2 \to M$, as $\mathbb{N}$ is closed under the semiring operations, and $\mu^D : D^2 \to D$, as a convex function of convex functions flattens to a convex function.

We will sometimes annotate η and μ with superscripts M and D to emphasize that we are interpreting them in the respective submonads, but the definitions are the same. In more conventional notation, $\eta_X^D(a) = \delta_a$, the Dirac (point-mass) distribution on $a \in X$, and distributions can be expressed as weighted formal sums with real coefficients as in [8].

The Eilenberg-Moore algebras for F are the $\mathbb{R}_+$-semimodules, although we do not need to know this for our study.

In the following sections, we will give a construction of a distributive law $\otimes^M : MF \to FM$, which contains the desired distributive law $MD \to DM$ as a special case.

3   Lists and Distributions

In this section we set the stage by giving an intuitive account of our approach. It will be evident that it is possible to obtain the desired distributive law in a piecemeal fashion using only arguments in this section; but in Sect. 4 we will show how to encapsulate these arguments in a general theorem to obtain them all at once. Nevertheless, the development of this section is essential motivation for understanding the more abstract version of Sect. 4.

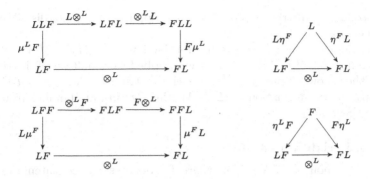

Fig. 1. Axioms for the distributive law $\otimes^L : LF \to FL$

3.1 A Distributive Law $\otimes^L : LF \to FL$

Let (L, μ^L, η^L) be the list monad. Define $\otimes^L_X : LFX \to FLX$,

$$\otimes^L_X ([f_1, \ldots, f_n])$$
$$= \lambda[a_1, \ldots, a_m] \in LX . \begin{cases} \prod_{i=1}^n f_i(a_i) & \text{if } m = n \\ 0 & \text{otherwise.} \end{cases}$$

where $[a_1, \ldots, a_n]$ denotes a list of n elements[1]

We abbreviate this by writing

$$\otimes^L_X([f_1, \ldots, f_n]) = \lambda[a_1, \ldots, a_n] \in X^n . \prod_{i=1}^n f_i(a_i)$$

with the convention that the right-hand side, while defined on all of LX, vanishes outside X^n. The support $\text{supp} \otimes^L_X([f_1, \ldots, f_n])$ is the cartesian product $\prod_{i=1}^n \text{supp } f_i$.

Theorem 1. $\otimes^L : LF \to FL$ *is a distributive law for* (L, μ^L, η^L) *over* (F, μ, η). *That is, it is a natural transformation and satisfies the axioms of Fig. 1.*

Proof. The proof uses only elementary equational reasoning and is given in its entirety in Sect. A.1.

Applied to a list of n distributions over X, $\otimes^L_X$ gives the joint distribution on the cartesian product X^n obtained by sampling the component distributions independently.

$$\otimes^L_X([d_1, \ldots, d_n])([a_1, \ldots, a_n]) = \prod_{i=1}^n d_i(a_i).$$

---

[1] This is not to be confused with the Iverson bracket $[\varphi]$. We trust the reader will be able to distinguish them by the difference in fonts and by context.

As $\otimes^L$ produces a distribution on LX when applied to a list of distributions on X, it specializes to $\otimes^L : LD \to DL$.

We would like to have a similar distributive law $\otimes^M : MF \to FM$, which will also specialize to $\otimes^M : MD \to DM$ when applied to a multiset of distributions. We will show how to derive $\otimes^M$ from the distributive law $\otimes^L : LF \to FL$ and a natural transformation $\# : L \to M$, the *Parikh map* familiar from formal language theory.

3.2   The Parikh Map $\# : L \to M$

Consider the monoid $(LX, \cdot, [])$, where $(\cdot)$ denotes list concatenation (often elided) and $[]$ is the empty list. This is the free monoid on generators X. Consider also the monoid $(MX, +, 0)$, where $+$ is pointwise sum of real-valued functions and 0 is the constant-zero function, representing multiset union and the empty multiset, respectively. This is the free commutative monoid on generators X. The two monoids are connected by the *Parikh map* $\#_X : LX \to MX$, where $\#_X(x)(a)$ gives the number of occurrences of $a \in X$ in the list $x \in LX$. This is a monoid homomorphism from the free monoid on generators X to its commutative image. Note that $\#_X$ is an epimorphism, as every multiset is the image of a list. Jacobs [8] refers to the map $\#_X$ as *accumulation* (*acc*).

More generally, we will show below that $\#$ is a morphism of monads

$$\# : (L, \mu^L, \eta^L) \to (M, \mu^M, \eta^M)$$

(Theorem 4). This means that $\# : L \to M$ is a natural transformation that commutes with μ and η; in other words, it is a monoid homomorphism, regarding monads as monoids over the category of endofunctors.

Because $\#_X$ is a monoid homomorphism, we have

$$\#_X([]) = 0 \qquad\qquad \#_X(xy) = \#_X(x) + \#_X(y).$$

On generators $[a]$,

$$\#_X([a]) = \lambda b \in X . [b = a] = \eta_X(a).$$

Using these facts, we can show that $\# : L \to M$ is natural transformation. For $f : X \to Y$,

$$M f(\#_X([a_1, \ldots, a_n]))$$

$$= M f(\sum_{i=1}^{n} \#_X([a_i])) = M f(\sum_{i=1}^{n} \lambda a \in X . [a = a_i])$$

$$= \lambda b \in Y . \sum_{f(a)=b} \sum_{i=1}^{n} [a = a_i] = \lambda b \in Y . \sum_{i=1}^{n} [b = f(a_i)]$$

$$= \sum_{i=1}^{n} \#_Y([f(a_i)]) = \#_Y([f(a_1), \ldots, f(a_n)])$$

$$= \#_Y(L f([a_1, \ldots, a_n])),$$

thus $M f \circ \#_X = \#_Y \circ L f$.

3.3   A Distributive Law $\otimes^M : MF \to FM$

We first define the natural transformation $\otimes^M : MF \to FM$, then show that it satisfies the requisite properties of a distributive law. For any $x \in LX$, define

$$S(x) = \{y \in LX \mid \#_X(y) = \#_X(x)\} = \#_X^{-1}(\#_X(x)).$$

Then $y \in S(x)$ iff y and x have the same length, say n, and y is a permutation of x; that is,

$$S([b_1, \ldots, b_n]) = \{[b_{\sigma(1)}, \ldots, b_{\sigma(n)}] \mid \sigma \text{ is a permutation on } \{1, \ldots, n\}\}.$$

Lemma 1. *Let* $S = S([b_1, \ldots, b_n])$. *For any fixed permutation* π *on* $\{1, \ldots, n\}$, *the map*

$$[a_1, \ldots, a_n] \mapsto [a_{\pi(1)}, \ldots, a_{\pi(n)}] \tag{1}$$

is a bijection $S \to S$.

Proof. Since S is closed under permutations,

$$\{[a_{\pi(1)}, \ldots, a_{\pi(n)}] \mid [a_1, \ldots, a_n] \in S\} \subseteq S.$$

But the map (1) is injective, so

$$|\{[a_{\pi(1)}, \ldots, a_{\pi(n)}] \mid [a_1, \ldots, a_n] \in S\}| \geq |S|,$$

therefore equality holds, so (1) must be a bijection on S.   □

The desired natural transformation $\otimes^M : MF \to FM$ is defined in the following lemma. This is essentially the same as Jacob's third characterization [8, Equation (5)].

Lemma 2. *There is a unique natural transformation* $\otimes^M : MF \to FM$ *such that the following diagram commutes:*

$$
\begin{array}{ccc}
LF & \xrightarrow{\ \otimes^L\ } & FL \\
{\scriptstyle \#F}\downarrow & & \downarrow{\scriptstyle F\#} \\
MF & \xrightarrow[\ \otimes^M\ ]{} & FM
\end{array}
\tag{2}
$$

Proof. Let

$$X^{(n)} = \{m \in MX \mid \sum_{a \in X} m(a) = n\},$$

the multisets over X of size n. For a given $[f_1, \ldots, f_n] \in LFX$, define

$$\otimes_X^M (\#_{FX}([f_1, \ldots, f_n]))$$

$$= \lambda y \in X^{(n)}. \sum_{[a_1, \ldots, a_n] \in \#_X^{-1}(y)} \prod_{i=1}^{n} f_i(a_i).$$

Again by convention, it is assumed that the function on the right-hand side is defined on all of MX, but vanishes on $MX \setminus X^{(n)}$. This defines $\otimes_X^M$ on the multiset $\#_{FX}([f_1, \ldots, f_n])$; we must show that the definition is independent of the choice of $[f_1, \ldots, f_n]$.

For any $[b_1, \ldots, b_n] \in LX$, we have $\#_X([b_1, \ldots, b_n]) \in MX$, and

$$\otimes_X^M (\#_{FX}([f_1, \ldots, f_n]))(\#_X([b_1, \ldots, b_n]))$$

$$= \sum_{[a_1,\ldots,a_n] \in \#_X^{-1}(\#_X([b_1,\ldots,b_n]))} \prod_{i=1}^n f_i(a_i)$$

$$= \sum_{[a_1,\ldots,a_n] \in S([b_1,\ldots,b_n])} \prod_{i=1}^n f_i(a_i).$$

Then for any permutation $\sigma : \{1, \ldots, n\} \to \{1, \ldots, n\}$,

$$\otimes_X^M (\#_{FX}([f_{\sigma(1)}, \ldots, f_{\sigma(n)}]))(\#_X([b_1, \ldots, b_n]))$$

$$= \sum_{[a_1,\ldots,a_n] \in S([b_1,\ldots,b_n])} \prod_{i=1}^n f_{\sigma(i)}(a_i)$$

$$= \sum_{[a_1,\ldots,a_n] \in S([b_1,\ldots,b_n])} \prod_{i=1}^n f_i(a_{\sigma^{-1}(i)}) \tag{3}$$

$$= \sum_{[a_{\sigma^{-1}(1)},\ldots,a_{\sigma^{-1}(n)}] \in S([b_1,\ldots,b_n])} \prod_{i=1}^n f_i(a_{\sigma^{-1}(i)}) \tag{4}$$

$$= \sum_{[a_1,\ldots,a_n] \in S([b_1,\ldots,b_n])} \prod_{i=1}^n f_i(a_i). \tag{5}$$

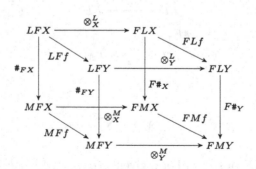

Fig. 2. Naturality of $\otimes^M$ from $\otimes^L$

The inference (3) is from the commutativity of multiplication, (4) is from Lemma 1, and (5) comes from reindexing. It follows that $\otimes_X^M (\#_{FX}([f_1, \ldots, f_n]))$

and $\otimes_X^M(\#_{FX}([f_{\sigma(1)},\dots,f_{\sigma(n)}]))$ agree on all elements of the form $\#_X([b_1,\dots,b_n]) \in MX$, and both take value 0 on elements of the form $\#_X([b_1,\dots,b_m])$ for $m \neq n$. But this is all of MX, as $\#_X$ is surjective.

The Eq. (2) now follows from the definitions of $F\#_X$ and $\otimes_X^L$:

$$\otimes_X^M(\#_{FX}([f_1,\dots,f_n]))$$

$$= \lambda y \in X^{(n)}. \sum_{[a_1,\dots,a_n]\in\#_X^{-1}(y)} \prod_{i=1}^{n} f_i(a_i)$$

$$= \lambda y \in X^{(n)}. \sum_{\#_X([a_1,\dots,a_n])=y} \prod_{i=1}^{n} f_i(a_i)$$

$$= F\#_X(\otimes_X^L([f_1,\dots,f_n])).$$

It remains to show the naturality of $\otimes^M$. That is, for $f : X \to Y$,

$$FMf \circ \otimes_X^M = \otimes_Y^M \circ MFf. \tag{6}$$

We will take this opportunity to illustrate the general observation that allows us to map properties interest from L down to M via $\#$. Suppose we know that $\otimes^L$ is a natural transformation. In Fig. 2, the commutativity of the top face asserts this fact, and we wish to derive the same for the bottom face. The front and back vertical faces commute by (2). The left and right vertical faces commute by the naturality of $\#$. Using these facts, we can show the following paths in Fig. 2 are equivalent:

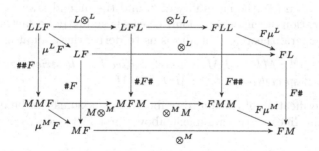

Fig. 3. An axiom of distributive laws

$$LFX \to MFX \to FMX \to FMY$$

$$= LFX \to FLX \to FLY \to FMY \tag{7}$$

$$= LFX \to LFY \to FLY \to FMY \tag{8}$$

$$= LFX \to MFX \to MFY \to FMY \tag{9}$$

The inference (7) is from the back and right vertical faces, (8) is from the top face, and (9) is from the left and front vertical faces. Thus

$$FMf \circ \otimes_X^M \circ \#_{FX} = \otimes_Y^M \circ MFf \circ \#_{FX}.$$

But since $\#_{FX} : LFX \to MFX$ is an epimorphism, we can conclude that

$$FMf \circ \otimes_X^M = \otimes_Y^M \circ MFf.$$

□

We can apply the same technique to show that $\otimes^M$ satisfies the axioms of distributive laws. These are the same as those of Fig. 1 with M in place of L. Assuming these diagrams hold for L (which we still have to prove—see Sect. A.1), we can derive them for M. For example, to transfer the upper left property of Fig. 1 from L to M, we use the diagram of Fig. 3, where

$$\#\# = M\# \circ \#L = \#M \circ L\#$$

$$\#F\# = MF\# \circ \#FL = \#FM \circ LF\#.$$

The top face of Fig. 3 asserts the property of interest for L, and we wish to derive the same for the bottom face. We obtain an equivalence of paths

$$LLF \to MMF \to MFM \to FMM \to FM$$
$$= LLF \to MMF \to MF \to FM$$

using the commutativity of the top and vertical side faces, then use the fact that $\#\#F : LLF \to MMF$ is an epimorphism to conclude that the bottom face commutes. The commutativity of the top face is proved below in Sect. A.1; the side faces are from (2), the naturality of #, and the monad laws.

The verification of the other axioms proceeds similarly. In fact, in Sect. 4 we will prove a general theorem that allows us to derive these all at once.

Theorem 2. $\otimes^M : MF \to FM$ *is a distributive law. Restricted to multisets of distributions, it specializes to* $\otimes^M : MD \to DM$.

Proof. The verification of the four distributive law axioms can be done individually along the lines of the argument above, or all at once using Corollary 1 below.

For the last statement, we observe that for distributions $d_1, \ldots, d_n$,

$$\sum_{m \in MX} \otimes_X^M(\#_{FX}([d_1, \ldots, d_n]))(m)$$

$$= \sum_{m \in MX} \sum_{\#_X([a_1, \ldots, a_n]) = m} \prod_{i=1}^n d_i(a_i)$$

$$= \sum_{[a_1, \ldots, a_n] \in LX} \prod_{i=1}^n d_i(a_i) = \prod_{i=1}^n \sum_{a \in X} d_i(a) = 1.$$

□

4   A General Result

In this section we show how results of the previous section can be obtained as a consequence of a general theorem on 2-categories to the effect that *equations between 2-cells are preserved under epic 2-natural transformations*. In our application, this theorem allows equations involving L and F to be mapped down to equations involving M and F via a 2-natural transformation defined in terms of #.

Recall that in a 2-category, composition of 1-cells and composition of 2-cells are called *horizontal* and *vertical composition*, respectively. We will write $\circ$ for vertical composition and denote horizontal composition by juxtaposition. Horizontal composition also acts on 2-cells and satisfies the property: If $F, F', F'' : \mathcal{C} \to \mathcal{D}$ and $G, G', G'' : \mathcal{D} \to \mathcal{E}$ are 1-cells and $\varphi : F \to F'$, $\varphi' : F' \to F''$, $\psi : G \to G'$, and $\psi' : G' \to G''$ are 2-cells, then

$$(\varphi' \circ \varphi)(\psi' \circ \psi) = (\varphi'\psi') \circ (\varphi\psi) : FG \to F''G''. \tag{10}$$

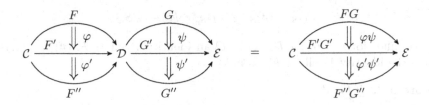

Categories, functors, and natural transformations form a 2-category Cat, in which the 0-cells are categories, the 1-cells are functors, and the 2-cells are natural transformations. In Cat, horizontal composition on natural transformations is defined by: for $\varphi : F \to G$ and $\psi : H \to K$,

$$\varphi\psi : FH \to GK = \varphi K \circ F\psi = G\psi \circ \varphi H \tag{11}$$

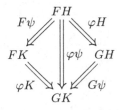

and conversely, $\varphi H = \varphi \, \mathrm{id}_H$ and $F\psi = \mathrm{id}_F \psi$.

A *2-natural transformation* between 2-functors is like a natural transformation between functors, but one level up. Formally, let $\mathcal{F}$ and $\mathcal{C}$ be 2-categories and let $\Phi, \Psi : \mathcal{F} \to \mathcal{C}$ be 2-functors such that $\Phi O = \Psi O$ for each 0-cell O of $\mathcal{F}$.

A *2-natural transformation* $\varphi : \Phi \to \Psi$ is a collection of 2-cells $\varphi_S : \Phi S \to \Psi S$ indexed by 1-cells S such that for each 2-cell $\sigma : S \to T$, the following diagram commutes:

$$
\begin{array}{ccc}
\Phi S & \xrightarrow{\Phi\sigma} & \Phi T \\
{\scriptstyle\varphi_S}\downarrow & & \downarrow{\scriptstyle\varphi_T} \\
\Psi S & \xrightarrow[\Psi\sigma]{} & \Psi T
\end{array}
\tag{12}
$$

The restriction that $\Phi O = \Psi O$ for each 0-cell O of $\mathcal{F}$ allows φ_S to be a 2-cell. A 2-natural transformation is *epic* if all components are epimorphisms.

Now let $\mathcal{F}$ be the free 2-category on finite sets of generators $\mathcal{O}$, $\mathcal{T}$, and $\mathcal{R}$ for the 0-, 1-, and 2-cells, respectively. Let $\mathcal{C}$ be another 2-category, and let $\Phi, \Psi : \mathcal{F} \to \mathcal{C}$ be 2-functors such that $\Phi O = \Psi O$ for each $O \in \mathcal{O}$.

Suppose that to each 1-cell generator $S \in \mathcal{T}$ there is associated a 2-cell $(\#)_S : \Phi S \to \Psi S$ of $\mathcal{C}$. If $S = S_1 \cdots S_n$ is a 1-cell of $\mathcal{F}$, define $(\#)_S$ to be the horizontal composition

$$
(\#)_S = (\#)_{S_1} \cdots (\#)_{S_n} : \Phi S \to \Psi S.
$$

In particular, for $n = 0$, for $I_O : O \to O$ an identity 1-cell of $\mathcal{F}$, $(\#)_{I_O} : \Phi I_O \to \Psi I_O$ is the identity 1-cell on $\Phi O = \Psi O$.

Theorem 3. *If the diagram*

$$
\begin{array}{ccc}
\Phi S & \xrightarrow{\Phi\sigma} & \Phi T \\
{\scriptstyle(\#)_S}\downarrow & & \downarrow{\scriptstyle(\#)_T} \\
\Psi S & \xrightarrow[\Psi\sigma]{} & \Psi T
\end{array}
\tag{13}
$$

commutes for each 2-cell generator $\sigma : S \to T$ in $\mathcal{R}$, then it commutes for all 2-cells σ generated by $\mathcal{R}$. Consequently, $(\#)$ is a 2-natural transformation $\Phi \to \Psi$ with components $(\#)_S$.

Proof. For identities,

$$
\begin{aligned}
(\#)_S \circ \Phi(\mathrm{id}_S) &= (\#)_S \circ \mathrm{id}_{\Phi S} & &\text{since } \Phi \text{ is a 2-functor} \\
&= \mathrm{id}_{\Psi S} \circ (\#)_S \\
&= \Psi(\mathrm{id}_S) \circ (\#)_S & &\text{since } \Psi \text{ is a 2-functor.}
\end{aligned}
$$

For vertical composition $\tau \circ \sigma : S \to U$ with $\sigma : S \to T$ and $\tau : T \to U$, if (13) commutes for σ and τ, that is,

$$
(\#)_T \circ \Phi\sigma = \Psi\sigma \circ (\#)_S \qquad\qquad (\#)_U \circ \Phi\tau = \Psi\tau \circ (\#)_T,
\tag{14}
$$

then

$$(\#)_U \circ \Phi(\tau \circ \sigma)$$

$$= (\#)_U \circ \Phi\tau \circ \Phi\sigma \qquad \text{since } \Phi \text{ is a 2-functor}$$

$$= \Psi\tau \circ (\#)_T \circ \Phi\sigma \qquad \text{by (14), right-hand equation}$$

$$= \Psi\tau \circ \Psi\sigma \circ (\#)_S \qquad \text{by (14), left-hand equation}$$

$$= \Psi(\tau \circ \sigma) \circ (\#)_S \qquad \text{since } \Psi \text{ is a 2-functor.}$$

For horizontal composition $\sigma\tau : SU \to TV$ with $\sigma : S \to T$ and $\tau : U \to V$, if (13) commutes for σ and τ, that is,

$$(\#)_T \circ \Phi\sigma = \Psi\sigma \circ (\#)_S \qquad\qquad (\#)_V \circ \Phi\tau = \Psi\tau \circ (\#)_U, \qquad (15)$$

then

$$(\#)_{TV} \circ \Phi(\sigma\tau)$$

$$= ((\#)_T (\#)_V) \circ ((\Phi\sigma)(\Phi\tau)) \qquad \text{since } \Phi \text{ is a 2-functor}$$

$$= ((\#)_T \circ \Phi\sigma)((\#)_V \circ \Phi\tau) \qquad \text{by (10)}$$

$$= (\Psi\sigma \circ (\#)_S)(\Psi\tau \circ (\#)_U) \qquad \text{by (15)}$$

$$= ((\Psi\sigma)(\Psi\tau)) \circ ((\#)_S(\#)_U) \qquad \text{by (10)}$$

$$= \Psi(\sigma\tau) \circ (\#)_{SU} \qquad \text{since } \Psi \text{ is a 2-functor.}$$

$$\square$$

Corollary 1. *Suppose the conditions of Theorem 3 hold. If in addition* $(\#)_S$ *is an epimorphism for every 1-cell S of $\mathcal{F}$, then every 2-cell equation that holds under the interpretation Φ also holds under the interpretation Ψ. That is, for all 2-cells $\sigma, \tau : S \to T$ of $\mathcal{F}$, if $\Phi\sigma = \Phi\tau$, then $\Psi\sigma = \Psi\tau$.*

Proof. We have

$$\Psi\sigma \circ (\#)_S = (\#)_T \circ \Phi\sigma \qquad \text{by Theorem 3}$$

$$= (\#)_T \circ \Phi\tau \qquad \text{since } \Phi\sigma = \Phi\tau$$

$$= \Psi\tau \circ (\#)_S \qquad \text{by Theorem 3.}$$

Since $(\#)_S$ is an epimorphism, $\Psi\sigma = \Psi\tau$.

In our application, $\mathcal{C}$ is Cat, the 2-category of categories, functors, and natural transformations, and

$$
\begin{aligned}
&- \mathcal{O} = \{O\}, \\
&- \mathcal{T} = \{P : O \to O, Q : O \to O\}, \\
&- \mathcal{R} = \{\mu : PP \to P, \eta : I \to P, \\
&\qquad\quad \mu' : QQ \to Q, \eta' : I \to Q, \otimes : PQ \to QP\}, \\
&- \Phi = \{O \mapsto \mathsf{Set}, P \mapsto L, Q \mapsto F, \mu \mapsto \mu^L, \eta \mapsto \eta^L, \\
&\qquad\quad \mu' \mapsto \mu^F, \eta' \mapsto \eta^F, \otimes \mapsto \otimes^L\},
\end{aligned}
$$

$- \Psi = \{O \mapsto \mathsf{Set}, P \mapsto M, Q \mapsto F, \mu \mapsto \mu^M, \eta \mapsto \eta^M,$
   $\mu' \mapsto \mu^F, \eta' \mapsto \eta^F, \otimes \mapsto \otimes^M\}\,,$
$- (\#)_P = \#, (\#)_Q = \mathsf{id}_F.$

We need to verify the conditions of Theorem 3 and Corollary 1. For the condition (13) of Theorem 3, we must show that (13) commutes for each of the 2-cell generators in $\mathcal{R}$. These conditions are

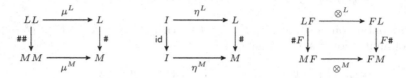

plus two more conditions for μ^F and η^F, which only involve identities and are trivial. The rightmost diagram is just (2), which was verified in Lemma 2. A complete proof for the remaining two is given in Sect. A.2.

For Corollary 1, we must show that all $(\#)_S$ are epimorphisms. In light of (11), we need only show that # is an epimorphism, and that the functors L, F, and M preserve epimorphisms. For any set X, $\#_X : LX \to MX$ takes a list of elements of X to the multiset of its elements and is clearly surjective. That the functors L, F, and M preserve epimorphisms is evident after a few moments' thought; but we provide an explicit proof in Sect. A.2.

These results lead to an alternative proof of Theorem 2 using a general method to transfer properties involving L to properties involving M. Since our application satisfies the conditions of Theorem 3 and Corollary 1, we have that any 2-cell equation that holds under the interpretation Φ also holds under Ψ. This allows us to derive several needed equations involving M all at once from the corresponding equations involving L: the monad equations for $\mu^M : MM \to M$ and $\eta^M : I \to M$, the naturality of $\otimes^M$, and the equations for the distributive law $\otimes^M : MF \to FM$.

To see how this works, observe that the diagrams for these properties (Figs. 2 and 3 for example) all have the same character: they are all cubical diagrams consisting of a diagram involving L on the top face, the same diagram on the bottom face with L replaced by M, and vertical arrows connecting them consisting of components of the 2-natural transformation (#). These components just replace L's with M's and are epimorphisms. The vertical faces all commute, because (#) is a 2-natural transformation. One can then conclude that the bottom face commutes whenever the top face does by the same argument as in Lemma 2.

5   Conclusion

We have provided a simplified proof of a distributive law between the finite distribution and finite multiset monads, first proved by Jacobs [8]. Unlike Jacobs's proof, we do not require any combinatorial machinery, which we view as not germane. The proof consists of two parts: (a) a distributive law between the distribution and list monads, which is more intuitive and easier to prove equationally due to the ordering of lists, and (b) a general theorem on 2-categories to the effect that category-theoretic laws, such as monad and distributive laws, are preserved by epic 2-natural transformations. In this case, laws involving lists are mapped to laws involving multisets via the Parikh map that forgets order, taking a list to its multiset of elements.

We believe that Jacobs's result has the potential to significantly advance the understanding of the interaction of nondeterminism and probability, especially in light of previous well-known difficulties combining the distribution and powerset monads. We did not review Jacobs's proof in this paper, as it is heavily technical and not related to our approach in any meaningful way. However, we do believe that important results deserve more than one proof, and that some readers may find our approach a more accessible alternative.

An interesting and novel aspect of our work is the general 2-categorical formulation of the problem, in which 2-categories are presented as a logical vehicle for the equational logic of natural transformations. There are several innovations here that we believe deserve further study and that may find further application: the use of free 2-categories as logical syntax, the use of 2-functors as algebraic interpretations, and the observation that epic 2-natural transformations preserve commutative diagrams involving natural transformations. We note explicitly the elegant way the proof plays out with both horizontal and vertical composition, leading to the preservation of an entire class of equations between natural transformations as special cases. We suspect that there may be potential for further applications of this technique involving other classes of coherence diagrams.

For simplicity, we have derived the distributive law for the finite multiset monad and finite distribution monad only; however, we are confident that the proof generalizes to the full Giry monad of measures on an arbitrary measurable spaces, as claimed in [5], using the same technique.

Acknowledgments. The support of the National Science Foundation under grant CCF-2008083 is gratefully acknowledged.

A   Extra Proofs

A.1 $\otimes^L : LF \to FL$ Is a Distributive Law

In this section we verify that $\otimes^L : LF \to FL$ is a distributive law. Because we are working with lists instead of multisets, the task is notationally much less cumbersome and involves only elementary equational reasoning. This is one major benefit of our approach.

Proof (Proof of Theorem 1). We first argue that the diagram

$$
\begin{array}{ccc}
LFX & \xrightarrow{\otimes^L_X} & FLX \\
{\scriptstyle LFh}\downarrow & & \downarrow{\scriptstyle FLh} \\
LFY & \xrightarrow[\otimes^L_Y]{} & FLY
\end{array}
$$

commutes, thus $\otimes^L : LF \to FL$ is a natural transformation.

Let $\bar{f} = [f_1, \ldots, f_n] \in LFX$, $f_i \in FX$. By definition, for $\bar{a} = [a_1, \ldots, a_n] \in X^n$,

$$
\otimes^L_X (\bar{f})(\bar{a}) = \prod_{i=1}^{n} f_i(a_i) \tag{16}
$$

and $\otimes^L_X(\bar{f})(\bar{a}) = 0$ if the lists $\bar{f}$ and $\bar{a}$ are not the same length. The support of $\otimes^L_X(\bar{f})$ is $\mathsf{supp}\,\otimes^L_X(\bar{f}) = \prod_{i=1}^{n} \mathsf{supp}\, f_i$.

If $h : X \to Y$, then $Fh : FX \to FY$. If $f \in FX$ and $b \in Y$, then $Fh(f) \in FY$ and

$$
Fh(f) = \lambda b \in Y . \sum_{\substack{a \in X \\ h(a)=b}} f(a)
$$

$$
Fh(f)(b) = \sum_{\substack{a \in X \\ h(a)=b}} f(a). \tag{17}
$$

Although there may be infinitely many $a \in X$ with $h(a) = b$, since f is of finite support, at most finitely many of these contribute to the sum.

Using (17) and (16),

$$
FLh(\otimes^L_X(\bar{f}))(\bar{b}) = \sum_{\substack{\bar{a} \in X^n \\ Lh(\bar{a})=\bar{b}}} \otimes^L_X(\bar{f})(\bar{a}) = \sum_{\substack{\bar{a} \in X^n \\ Lh(\bar{a})=\bar{b}}} \prod_{i=1}^{n} f_i(a_i). \tag{18}
$$

On the other hand, using (16) and (17),

$$
\otimes^L_Y(LFh(\bar{f}))(\bar{b}) = \prod_{i=1}^{n} Fh(f_i)(b_i) = \prod_{i=1}^{n} \sum_{\substack{a \in X \\ h(a)=b_i}} f_i(a). \tag{19}
$$

The right-hand sides of (18) and (19) are equal by the distributive law of semirings, thus

$$
\otimes^L_Y(LFh(\bar{f}))(\bar{b}) = FLh(\otimes^L_X(\bar{f}))(\bar{b}).
$$

As $\bar{f}$ and $\bar{b}$ were arbitrary,

$$
\otimes^L_Y \circ LFh = FLh \circ \otimes^L_X.
$$

We must show that the diagrams of Fig. 1 commute. Let us start with the easy ones. For $f \in FX$,

$$\otimes_X^L (\eta_{FX}^L(f))$$
$$= \otimes_X^L([f]) = \lambda[a] . f(a)$$
$$= \lambda y . \sum_{\eta_X^L(a)=y} f(a) = F\eta_X^L(f).$$

For $\bar{a} = [a_1, \ldots, a_n] \in LX$,

$$\otimes_X^L (L\eta_X^F(\bar{a}))$$
$$= \otimes_X^L([\eta_X^F(a_1), \ldots, \eta_X^F(a_n)])$$
$$= \otimes_X^L([\lambda b . [b = a_1], \ldots, \lambda b . [b = a_n]])$$
$$= \lambda[b_1, \ldots, b_n] \in X^n . \prod_{i=1}^{n}(\lambda b . [b = a_i])(b_i)$$
$$= \lambda[b_1, \ldots, b_n] \in X^n . \prod_{i=1}^{n} [b_i = a_i]$$
$$= \lambda\bar{b} \in X^n . [\bar{b} = \bar{a}]$$
$$= \eta_{LX}^F(\bar{a}).$$

For the top-left diagram of Fig. 1, we need to show

$$F\mu_X^L \circ \otimes_{LX}^L \circ L\otimes_X^L = \otimes_X^L \circ \mu_{FX}^L.$$

Let $[\bar{f}_1, \ldots, \bar{f}_n] \in LLFX$, where $\bar{f}_i = [f_{i1}, \ldots, f_{ik_i}]$, $1 \le i \le n$. Then

$$F\mu_X^L(\otimes_{LX}^L(L \otimes_X^L ([\bar{f}_1, \ldots, \bar{f}_n])))$$
$$= F\mu_X^L(\otimes_{LX}^L([\otimes_X^L (\bar{f}_1), \ldots, \otimes_X^L(\bar{f}_1)]))$$
$$= F\mu_X^L(\otimes_{LX}^L([\lambda\bar{a}_1 . \prod_{i=1}^{k_1} f_{1i}(a_{1i}), \ldots, \lambda\bar{a}_n . \prod_{i=1}^{k_n} f_{1i}(a_{1i})])))$$
$$= F\mu_X^L(\lambda[\bar{a}_1, \ldots, \bar{a}_n] . \prod_{i=1}^{n}\prod_{j=1}^{k_i} f_{ij}(a_{ij}))$$
$$= \lambda y . \sum_{\mu_X^L(x)=y} (\lambda[\bar{a}_1, \ldots, \bar{a}_n] . \prod_{i=1}^{n}\prod_{j=1}^{k_i} f_{ij}(a_{ij}))(x)$$
$$= \lambda[a_{11}, \ldots, a_{1k_1}, \ldots, a_{n1}, \ldots, a_{nk_n}] . \prod_{i=1}^{n}\prod_{j=1}^{k_i} f_{ij}(a_{ij})$$
$$= \otimes_X^L([f_{11}, \ldots, f_{1k_1}, \ldots, f_{n1}, \ldots, f_{nk_n}])$$
$$= \otimes_X^L(\mu_{FX}^L([\bar{f}_1, \ldots, \bar{f}_n])).$$

Finally, for the bottom-left diagram of Fig. 1, we need to show

$$\mu_{LX}^F \circ F \otimes_X^L \circ \otimes_{FX}^L = \otimes_X^L \circ L\mu_X^F.$$

Let $\bar{f} = [f_1, \ldots, f_n] \in LFFX$, $f_i \in FFX = \mathbb{R}^{\mathbb{R}^X}$. One path in the diagram gives

$$\mu_{LX}^F(F \otimes_X^L (\otimes_{FX}^L(\bar{f})))$$

$$= \mu_{LX}^F(F \otimes_X^L (\lambda \bar{g}. \prod_{i=1}^n f_i(g_i)))$$

$$= \mu_{LX}^F(\lambda y. \sum_{\otimes_X^L(x)=y} (\lambda \bar{g}. \prod_{i=1}^n f_i(g_i))(x))$$

$$= \mu_{LX}^F(\lambda y. \sum_{\otimes_X^L(\bar{g})=y} \prod_{i=1}^n f_i(g_i))$$

$$= \lambda a \in LX. \sum_{g \in FLX} (\lambda y. \sum_{\otimes_X^L(\bar{g})=y} \prod_{i=1}^n f_i(g_i))(g) \cdot g(a)$$

$$= \lambda a \in LX. \sum_{g \in FLX} (\sum_{\otimes_X^L(\bar{g})=g} \prod_{i=1}^n f_i(g_i)) \cdot g(a)$$

$$= \lambda a \in LX. \sum_{\bar{g} \in (FX)^n} \prod_{i=1}^n f_i(g_i) \cdot (\otimes_X^L(\bar{g}))(a)$$

$$= \lambda [a_1, \ldots, a_n] \in LX. \sum_{\bar{g} \in (FX)^n} \prod_{i=1}^n f_i(g_i) \cdot \prod_{i=1}^n g_i(a_i)$$

$$= \lambda [a_1, \ldots, a_n]. \sum_{\bar{g} \in (FX)^n} \prod_{i=1}^n f_i(g_i) \cdot g_i(a_i). \tag{20}$$

The other path gives

$$\otimes_X^L (L\mu_X^F(\bar{f}))$$

$$= \otimes_X^L([\mu_X^F(f_1), \ldots, \mu_X^F(f_n)])$$

$$= \otimes_X^L([\lambda a \in X. \sum_{g \in FX} f_1(g) \cdot g(a), \ldots,$$

$$\lambda a \in X. \sum_{g \in FX} f_n(g) \cdot g(a)]))$$

$$= \lambda [a_1, \ldots, a_n]. \prod_{i=1}^n \sum_{g \in FX} f_i(g) \cdot g(a_i), \tag{21}$$

and (20) and (21) are equivalent by distributivity of multiplication over addition.

A.2   Conditions of Theorem 3

We must also establish the remaining conditions (13) required for the application of Theorem 3. Again, the proofs involve only elementary equational reasoning.

Theorem 4. *The following diagrams commute:*

$$
\begin{array}{ccc}
LL & \xrightarrow{\ \mu^L\ } & L \\
{\scriptstyle \#\#}\downarrow & & \downarrow{\scriptstyle \#} \\
MM & \xrightarrow{\ \mu^M\ } & M
\end{array}
\qquad\qquad
\begin{array}{ccc}
I & \xrightarrow{\ \eta^L\ } & L \\
{\scriptstyle id}\downarrow & & \downarrow{\scriptstyle \#} \\
I & \xrightarrow{\ \eta^M\ } & M
\end{array}
$$

Proof. For the left-hand diagram, let $\ell_i = [a_{i1}, \ldots, a_{ik_i}]$, $a_{ij} \in X$, $1 \le i \le n$, and let

$$
f_i = \#_X(\ell_i) = \sum_{j=1}^{k_i} \#_X([a_{ij}])
$$

$$
= \sum_{j=1}^{k_i} \lambda a \in X.[a = a_{ij}], \quad 1 \le i \le n.
$$

Recall that $\#\# = \#M \circ L\#$. Then

$$
\#_X(\mu_X^L([\ell_1, \ldots, \ell_n]))
$$
$$
= \#_X([a_{11}, \ldots, a_{1k_1}, \ldots, a_{n1}, \ldots, a_{nk_n}])
$$
$$
= \sum_{i=1}^{n}\sum_{j=1}^{k_i} \#_X([a_{ij}]) = \sum_{i=1}^{n} f_i,
$$

$$
\mu_X^M(\#_{MX}(L\#_X([\ell_1, \ldots, \ell_n])))
$$
$$
= \mu_X^M(\#_{MX}([\#_X(\ell_1), \ldots, \#_X(\ell_n)]))
$$
$$
= \mu_X^M(\#_{MX}([f_1, \ldots, f_n])) = \mu_X^M(\sum_{i=1}^{n} \#_{MX}([f_i]))
$$

$$
= \mu_X^M(\sum_{i=1}^{n} \lambda h \in FX.[h = f_i])
$$

$$
= \lambda a \in X.\sum_{h \in FX}(\sum_{i=1}^{n} \lambda h \in FX.[h = f_i])(h) \cdot h(a)
$$

$$
= \lambda a \in X.\sum_{h \in FX}\sum_{i=1}^{n}[h = f_i] \cdot h(a)
$$

$$
= \lambda a \in X.\sum_{i=1}^{n} f_i(a) = \sum_{i=1}^{n} f_i.
$$

Finally, for the right-hand diagram, for $a \in X$,

$$\#_X(\eta_X^L(a)) = \#_X([a]) = \lambda b \in X . [b = a] = \eta^M(a).$$

$\square$

Lemma 3. *The functors L, F, and M preserve epimorphisms.*

Proof. Actually, all functors defined on Set preserve epimorphisms. By the axiom of choice, every surjection $f : X \to Y$ has a right inverse $f' : Y \to X$, that is, $f \circ f' = \mathsf{id}_Y$. This says that f is a split epimorphism, and all functors G in any category preserve split epimorphisms:

$$Gf \circ Gf' = G(f \circ f') = G\mathsf{id}_Y = \mathsf{id}_{GY}.$$

A split epimorphism f in any category is an epimorphism, since

$$e_1 \circ f = e_2 \circ f \Rightarrow e_1 \circ f \circ f' = e_2 \circ f \circ f' \Rightarrow e_1 = e_2.$$

$\square$

References

1. Affeldt, R., Garrigue, J., Nowak, D., Saikawa, T.: A trustful monad for axiomatic reasoning with probability and nondeterminism. J. Funct. Program. **31**, e17 (2021). https://doi.org/10.1017/S0956796821000137
2. Beck, J.: Distributive laws. In: Eckman, B. (ed.) Seminar on Triples and Categorical Homology Theory. LNM, vol. 80, pp. 119–140. Springer, Heidelberg (1969). https://doi.org/10.1007/BFb0083084
3. Chen, Y., Sanders, J.W.: Unifying probability with nondeterminism. In: Cavalcanti, A., Dams, D. (eds.) FM 2009. LNCS, vol. 5850, pp. 467–482. Springer, Heidelberg (2009). https://doi.org/10.1007/978-3-642-05089-3_30
4. Dahlqvist, F., Parlant, L., Silva, A.: Layer by layer - combining monads. In: Fischer, B., Uustalu, T. (eds.) ICTAC 2018. LNCS, vol. 11187, pp. 153–172. Springer, Cham (2018). https://doi.org/10.1007/978-3-030-02508-3_9
5. Dash, S., Staton, S.: A monad for probabilistic point processes. In: Electronic Proceedings in Theoretical Computer Science, vol. 333, pp. 19–32 (2021). https://doi.org/10.4204/eptcs.333.2
6. den Hartog, J., de Vink, E.P.: Mixing up nondeterminism and probability: a preliminary report. In: Baier, C., Huth, M., Kwiatkowska, M.Z., Ryan, M. (eds.) First International Workshop on Probabilistic Methods in Verification, PROBMIV 1998, Indianapolis, Indiana, USA, 19–20 June 1998 (1998). Electron. Notes Theor. Comput. Sci. **22**, 88–110. https://doi.org/10.1016/S1571-0661(05)82521-6
7. Goy, A., Petrisan, D.: Combining probabilistic and non-deterministic choice via weak distributive laws. In: Hermanns, H., Zhang, L., Kobayashi, N., Miller, D. (eds.) LICS 2020: 35th Annual ACM/IEEE Symposium on Logic in Computer Science, Saarbrücken, Germany, 8–11 July 2020, pp. 454–464 (2020). https://doi.org/10.1145/3373718.3394795

8. Jacobs, B.: From multisets over distributions to distributions over multisets. In: 2021 36th Annual ACM/IEEE Symposium Logic in Computer Science (LICS), pp. 1–13 (2021)

9. Keimel, K., Plotkin, G.D.: Mixed powerdomains for probability and nondeterminism. Log. Methods Comput. Sci. **13** (2017). https://doi.org/10.23638/LMCS-13(1: 2)2017

10. Mislove, M.W.: Nondeterminism and probabilistic choice: obeying the laws. In: Palamidessi, C. (ed.) CONCUR 2000. LNCS, vol. 1877, pp. 350–364. Springer, Heidelberg (2000). https://doi.org/10.1007/3-540-44618-4_26

11. Mislove, M., Ouaknine, J., Worrell, J.: Axioms for probability and nondeterminism. In: Corradini, F., Nestmann, U. (eds.) Proceedings of the 10th International Workshop on Expressiveness in Concurrency, EXPRESS 2003, Marseille, France, 2 September 2003 (2003). Electron. Notes Theor. Comput. Sci. **96**, 7–28. https:// doi.org/10.1016/j.entcs.2004.04.019

12. Varacca, D.: Probability, nondeterminism and concurrency: two denotational models for probabilistic computation. Ph.D. thesis, Aarhus University (2003)

13. Varacca, D., Winskel, G.: Distributing probability over non-determinism. Math. Struct. Comput. Sci. **16**, 87–113 (2006). https://doi.org/10.1017/ S0960129505005074

14. Wang, D., Hoffmann, J., Reps, T.W.: A denotational semantics for low-level probabilistic programs with nondeterminism. In: König, B. (ed.) Proceedings of the Thirty-Fifth Conference on the Mathematical Foundations of Programming Semantics, MFPS 2019, London, UK, 4–7 June 2019 (2019). Electron. Notes Theor. Comput. Sci. **347**, 303–324. https://doi.org/10.1016/j.entcs.2019.09.016

15. Zwart, M.: On the non-compositionality of monads via distributive laws. Ph.D. thesis, Oxford University (2020)

16. Zwart, M., Marsden, D.: No-go theorems for distributive laws. Log. Methods Comput. Sci. **18** (2022). https://doi.org/10.46298/lmcs-18(1:13)2022

Minimal Depth Distinguishing Formulas Without Until for Branching Bisimulation

Jan Martens[✉][ID] and Jan Friso Groote[ID]

Eindhoven University of Technology, Eindhoven, The Netherlands
{j.j.m.martens,j.f.groote}@tue.nl

Abstract. In [16] an algorithm for distinguishing formulas for branching bisimulation is proposed. This algorithm has two shortcomings. First, it uses a dedicated until operator, and second, the generated formulas are in no sense minimal. Here we propose a method that generates formulas fitting in the modal mu-calculus, or more precisely, in Hennessy-Milner logic with one regular modality. We provide a polynomial-time algorithm that generates a distinguishing formula that is guaranteed to have minimal depth. Our technical exposition heavily relies on branching apartness, the dual of branching bisimulation.

1 Introduction

Behavioural equivalences are useful to establish whether two systems behave the same. For instance an implementation of a system must behave the same as its specification. Well known behavioural equivalences are strong bisimulation [19] and branching bisimulation [11], but there are many others [10].

When the behaviour of two systems are not equivalent, i.e., when they are apart, it can be very difficult to figure out what the reason is, as in realistic systems it is not uncommon that the difference is only exposed after performing hundreds of equal steps. Fortunately, modal formulas come to the rescue. Following [14] if two systems are not strongly bisimilar, there is a modal formula that is valid in one and not in the other, exemplifying what the reason for inequality is. This formula is called a *distinguishing formula*, or in case of apartness, a *witness*. In [4] an algorithm is given to calculate distinguishing formulas for strong bisimulation.

In order to make a distinguishing formula extra useful it should be minimal in size. Unfortunately, generating a minimal distinguishing formula for strong bisimulation is NP-hard [17], and as a consequence it is NP-hard for many other equivalences as well. But fortunately, it turns out that generating distinguishing formulas that have a minimal depth or have a minimal number of nested negations is computationally tractable [17]. This also holds for the rather weak notion of minimality proposed in [4]. If modal formulas are generated that are not at least minimal depth, they tend to become much larger than needed, and this often renders them effectively unusable.

Algorithms or generation schemes for distinguishing formulas for branching bisimulation are proposed in [7,16]. These formulas contain the until operator

V. Capretta et al. (Eds.): *Logics and Type Systems in Theory and Practice*, LNCS 14560, pp. 188–202, 2024.
https://doi.org/10.1007/978-3-031-61716-4_12

to explicitly deal with the internal action τ [5]. The until operator is not part of the common mu-calculus [2,13,18]. We want to generate distinguishing formulas that fit the modal mu-calculus, and only require a minimal extension of Hennessy-Milner logic. As the modal mu-calculus is very expressive, the proposed until operators can of course be expressed in the modal mu-calculus, but this makes the formulas unduly complex. Our extension only consists of the regular modality $\langle \tau^* \rangle \phi$ with its normal classical semantic interpretation. Besides that we use the abbreviation $\langle \hat{\tau} \rangle \phi$ standing for $\langle \tau \rangle \phi \vee \phi$, but this is not a real extension. This results in a class of formulas preserved under branching bisimulation and corresponds to the modalities used in [6].

Geuvers and Jacobs [9] studied the dual of bisimulation relations, namely the inductive notion of apartness relations. A derivation that proves an apartness relation can be related to a distinguishing formula, proving the characterization with the corresponding logic. This is shown in [7] for strong and weak bisimulation, but for branching bisimulation obtaining the derivation that proves the apartness relation from a distinguishing formula was not straightforward and this question was left open. A possible solution is given by Geuvers and Golov [8] by studying a directed notion of branching bisimilarity. Theorem 12 below provides an alternative answer where no directed notion of apartness is needed and a slightly modified definition of branching apartness makes a direct inductive proof possible.

In this article we propose an algorithm to generate minimal depth witnesses for branching apartness, or minimal depth distinguishing formulas for branching bisimulation. The syntax of the formulas fits Hennessy-Milner formulas with the additional use of the regular modality $\langle \tau^* \rangle \phi$, which, unlike the until operator, is part of the standard syntax of the modal mu-calculus as supported by standard behavioural analysis tools [3,18].

2 Preliminaries

Branching bisimilarity is a behavioural equivalence on labelled transition systems that accounts for internal actions. Its dual, branching apartness, formalises that two processes are not branching bisimilar. We use the set of action labels $Act_\tau = Act \cup \{\tau\}$ where Act is an alphabet and $\tau \notin Act$ is the label for internal steps.

Definition 1 (Labelled transition systems). *A labelled transition system (LTS) is given by a 3-tuple $L = (S, Act_\tau, \rightarrow)$ consisting of:*

- *a finite set of states S,*
- *a finite set of action labels Act_τ containing a silent action $\tau \in Act_\tau$, and*
- *a transition relation $\rightarrow \subseteq S \times Act_\tau \times S$.*

Given a transition relation $\rightarrow$, we write $s \xrightarrow{a} s'$ iff $(s, a, s') \in \rightarrow$. We write $s_0 \twoheadrightarrow s_n$ iff there is a silent path from s_0 to s_n, i.e., there is a sequence of states $s_0, \ldots, s_n \in S$, such that $s_0 \xrightarrow{\tau} s_1 \cdots s_{n-1} \xrightarrow{\tau} s_n$. For states $s, s' \in S$ and an action label $a \in Act_\tau$, we write $s \xrightarrow{(a)} s'$ iff either $s \xrightarrow{a} s'$ or $a = \tau$ and $s = s'$, to indicate silent or non-behaviour.

We define branching bisimilarity using branching bisimulation relations in the standard way.

Definition 2. *(Cf. [13, Def. 2.4.2]) Given an LTS $L = (S, Act_\tau, \rightarrow)$. A symmetric relation $R \subseteq S \times S$ is called a branching bisimulation, iff for all sRt and $s \xrightarrow{a} s'$, then either*

1. $a = \tau$ *and* $s'Rt$*, or*
2. *there are* $t', t'' \in S$ *such that* $t \twoheadrightarrow t' \xrightarrow{a} t''$*,* sRt'*, and* $s'Rt''$*.*

Two states $s, t \in S$ *are said to be branching bisimilar, written as* $s \underset{b}{\leftrightarrow} t$*, iff there is a branching bisimulation R such that sRt.*

Next, we define branching apartness #, the dual of branching bisimilarity. Branching apartness is defined as the union of all i-apartness relations.

Definition 3. *Given an LTS $L = (S, Act_\tau, \rightarrow)$. We characterise branching apartness $\#_i$ for all $i \in \mathbb{N}$ inductively. We define $\#_0 = \emptyset$, and for each $s, t \in S$ we let $s \#_{i+1} t$ iff $s \#_i t$ or:*

1. *there is a path* $s \twoheadrightarrow s' \xrightarrow{a} s''$ *such that for all paths* $t \twoheadrightarrow t' \xrightarrow{(a)} t''$*, either* $s' \#_i t'$ *ór* $s'' \#_i t''$*, or*
2. *symmetrically, there is a path* $t \twoheadrightarrow t' \xrightarrow{a} t''$ *such that for all paths* $s \twoheadrightarrow s' \xrightarrow{(a)} s''$*, either* $t' \#_i s'$ *ór* $t'' \#_i s''$*.*

The relation $\# \subseteq S \times S$*, called the branching apartness relation, is defined by* $\# = \bigcup_{i \in \mathbb{N}} \#_i$*.*

As our LTSs are finite, # is also the fixed point of the chain $\#_0, \#_1, \ldots$. In other words, there is an $i \in \mathbb{N}$ for which $\#_{i+1} = \#_i$ and $\# = \#_i$ for any such i. We first prove that branching apartness is a *proper apartness* relation conform [9].

Theorem 4. *Given an LTS $L = (S, Act, \rightarrow)$. For all $k \in \mathbb{N}$ the relation $\#_k$ is a proper apartness relation, i.e. the relation $\#_k$ is:*

- *irreflexive:* $\neg(s \#_k s)$ *for all* $s \in S$*,*
- *symmetric:* $s \#_k t \implies t \#_k s$ *for all* $s, t \in S$*, and*
- *co-transitive:* $s \#_k t \implies (t \#_k u \lor s \#_k u)$ *for all* $s, t, u \in S$*.*

Proof. The only property which is non-trivial is co-transitivity. We prove the property with induction on k. Assume $\#_i$ is a proper apartness relation for some $i \in \mathbb{N}$, and let $s, t \in S$ be two states such that $s \#_{i+1} t$. In order to prove co-transitivity we assume a state $u \in S$ and show $\neg(s \#_{i+1} u) \implies t \#_{i+1} u$.

By definition, and without loss of generality we can assume that there is a path $s \twoheadrightarrow s' \xrightarrow{a} s''$ such that for all $t \twoheadrightarrow t' \xrightarrow{(a)} t''$ it holds that either $s' \#_i t'$ or $s'' \#_i t''$. Assuming $\neg(s \#_{i+1} u)$, we know there is a path $u \twoheadrightarrow u' \xrightarrow{(a)} u''$ such that $\neg(s' \#_i u')$ and $\neg(s'' \#_i u'')$.

From this path we reason towards the conclusion that $t \#_{i+1} u$. We distinguish the following two cases.

- If $a \neq \tau$ or $u' \neq u''$ then $u' \xrightarrow{a} u''$. For any path $t \twoheadrightarrow t' \xrightarrow{(a)} t''$ it holds that $t' \#_i s'$ or $t'' \#_i s''$. Since we assumed that $\neg(s' \#_i u')$ and $\neg(s'' \#_i u'')$ we know by applying the induction hypothesis that $t' \#_i u'$ or $t'' \#_i u''$. This witnesses $t \#_{i+1} u$.
- If $a = \tau$ and $u' = u''$, then $\neg(s' \#_i u')$ and $\neg(s'' \#_i u')$. Now distinguish:
  - If $u \neq u'$ then $u \twoheadrightarrow u_p \xrightarrow{\tau} u'$ for some $u_p \in S$. For any $t \twoheadrightarrow t'$ it holds that $t' \#_i s'$ or $t' \#_i s''$ since $t' \xrightarrow{(\tau)} t'$. By applying the induction hypothesis and since $\neg(s' \#_i u')$ and $\neg(s'' \#_i u')$ it holds that $t' \#_i u'$. This witnesses $t \#_{i+1} u$.
  - If $u = u'$ then also $\neg(s' \#_i u)$. Since $t \twoheadrightarrow t \xrightarrow{(a)} t$ applying the induction hypothesis gives $t \#_i u$. By definition this also means $t \#_{i+1} u$.

This covers all cases and establishes that $t \#_{i+1} u$, which means that $\#_{i+1}$ is co-transitive. Since the base case $\#_0 = \emptyset$ is trivially co-transitive the induction holds. $\qquad\square$

Next, we prove that branching apartness $\#$ is the dual of branching bisimilarity.

Theorem 5. *Given an LTS $L = (S, Act_\tau, \rightarrow)$ and states $s, t \in S$. Then*

$$s \# t \iff s \not\Leftrightarrow_b t.$$

The proof of this theorem employs the co-transitivity property of apartness relations, and the following lemma, cf. [1, Lemma 6].

Lemma 6. *Given an LTS $L = (S, Act_\tau, \rightarrow)$, states $s, t \in S$, and a branching bisimulation relation R such that sRt. For all states $s' \in S$ if $s \twoheadrightarrow s'$ then there is a state $t \in S$ such that $t \twoheadrightarrow t'$ and $s'Rt'$.*

Proof. Assume $s \twoheadrightarrow s'$, then there is a path $s_0 \xrightarrow{\tau} \ldots \xrightarrow{\tau} s_n$ such that $s_0 = s$ and $s_n = s'$. We prove by induction that for every $i \in [0, n]$ there is a state $t_i \in S$ such that $t \twoheadrightarrow t_i$ and $s_i R t_i$.

For the base case, $i = 0$, the property trivially holds since $t \twoheadrightarrow t$ and sRt. For the induction step $i + 1$ we distinguish two cases conform Definition 2. Since $s_i \xrightarrow{\tau} s_{i+1}$, either

1. $s_{i+1} R t_i$, and since $t_i \twoheadrightarrow t_i$ the property holds, or
2. there are $t', t'' \in S$ such that $t_i \twoheadrightarrow t' \xrightarrow{\tau} t''$, $s_i R t'$ and $s_{i+1} R t''$. So, $t \twoheadrightarrow t''$ and $s_{i+1} R t''$ as had to be shown.

This finishes the induction step and proves the lemma. $\qquad\square$

This lemma allows us to prove Theorem 5.

Proof of Theorem 5. We show this in both directions separately.

$\Rightarrow$ We prove by induction on $i \in \mathbb{N}$ that if $s \#_i t$ then $s \not\Leftrightarrow_b t$. This property trivially holds for $i = 0$. For the inductive case, we assume $s \#_{i+1} t$. This means there is a path $s \twoheadrightarrow s' \xrightarrow{a} s''$ that witnesses for any $t \twoheadrightarrow t' \xrightarrow{(a)} t''$ either $s' \#_i t'$ or $s'' \#_i t''$.

To arrive at a contradiction we assume that $R \subseteq S \times S$ is a branching bisimulation relation such that sRt. By Lemma 6 there is a $t \twoheadrightarrow t'$ such that $s'Rt'$. We distinguish both cases of Definition 2 with respect to the transition $s' \xrightarrow{a} s''$ and show that both cases lead to a contradiction.

1. In the first case $a = \tau$ and $s''Rt'$. In this case both $s'Rt'$ and $s''Rt'$. By applying the induction hypothesis, this means $\neg(s' \#_i t')$, and $\neg(s'' \#_i t')$ contradicting our assumption since neither $t' \#_i s'$ or $t'' \#_i s''$.

2. In the second case there is a path $t' \twoheadrightarrow \tilde{t}' \xrightarrow{a} t''$ such that $s'R\tilde{t}'$, and $s''Rt''$. This means by the induction hypothesis that $\neg(s' \#_i \tilde{t}')$ and $\neg(s'' \#_i t'')$. This also contradicts the assumption.

We showed that both cases lead to a contradiction which means there is no branching bisimulation relation R such that $(s,t) \in R$. This completes the induction, and hence if $s \# t$ then $s \not\Leftrightarrow_b t$.

$\Leftarrow$ We prove this by showing that if $\neg(s \# t)$ then $s \Leftrightarrow_b t$. For this purpose assume $\neg(s \# t)$, and consider the relation

$$R = \{\langle u, v \rangle \mid \neg(u \#_{i+1} v)\}$$

for some $i \in \mathbb{N}$ such that $\#_{i+1} = \#_i$.

We show that R is a branching bisimulation relation. Note that this is enough to prove this part of this Theorem, as if $\neg(s \# t)$ then sRt, and as R is a branching bisimulation relation $s \Leftrightarrow_b t$.

So, consider a pair of states $u, v \in S$ such that uRv. Given a transition $u \xrightarrow{a} u'$, since $\neg(u \#_{i+1} v)$ and $u \twoheadrightarrow u \xrightarrow{a} u'$, by contraposition of Definition 3, there is a $v \twoheadrightarrow v' \xrightarrow{(a)} v''$ such that both $\neg(u \#_i v')$ and $\neg(u' \#_i v'')$. Since $\#_i = \#_{i+1}$ this implies $\langle u, v' \rangle \in R$ and $\langle u', v'' \rangle \in R$. Either $a \neq \tau$ and this is exactly conform item 2 in Definition 2. Otherwise, if $a = \tau$, then either

1. $v' = v''$, then since uRv' and $u'Rv''$, by the fact that R is transitive (Theorem 4) we conclude that $u'Ru$, and since uRv this means $u'Rv$ conform case 1 of Definition 2. Or

2. $v' \neq v''$, in which case $v' \xrightarrow{\tau} v''$, conform case 2 of Definition 2.

The relation R is clearly symmetric and hence, this completes the proof that R is a branching bisimulation relation. $\square$

3   A Hennessy-Milner Theorem

We want to exemplify states that are not branching bisimilar by modal formulas. For strong bisimulation this can be performed using formulas in Hennessy-Milner logic [14]. A nice and natural property is that the distinguishing formulas that we generate stem from a class of formulas that is respected by branching bisimulation. The modality $\langle \tau \rangle \phi$ can therefore not be used, as $\langle a \rangle \langle \tau \rangle true$ is valid in the transition system at the left in Fig. 1 but not in the branching bisimilar transition system at the right. So, some adaptation of Hennessy-Milner logic is required.

Fig. 1. Two simple branching bisimilar transition systems

A possible solution is provided by allowing dedicated until operators [5, 16]. But we do not like this, because until operators are not common within the modal mu-calculus. However, regular formulas are commonly used within the modal mu-calculus [3, 18] and therefore we decide to use $\langle \tau^* \rangle \phi$, which is equivalent to $\mu X.\langle \tau \rangle X \vee \phi$. This formula expresses that zero or more internal actions τ can be performed after which ϕ holds.

This leads to the following logic HML_τ. This logic corresponds with the modalities of the logic introduced in [6].

Definition 7. *(HML$_\tau$) Hennessy-Milner Logic with silent closure is defined over the following syntax:*

$$\phi ::= true \mid \phi \wedge \phi \mid \neg\phi \mid \langle \tau^* \rangle \phi \mid \langle a \rangle \phi,$$

where $a \in Act_\tau$. We write HML$_\tau$ for the set of all formulas over this syntax.

Given an LTS $L = (S, Act_\tau, \rightarrow)$, the semantics $[\![\phi]\!] \subseteq S$ of a formula $\phi \in \mathrm{HML}_\tau$ is the set of states where ϕ holds, defined by:

$$[\![true]\!] = S,$$
$$[\![\phi_1 \wedge \phi_2]\!] = [\![\phi_1]\!] \cap [\![\phi_2]\!],$$
$$[\![\neg\phi]\!] = S \setminus [\![\phi]\!],$$
$$[\![\langle a \rangle \phi]\!] = \{s \mid s \xrightarrow{a} s' \text{ such that } s' \in [\![\phi]\!]\}, \text{ and}$$
$$[\![\langle \tau^* \rangle \phi]\!] = \{s \mid s \twoheadrightarrow s' \text{ such that } s' \in [\![\phi]\!]\}.$$

Other modalities can be expressed in this syntax, e.g., $[a]\phi := \neg\langle a \rangle \neg\phi$ and $\phi_1 \vee \phi_2 := \neg(\neg\phi_1 \wedge \neg\phi_2)$.

Given a formula $\phi \in \mathrm{HML}_\tau$, and two states $s, t \in S$, we write $s \sim_\phi t$, iff $s \in [\![\phi]\!] \iff t \in [\![\phi]\!]$. For a set of formulas $\Phi \subseteq \mathrm{HML}_\tau$ we use the notation $s \sim_\Phi t$ iff $s \sim_\phi t$ for all $\phi \in \Phi$.

If formulas of the shape of $\langle \tau \rangle \phi$ are allowed, branching bisimilar states can be distinguished. Therefore, we instead use the modality $\langle \hat{\tau} \rangle$ which is defined as $\langle \hat{\tau} \rangle \phi := \langle \tau \rangle \phi \vee \phi$. Similar to [6], we define a syntactic restriction in order to characterize exactly branching bisimilar states in a Hennessy-Milner-like theorem.

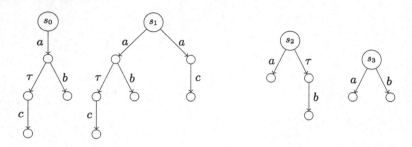

Fig. 2. Two pairs of non branching bisimilar transition systems

Definition 8. *For every $i \in \mathbb{N}$, we define the set $\mathcal{F}_{br}^i \subseteq HML_\tau$ inductively. As base case we take $\mathcal{F}_{br}^0 = \{true\}$, and we define $\mathcal{F}_{br}^{i+1}$ to be the smallest set containing:*

- $\phi \in \mathcal{F}_{br}^{i+1}$, *if* $\phi \in \mathcal{F}_{br}^i$,
- $\neg\phi \in \mathcal{F}_{br}^{i+1}$, *if* $\phi \in \mathcal{F}_{br}^{i+1}$,
- $\phi_1 \wedge \phi_2 \in \mathcal{F}_{br}^{i+1}$, *if* $\phi_1, \phi_2 \in \mathcal{F}_{br}^{i+1}$, *and*
- $\langle\tau^*\rangle(\langle a\rangle\phi_1 \wedge \phi_2) \in \mathcal{F}_{br}^{i+1}$, *where* $a \in Act \cup \{\hat{\tau}\}$, *and* $\phi_1, \phi_2 \in \mathcal{F}_{br}^i$.

Note that with this restriction, the logic is equal to a depth-indexed version of the Hennessy-Milner with Until (HMLU) from [5]. The binary until modality in HMLU can be seen as an abbreviation for $\langle\tau^*\rangle(\langle a\rangle\phi_1 \wedge \phi_2)$.

Example 9. In Fig. 2 two pairs of labelled transition systems are depicted that are not branching bisimilar, but the two LTSs at the left are weakly bisimilar [19]. The formula

$$\langle\tau^*\rangle\langle a\rangle\neg\langle\tau^*\rangle\langle b\rangle true$$

is valid in s_0 but not in s_1, whereas the formula

$$\langle\tau^*\rangle(\langle b\rangle true \wedge \neg\langle\tau^*\rangle\langle a\rangle true)$$

is valid in s_2 but not in s_3. We simplified the distinguishing formulas using $\phi \wedge true = \phi$. Before this simplification they both fit in $\mathcal{F}_{br}^2$.

Fig. 3. The formula $\langle\tau^*\rangle(\langle\tau\rangle true \wedge true)$ holds in s_0 but not in s_1.

Example 10. As already indicated above, we explicitly disallow the $\langle\tau\rangle$ modality and instead use the modality $\langle\hat{\tau}\rangle$. If formulas of the shape of $\langle\tau^*\rangle(\langle\tau\rangle\phi_1 \wedge \phi_2)$ are allowed, branching bisimilar states can be distinguished.

Consider the LTS $\mathcal{A}$ shown in Fig. 3. In this example we see that $s_0 \leftrightarrow_b s_1$. The formula $\phi = \langle\tau^*\rangle(\langle\tau\rangle true \wedge true)$ is a distinguishing formula for s_0 and s_1, since $s_0 \in [\![\phi]\!]$ and $s_1 \notin [\![\phi]\!]$. After restricting to formulas in $\mathcal{F}_{br}^i$ only allowing $\langle\hat{\tau}\rangle$ modalities, there is no distinguishing formula. The modified formula $\phi' = \langle\tau^*\rangle(\langle\hat{\tau}\rangle true \wedge true)$ does not distinguish s_0 and s_1.

Lemma 11. *Let $L = (S, Act_\tau, \rightarrow)$ be an LTS, $s,t \in S$ two states, and $i \in \mathbb{N}$. If $s \not\sim_{\mathcal{F}_{br}^i} t$, then there is a formula $\phi = \langle\tau^*\rangle(\langle a\rangle\phi_1 \wedge \phi_2) \in \mathcal{F}_{br}^i$ such that $s \not\sim_\phi t$.*

Proof. As $s \not\sim_{\mathcal{F}_{br}^i} t$, there is a formula $\phi \in \mathcal{F}_{br}^i$ such that $s \not\sim_\phi t$. Assume there is no formula smaller than ϕ with the same property.

- It is not possible that $\phi = true$ as it cannot distinguish s and t.
- If $\phi = \neg\phi'$ for some $\phi' \in \mathcal{F}_{br}^i$, then also $s \not\sim_{\phi'} t$, and so, ϕ' distinguishes s and t meaning that ϕ was not smallest.
- If $\phi = \phi_1 \wedge \phi_2$, then at least $s \not\sim_{\phi_1} t$ or $s \not\sim_{\phi_2} t$. In either case ϕ was not the smallest formula to distinguish s and t.
- The only option that remains is that $\phi = \langle\tau^*\rangle(\langle a\rangle\phi_1 \wedge \phi_2)$.

$\square$

Theorem 12. *Given an LTS $L = (S, Act_\tau, \rightarrow)$, and states $s,t \in S$. For every $i \in \mathbb{N}$:*

$$s \mathrel{\#_i} t \iff s \not\sim_{\mathcal{F}_{br}^i} t.$$

Proof. We prove this property with induction. The property holds trivially for $i = 0$. Assuming the property holds for some $i \in \mathbb{N}$, we show both directions for $i + 1$.

($\Rightarrow$) Under the assumption $s \mathrel{\#_{i+1}} t$, we build a formula $\phi_{dist} \in \mathcal{F}_{br}^{i+1}$ such that $s \not\sim_{\phi_{dist}} t$. By definition of $s \mathrel{\#_{i+1}} t$, we can without loss of generality assume that there is a path $s \twoheadrightarrow s' \xrightarrow{a} s''$ such that for all $t \twoheadrightarrow t' \xrightarrow{(a)} t''$ it holds that $s' \mathrel{\#_i} t'$ or $s'' \mathrel{\#_i} t''$. Consider the sets of states:

$$T = \{t'' \mid t \twoheadrightarrow t' \xrightarrow{(a)} t'' \text{ where } t'' \mathrel{\#_i} s''\} \text{ and}$$
$$T_\tau = \{t' \mid t \twoheadrightarrow t' \xrightarrow{(a)} t'' \text{ where } t' \mathrel{\#_i} s'\}.$$

By the induction hypothesis there is for each $u \in T$ a formula $\phi_u \in \mathcal{F}_{br}^i$ such that $s'' \not\sim_{\phi_u} u$. Define $\phi'_u = \phi_u$ if $s'' \in [\![\phi_u]\!]$, and $\phi'_u = \neg\phi_u$ otherwise. Now by construction, $\phi'_u \in \mathcal{F}_{br}^i$ and also $s'' \in [\![\phi'_u]\!]$ and $u \notin [\![\phi'_u]\!]$. We define $\Phi_T = \bigwedge_{u \in T} \phi'_u$. Note that $\Phi_T \in \mathcal{F}_{br}^i$, $s'' \in [\![\Phi_T]\!]$ and $[\![\Phi_T]\!] \cap T = \emptyset$. Similarly, for the set T_τ we construct the formula

$$\Phi_{T_\tau} = \bigwedge_{u \in T_\tau, s' \in [\![\phi_u]\!], u \notin [\![\phi_u]\!]} \phi_u,$$

which is the conjunction containing a formula $\phi_u \in \mathcal{F}_{br}^i$ for each $u \in T_\tau$ such that $s' \in [\![\phi_u]\!]$ and $u \notin [\![\phi_u]\!]$. By construction, $\phi_{T_\tau} \in \mathcal{F}_{br}^i$, while $s' \in [\![\Phi_{T_\tau}]\!]$ and $[\![\Phi_{T_\tau}]\!] \cap T_\tau = \emptyset$.

Using the constructed conjuncts Φ_{T_τ} and Φ_T we construct $\phi_{dist} = \langle \tau^* \rangle \Phi$ as follows:

$$\phi_{dist} = \begin{cases} \langle \tau^* \rangle (\langle \hat{\tau} \rangle \Phi_T \wedge \Phi_{T_\tau}) & \text{if } a = \tau, \\ \langle \tau^* \rangle (\langle a \rangle \Phi_T \wedge \Phi_{T_\tau}) & \text{otherwise.} \end{cases}$$

Now we confirm that $s \not\sim_{\phi_{dist}} t$. The path $s \twoheadrightarrow s' \xrightarrow{a} s''$ witnesses that $s \in [\![\phi_{dist}]\!]$, since $s' \in [\![\Phi_{T_\tau}]\!]$ and $s'' \in [\![\Phi_T]\!]$. To see that $t \notin [\![\phi_{dist}]\!]$ we must understand, given a path $t \twoheadrightarrow t'$, that $t' \notin [\![\Phi]\!]$. To do so we case distinguish on whether $t' \#_i s'$:

1. In the first case if $t' \#_i s'$, then $t' \in T_\tau$ and $t' \notin \Phi$.
2. In the second case, necessarily $t'' \#_i s''$ for all $t' \xrightarrow{(a)} t''$, and thus $t'' \notin [\![\Phi_T]\!]$. If $a \neq \tau$ this means that $t' \notin [\![\langle a \rangle \Phi_T]\!]$. If $a = \tau$ then by definition $t' \xrightarrow{(\tau)} t'$ and thus also $t' \notin [\![\Phi_T]\!]$. Hence, also in this case $t' \notin [\![\langle \hat{\tau} \rangle \Phi_T]\!]$.

This covers that in both cases $t' \notin \Phi$, and we conclude that $t \notin [\![\phi_{dist}]\!]$.

($\Leftarrow$) Assume $s \not\sim_{\mathcal{F}_{br}^{i+1}} t$ for some $s, t \in S$. By Lemma 11 there is a formula $\phi = \langle \tau^* \rangle (\langle a \rangle \phi_1 \wedge \phi_2)$ for some $\phi_1, \phi_2 \in \mathcal{F}_{br}^i$ and $a \in Act \cup \{\hat{\tau}\}$, such that $s \not\sim_\phi t$. Since s and t are chosen arbitrarily, we can safely assume that $s \in [\![\phi]\!]$ and $t \notin [\![\phi]\!]$. Since $s \in [\![\phi]\!]$, there is a path $s \twoheadrightarrow s'$ such that $s' \in [\![\langle a \rangle \phi_1 \wedge \phi_2]\!]$. We distinguish on whether $a \in Act$ or $a = \hat{\tau}$.

- In the first case if $a \in Act$, there is a transition $s' \xrightarrow{a} s''$ such that $s'' \in [\![\phi_1]\!]$. We show that the path $s \twoheadrightarrow s' \xrightarrow{a} s''$ witnesses $s \#_{i+1} t$. Since $t \notin [\![\phi]\!]$, we know that for all $t \twoheadrightarrow t'$ it holds that $t' \notin [\![\langle a \rangle \phi_1 \wedge \phi_2]\!]$. Hence, if $t \twoheadrightarrow t' \xrightarrow{(a)} t''$ either $t'' \notin [\![\phi_1]\!]$ or $t' \notin [\![\phi_2]\!]$. We apply our induction hypothesis on $\phi_1, \phi_2 \in \mathcal{F}_{br}^i$ and derive that either $t' \#_i s'$ or $t'' \#_i s''$.
- In the other case if $a = \hat{\tau}$ either $s' \xrightarrow{\tau} s''$ such that $s'' \in [\![\phi_1]\!]$ or $s' \in [\![\phi_1]\!]$. In the first case the path $s \twoheadrightarrow s' \xrightarrow{\tau} s''$ witnesses $s \#_{i+1} t$ as in the case above. In the latter case, when $s' \in [\![\phi_1]\!]$, then for all $t \twoheadrightarrow t'$ since it holds that $t' \notin [\![\phi_1]\!]$ or $t' \notin [\![\phi_2]\!]$, by induction $s' \#_i t'$. If $s = s'$ then because $t \twoheadrightarrow t$ this means $s \#_i t$ and otherwise if $s \neq s'$ then since $s \twoheadrightarrow s'$ there is a path $s \twoheadrightarrow s_p \xrightarrow{\tau} s'$ for some $s_p \in S$. This path $s \twoheadrightarrow s_p \xrightarrow{\tau} s'$ now witnesses $s \#_{i+1} t$.

This completes the inductive case, and proves the property. $\square$

From Theorem 12 and 5 it follows that if two states are not branching bisimilar there is a distinguishing formula in Hennessy-Milner logic with silent closure. This reproves the correspondence as it was also proven in [6], where the same logic is used for a different purpose.

Corollary 13. *(cf. [6, Theorem 1.]) Let $L = (S, Act_\tau, \rightarrow)$ be a labelled transition system. Then for any two states $s, t \in S$:*

$$s \not\Leftrightarrow_b t \iff s \not\sim_\phi t \text{ for some } \phi \in \mathcal{F}_{br}^i \text{ and } i \in \mathbb{N}.$$

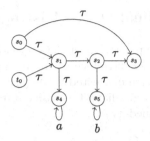

Fig. 4. The LTS $L = (\{s_0, s_1, s_2, s_3, s_4, s_5, t_0\}, \{a, b, \tau\}, \rightarrow)$ in which s_0 and t_0 are branching apart.

A natural question that could arise is whether $\langle \hat{\tau} \rangle$ is required. In other words, whether there is a logic without $\langle \hat{\tau} \rangle \phi$ and with $\langle \tau^* \rangle \phi$ that characterises branching bisimilarity.

To make this statement formal we consider HML_{τ^*} containing the formulas over the syntax:

$$\psi ::= \mathit{true} \mid \psi \wedge \psi \mid \neg \psi \mid \langle \tau^* \rangle \psi \mid \langle a \rangle \psi,$$

where $a \in \mathit{Act}$ and $a \neq \tau$. We prove that formulas over this logic, or any subset thereof, does not characterise branching bisimilarity. This is witnessed by the fact that it cannot distinguish the non branching bisimilar states s_0 and t_0 in Fig. 4.

Observation 14. *The logic HML_{τ^*} does not distinguish all branching bisimilar states.*

Proof. Consider the LTS L shown in Fig. 4. There is no distinguishing formula for s_0 and t_0 as we show below for all formulas $\psi \in \mathrm{HML}_{\tau^*}$ that

$$s_0 \in [\![\psi]\!] \iff t_0 \in [\![\psi]\!].$$

We prove this with structural induction on the formula ψ. For the base case if $\psi = \mathit{true}$ this is trivial. For the induction assume $\psi \in \mathrm{HML}_{\tau^*}$ and the property holds for all smaller formulas $\psi_1 \in \mathrm{HML}_{\tau^*}$. We show that for ψ it holds that $s_0 \in [\![\psi]\!] \iff t_0 \in [\![\psi]\!]$ by distinguishing on the shape of ψ. The cases $\psi = \psi_1 \wedge \psi_2$ and $\psi = \neg \psi_1$ follow directly from the induction hypothesis.

If $\psi = \langle a \rangle \psi_1$ then since $a \neq \tau$ it follows that $s_0 \notin [\![\psi]\!]$ and $t_0 \notin [\![\psi]\!]$. Which leaves the only interesting case when $\psi = \langle \tau^* \rangle \psi_1$. To show $s_0 \in [\![\psi]\!] \implies t_0 \in [\![\psi]\!]$ assume $s_0 \in [\![\psi]\!]$. This means there is a path $s \twoheadrightarrow s'$ such that $s' \in \psi_1$. We distinguish on whether $s' = s$.

- In the case that $s' \neq s_0$ there is also a path $t \twoheadrightarrow s'$ and thus $t \in [\![\psi]\!]$.
- If it is the case that $s' = s_0$ then by induction $t_0 \in [\![\psi_1]\!]$, which means since $t_0 \twoheadrightarrow t_0$ also $t_0 \in [\![\psi]\!]$.

This covers all the cases and proves $s_0 \in [\![\psi]\!] \implies t_0 \in [\![\psi]\!]$. The proof for the other direction is identical, and completes the proof of the theorem. □

4   Computing Minimal Depth Distinguishing Formulas

The proof of Theorem 12 explicitly constructs a distinguishing formula in HML_τ. Based on this construction we introduce an algorithm that computes a minimal depth distinguishing HML_τ formula, listed as Algorithm 1.

The algorithm uses two auxiliary functions. First, it requires the function $\Delta : S \times S \to \mathbb{N} \cup \{\infty\}$, defined as:

$$\Delta(s,t) = \begin{cases} i & \text{if } s \not\#_i t \text{ and } \neg(s \not\#_{i-1} t), \\ \infty & \text{otherwise.} \end{cases}$$

Second, it uses the function δ_i to select suitable distinguishing observations, defined as:

$$\delta_i(s,t) = \{(a, s', s'') \mid s \twoheadrightarrow s' \xrightarrow{a} s'' \text{and for all } t \twoheadrightarrow t' \xrightarrow{(a)} t''$$
$$\text{it holds that } t' \not\#_{i-1} s' \text{ or } t'' \not\#_{i-1} s''\}.$$

Given an input $s, t \in S$ such that $s \not\#_i t$, the algorithm constructs a distinguishing formula as follows. First a pair $(a, s', s'') \in \delta_i(s,t)$ is selected. This pair encodes a path $s \twoheadrightarrow s' \xrightarrow{a} s''$ that witnesses the properties in Definition 3. If this path does not exist then for t such a path necessarily exists, and the negation is returned.

For all states $t', t'' \in S$ such that $t \twoheadrightarrow t' \xrightarrow{a} t''$, the algorithm recursively computes a distinguishing formula that distinguishes s'' from t'', or if that is not possible s' from t'. The conjunction of all these formulas is returned and guarantees by construction that $s \in \llbracket \phi(s,t) \rrbracket$ and $t \notin \llbracket \phi(s,t) \rrbracket$.

In next section we show how to compute Δ in polynomial time. Given a procedure for Δ, we know that this algorithm has polynomial runtime. Indeed, if we use techniques from dynamic programming to not recompute the function with the same arguments, then the function is called at most once for every combination of states.

4.1   Implementation Details and Example

We implemented this method. In this section we show how we compute the relations $\#_i$ for every $i \in \mathbb{N}$. We first compute the LTS modulo branching bisimulation with the more efficient algorithm introduced in [12,15]. On this reduced state space we compute the relations $\#_i$ for every $i \in \mathbb{N}$ by a naive partition refinement algorithm. In this section we introduce this simple partition refinement algorithm.

A partition $\pi \subseteq 2^S$ is a disjoint cover of S, i.e. every element $B \in \pi$ is nonempty, i.e., $B \neq \emptyset$, and for all $B' \in \pi$ it holds that $B' \cap B = \emptyset$ or $B' = B$. A partition π naturally describes an apartness relation $\#_\pi \subseteq S \times S$:

$$\#_\pi = \{(s,t) \mid \forall B \in \pi . s \notin B \vee t \notin B\}.$$

input : A state $s \in S$, and a set $T \subseteq S$ such that $s \# t$ for all $t \in T$.
output: A formula $\phi \in \mathcal{F}_{br}$ such that $s \in [\![\phi]\!]$ and $T \cap [\![\phi]\!] = \emptyset$.

```
1 Function Dist(s, T) is
        /* Return a formula Φ such that s ∈ [[Φ]] and T ∩ [[Φ]] = ∅.        */
2       while T ≠ ∅ do
3           Select t_max ∈ T such that Δ(s, t_max) ≥ Δ(s, t') for all t' ∈ T;
4           φ_{t_max} := φ(s, t_max);
5           Φ := Φ ∧ φ_{t_max};
6           T := T ∩ [[φ_{t_max}]];
7       end
8       return Φ;
9 end
```

input : Two states $s, t \in S$ such that $s \# t$.
output: A formula $\phi \in \mathcal{F}_{br}$ such that $s \in [\![\phi]\!]$ and $t \notin [\![\phi]\!]$.

```
10 Function φ(s, t) is
11      i := Δ(s, t);
12      if δ_i(s, t) = ∅ then
13          return ¬ φ(t, s)
14      Select (a, s', s'') ∈ δ_i(s, t);
15      â = a;
16      if a = τ then
17          â := τ̂;
18      T_τ := {t' | t ↠ t'};
19      T := {t'' | t' ∈ T_τ, t' (a)→ t'' and t'' #_{i-1} s''};
20      Φ_T := Dist(s'', T);
21      T_τ := T_τ ∩ [[⟨â⟩Φ_T]];
22      Φ_{T_τ} := Dist(s', T_τ);
23      return ⟨τ*⟩(⟨â⟩Φ_T ∧ Φ_{T_τ});
24 end
```

Algorithm 1: Minimal-depth distinguishing witnesses.

Given an LTS $L = (S, Act_\tau, \rightarrow)$, an action $a \in Act_\tau$ and two subsets $B, B' \subseteq S$ we define the set $split(B, a, B')$ as all states of S that can follow a silent path to a state in B which reaches B' with an a action:

$$split(B, a, B') = \{s \mid \exists s' \in B, s'' \in B'. s \twoheadrightarrow s' \xrightarrow{a} s'' \text{ such that } a \neq \tau \text{ or } B \neq B'\}.$$

A partition π' is a stable refinement of the partition π iff for every block $C \in \pi'$ it holds that

$$C \subseteq split(B, a, B') \text{ or } C \cap split(B, a, B') = \emptyset$$

for all actions $a \in Act_\tau$ and blocks $B, B' \in \pi$.

We establish the correctness of our partition refinement with the following theorem.

Theorem 15. *Given an LTS $L = (S, Act_\tau, \rightarrow)$, a partition π_i such that $\natural_{\pi_i} = \#_i$ for some $i \in \mathbb{N}$, and states $s, t \in S$ such that $\neg(s \#_i t)$. It holds that $s \#_{i+1} t$ if and only if there are blocks $B, B' \in \pi$ and an action $a \in Act_\tau$ such that $s \in split(B, a, B') \iff t \notin split(B, a, B')$.*

Proof. We prove this in both directions separately.

($\Rightarrow$) Assume $s \#_{i+1} t$. Additionally, without loss of generality assume there is a path $s \twoheadrightarrow s' \xrightarrow{a} s''$ such that for all $t \twoheadrightarrow t' \xrightarrow{(a)} t''$ either $s' \#_i t'$ or $s'' \#_i t''$. We distinguish two cases:

- In the first case assume $a \neq \tau$ or $s' \#_i s''$. Then by definition $s \in split(B, a, B')$ where $B, B' \in \pi$ are the blocks such that $s' \in B$ and $s'' \in B'$.
  Since we know $t' \notin B$ or $t'' \notin B'$, it follows that $t \notin split(B, a, B')$.
- For the remaining case assume $a = \tau$ and $\neg(s' \#_i s'')$. In this case the path $s \twoheadrightarrow s' \xrightarrow{\tau} s''$ does not directly give a pair of blocks B, B' that witnesses the split. In order to obtain this pair we first show that $s \#_i s'$. The path $t \twoheadrightarrow t \xrightarrow{(a)} t$ witnesses $t \#_i s'$, and since we assumed $\neg(s \#_i t)$ by co-transitivity $s' \#_i s$.
  Now, there is a path $s \twoheadrightarrow u \xrightarrow{\tau} u' \twoheadrightarrow s' \xrightarrow{\tau} s''$ such that $\neg(u' \#_i s')$ and $u \#_i u'$. For the blocks $B, B' \in \pi_i$ such that $u \in B$ and $u' \in B'$ it holds that $s \in split(B, \tau, B')$ and $t \notin split(B, \tau, B')$ similar to the first case of this case distinction.

($\Leftarrow$) Assume $s \in split(B, a, B')$ and $t \notin split(B, a, B')$ for some $B, B' \in \pi_i$ and $a \in Act_\tau$. Let $s \twoheadrightarrow s' \xrightarrow{a} s''$ be the path that witnesses $s \in split(B, a, B')$. Let $t \twoheadrightarrow t' \xrightarrow{(a)} t''$ be a path. Since $t \notin split(B, a, B')$, either $t' \notin B$ or $t'' \notin B'$. Distinguish the following cases,

- $a = \tau$ and $\neg(t' \#_i t'')$. Then since $B \neq B'$, either $t' \#_i s'$ or $t'' \#_i s''$.
- otherwise if $a \neq \tau$ or $t' \#_i t''$ then since $t \notin split(B, a, B')$ either $t' \notin B$ or $t'' \notin B'$. In other words $t' \#_i s'$ or $t'' \#_i s''$.

So, in both cases we can conclude $s \#_{i+1} t$, which had to be shown.   □

Example 16. In this example we apply our algorithm to the LTS in Fig. 4. In this LTS the distinguishing behaviour of the states s_0 and t_0 is a τ-action. This results in a bit more complicated distinguishing formula. First we compute the partitions π_0, π_1, π_2 such that $\natural_{\pi_i} = \#_i$.

$$\pi_0 = \{\{t_0, s_0, s_1, s_2, s_3, s_4, s_5\}\}$$
$$\pi_1 = \{\{t_0, s_0, s_1\}, \{s_2, s_5\}, \{s_4\}, \{s_3\}\}$$
$$\pi_2 = \{\{t_0, s_1\}, \{s_0\}, \{s_2\}, \{s_5\}, \{s_4\}, \{s_3\}\}$$

When computing $\phi(s_0, t_0)$, a path in $\delta_2(s_0, t_0)$ is computed. The path from s_0 that witnesses $s_0 \#_2 t_0$ is $s_0 \twoheadrightarrow s_0 \xrightarrow{\tau} s_3$. Now indeed, for all paths $t_0 \twoheadrightarrow t' \xrightarrow{(\tau)} t''$ it holds that either $s_0 \#_1 t'$ or $s_3 \#_1 t''$. The algorithm computes the sets of states reachable by t_0 that can be distinguished from s_3, which is $T = \{t_0, s_1, s_2, s_4, s_5\}$, which is captured by the formulas $\phi_1 = \phi(s_3, s_4) \wedge \phi(s_3, s_5)$. Now compute the set $T_\tau \cap [\![\langle \hat{\tau} \rangle \phi_1]\!] = \{s_2, s_3\}$, which is distinguished by the formula $\phi(s_0, s_2) = \phi(s_0, s_3) = \langle \tau^* \rangle \langle a \rangle \, true$. Hence, the formula computed results in:

$$\phi(s_0, t_0) = \langle \tau^* \rangle (\langle \hat{\tau} \rangle \phi_1 \wedge \phi_2)$$

$$\phi_1 = \phi(s_3, s_4) \wedge \phi(s_3, s_5) = \neg \langle \tau^* \rangle \langle a \rangle \, true \wedge \neg \langle \tau^* \rangle \langle b \rangle \, true$$

$$\phi_2 = \phi(s_0, s_3) = \langle \tau^* \rangle \langle a \rangle \, true.$$

References

1. Basten, A.A.: Branching bisimilarity is an equivalence indeed! Inf. Process. Lett. **58**(3), 141–147 (1996). https://doi.org/10.1016/0020-0190(96)00034-8
2. Bradfield, J.C., Stirling, C.: Modal mu-calculi. In: Blackburn, P., van Benthem, J.F.A.K., Wolter, F. (eds.) Handbook of Modal Logic. Studies in Logic and Practical Reasoning, vol. 3, pp. 721–756. North-Holland (2007). https://doi.org/10.1016/s1570-2464(07)80015-2
3. Bunte, O., et al.: The mCRL2 toolset for analysing concurrent systems. In: Vojnar, T., Zhang, L. (eds.) TACAS 2019. LNCS, vol. 11428, pp. 21–39. Springer, Cham (2019). https://doi.org/10.1007/978-3-030-17465-1_2
4. Cleveland, R.: On automatically explaining bisimulation inequivalence. In: Clarke, E.M., Kurshan, R.P. (eds.) CAV 1990. LNCS, vol. 531, pp. 364–372. Springer, Heidelberg (1991). https://doi.org/10.1007/BFb0023750
5. De Nicola, R., Vaandrager, F.W.: Three logics for branching bisimulation. J. ACM **42**(2), 458–487 (1995). https://doi.org/10.1145/201019.201032
6. Fokkink, W., van Glabbeek, R., de Wind, P.: Divide and congruence: from decomposition of modalities to preservation of branching bisimulation. In: de Boer, F.S., Bonsangue, M.M., Graf, S., de Roever, W.-P. (eds.) FMCO 2005. LNCS, vol. 4111, pp. 195–218. Springer, Heidelberg (2006). https://doi.org/10.1007/11804192_10
7. Geuvers, J.H.: Apartness and distinguishing formulas in Hennessy-Milner logic. In: Jansen, N., Stoelinga, M., van den Bos, P. (eds.) A Journey from Process Algebra via Timed Automata to Model Learning. Lecture Notes in Computer Science, vol. 13560, pp. 266–282. Springer, Cham (2022). https://doi.org/10.1007/978-3-031-15629-8_14
8. Geuvers, J.H., Golov, A.: Directed branching bisimulation via apartness and positive logic. arXiv preprint arXiv:2210.07380 (2022)
9. Geuvers, J.H., Jacobs, B.: Relating apartness and bisimulation. Log. Methods Comput. Sci. **17**(3) (2021). https://doi.org/10.46298/lmcs-17(3:15)2021
10. van Glabbeek, R.J.: The linear time - branching time spectrum I. In: Bergstra, J.A., Ponse, A., Smolka, S.A. (eds.) Handbook of Process Algebra, pp. 3–99. North-Holland/Elsevier (2001). https://doi.org/10.1016/b978-044482830-9/50019-9
11. van Glabbeek, R.J., Weijland, W.P.: Branching time and abstraction in bisimulation semantics. J. ACM **43**(3), 555–600 (1996). https://doi.org/10.1145/233551.233556
12. Groote, J.F., Jansen, D.N., Keiren, J.J.A., Wijs, A.J.: An phO(phmlogphn) algorithm for computing stuttering equivalence and branching bisimulation. ACM Trans. Comput. Log. **18**(2), 13:1–13:34 (2017). https://doi.org/10.1145/3060140
13. Groote, J.F., Mousavi, M.R.: Modeling and Analysis of Communicating Systems. MIT Press (2014). https://mitpress.mit.edu/books/modeling-and-analysis-communicating-systems

14. Hennessy, M., Milner, R.: On observing nondeterminism and concurrency. In: de Bakker, J., van Leeuwen, J. (eds.) ICALP 1980. LNCS, vol. 85, pp. 299–309. Springer, Heidelberg (1980). https://doi.org/10.1007/3-540-10003-2_79

15. Jansen, D.N., Groote, J.F., Keiren, J.J.A., Wijs, A.: An $O(m \log n)$ algorithm for branching bisimilarity on labelled transition systems. In: TACAS 2020. LNCS, vol. 12079, pp. 3–20. Springer, Cham (2020). https://doi.org/10.1007/978-3-030-45237-7_1

16. Korver, H.: Computing distinguishing formulas for branching bisimulation. In: Larsen, K.G., Skou, A. (eds.) CAV 1991. LNCS, vol. 575, pp. 13–23. Springer, Heidelberg (1992). https://doi.org/10.1007/3-540-55179-4_3

17. Martens, J.J.M., Groote, J.F.: Computing minimal distinguishing Hennessy-Milner formulas is NP-hard, but variants are tractable. In: Pérez, G.A., Raskin, J.-F. (eds.) Proceedings of Conference on Concurrency Theory, (CONCUR 2023). LIPIcs, vol. 279, pp. 32:1–32:17. Schloss Dagstuhl - Leibniz-Zentrum für Informatik (2023). https://doi.org/10.4230/LIPIcs.CONCUR.2023.32

18. Mateescu, R., Sighireanu, M.: Efficient on-the-fly model-checking for regular alternation-free mu-calculus. Sci. Comput. Program. **46**(3), 255–281 (2003). https://doi.org/10.1016/S0167-6423(02)00094-1

19. Milner, R.: A Calculus of Communicating Systems. LNCS, vol. 92. Springer, Heidelberg (1980). https://doi.org/10.1007/3-540-10235-3

Relating Apartness and Branching Bisimulation Games

Jurriaan Rot[1(✉)], Sebastian Junges[1], and Harsh Beohar[2]

[1] Institute for Computing and Information Sciences (iCIS), Radboud University, Nijmegen, The Netherlands
jurriaan.rot@ru.nl
[2] Department of Computer Science, University of Sheffield, Sheffield, UK

Abstract. Geuvers and Jacobs (LMCS 2021) formulated the notion of apartness relation on state-based systems modelled as coalgebras. In this context apartness is formally dual to bisimilarity, and gives an explicit proof system for showing that certain states are not bisimilar. In the current paper, we relate apartness to another classical element of the theory of behavioural equivalences: that of turn-based two-player games. Studying both strong and branching bisimilarity, we show that winning configurations for the Spoiler player correspond to apartness proofs, for transition systems that are image-finite (in the case of strong bisimilarity) and finite (in the case of branching bisimilarity).

1 Introduction

Bisimilarity is one of the fundamental notions of equivalence [12], encoding when two states of a labelled transition system (LTS) have the same behaviour. Bisimilarity is well studied in the literature from both logical and game-theoretic viewpoints. For instance, the classical Hennessy-Milner characterisation theorem [5] states that two states of an image-finite LTS are bisimilar if and only if they satisfy the same set of modal formulas. Similarly, the well-known result by Stirling [9] states that two states of an LTS are bisimilar if and only if Duplicator has a winning strategy from this pair of states in the Spoiler/Duplicator bisimulation game. These two viewpoints have almost become a standard in the sense that it is expected that similar characterisation results hold, whenever a new notion of behavioural equivalence is proposed.

Orthogonally to these logical and game-theoretic viewpoints, in the recent work of Geuvers and Jacobs [4], a dual approach to bisimilarity is postulated in terms of *apartness* in transition systems. Instead of describing when two states are behaviourally equivalent, as in bisimilarity, the motive of an apartness relation is in showing *differences* in behaviour. More formally, where bisimilarity is a coinductive characterisation of behavioural equivalence, apartness inductively

This research is partially supported by the Royal Society International Exchange grant (IES\R3\223092). The third author was also partially supported by EPSRC NIA grant EP/X019373/1.

provide a proof system for constructing witnesses of such differences. Geuvers and Jacobs propose a general coalgebraic formulation of apartness, and show how this yields concrete proof systems for deterministic automata, labelled transition systems and streams. They also develop versions of apartness for weak and branching bisimilarity.

This research strand allows us to study connections between modal logic, games and bisimilarity through the lens of apartness. In particular, the Hennessy-Milner theorem says that two states are apart if and only if there is a distinguishing formula, i.e., a formula that holds in one state but not the other. For games, a natural formulation is that two states are apart if and only if Spoiler has a winning strategy. Both results hold by simply observing that bisimilarity is the complement of apartness [4]. However, such an approach is rather implicit: it does not really show how to move between apartness proofs, distinguishing formulas and winning strategies for Spoiler.

The relation between apartness proofs and distinguishing formulas is studied in [3], and revisited in an abstract coalgebraic setting in [11]. In the current paper, we focus on the relation between apartness and bisimulation games. One of the main messages of this paper is the following dichotomy: bisimulations correspond to the winning strategies for Duplicator, while apartness relations correspond to winning strategies for Spoiler. We explicitly relate winning configurations for Spoiler to apartness proofs. We first develop the correspondence between apartness relations and Spoiler strategies for strong bisimilarity, and then move on to branching bisimilarity [13], following the game characterisation in [15]. Our proofs rely on the assumption that Duplicator has only finitely many possible moves; this is true under the assumption that the LTS is image-finite, in the case of strong bisimilarity. For branching bisimilarity, since Duplicator can answer with a sequence of τ moves, we make a stronger assumption for our proof strategy to work: that the LTS is finite.

2   Strong Bisimilarity

The objective of this section is to show that the winning strategy of Spoiler in a strong bisimulation game corresponds to an apartness relation. To this end, we first recall preliminaries on bisimulations [8], games [9,10], and (proper) apartness relations [3,4].

Throughout this section, we fix an LTS, which we take to be a tuple $(X, A, \rightarrow)$ consisting of a set X of states, a set A of actions, and a transition relation $\rightarrow \subseteq X \times A \times X$. We assume our LTS to be *image-finite*: for every state $x \in X$ and label $a \in A$, the set $\{x' \mid x \xrightarrow{a} x'\}$ is finite.

Definition 1. *A symmetric relation $R \subseteq X \times X$ is a (strong) bisimulation if for all $(x, y) \in R$:*

- *if $x \xrightarrow{a} x'$ then $\exists y'. y \xrightarrow{a} y' \wedge x' R y'$.*

Two states $x, y \in X$ are bisimilar, *denoted $x \leftrightarrow y$, iff there exists a bisimulation R such that $x R y$; i.e., $\leftrightarrow = \bigcup \{R \mid R \text{ is a bisimulation}\}$.*

The relation $\leftrightarrow$ is itself a bisimulation, and it is trivially the largest one.

Definition 2. *A symmetric relation $R \subseteq X \times X$ is called an* apartness relation *if it satisfies:*

$$\frac{x \xrightarrow{a} x' \quad \forall y'. y \xrightarrow{a} y' \text{ implies } x' \, R \, y'}{x \, R \, y}$$

Two states $x, y \in X$ are apart, *denoted $x \# y$, iff (x, y) are related in every apartness relation R, i.e., $\# = \bigcap\{R \mid R \text{ is an apartness relation}\}$. We refer to the relation $\#$ simply as* apartness.

Apartness $\#$ is an apartness relation itself. Since apartness is characterised as the *least* relation satisfying the above rule, it provides a proof technique: showing that two states are apart amounts to giving a proof using the rule. Moreover, it suffices to consider *finite* proofs, by the assumption that the LTS is image-finite.

Example 3. Consider the following LTS.

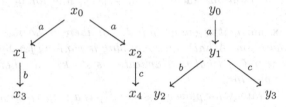

The states x_0 and y_0 are apart, which we can show with following proof tree, where we also explicitly use that the apartness is a symmetric relation:

$$\frac{\dfrac{y_1 \xrightarrow{c} y_3 \quad \neg(x_1 \xrightarrow{c})}{\dfrac{y_1 \# x_1}{x_1 \# y_1}}}{\dfrac{x_0 \xrightarrow{a} x_1 \quad \forall y'. y_0 \xrightarrow{a} y' \text{ implies } x_1 \# y'}{x_0 \# y_0}}$$

In the above proof, $\neg(x_1 \xrightarrow{c})$ means there is no transition of the form $x_1 \xrightarrow{c} x'$, that is, no outgoing c-transition from x_1. This means the universally quantified statement in the rule for apartness vacuously holds.

Theorem 4 (Geuvers and Jacobs [4]). *Apartness is dual to bisimilarity, i.e., $\# = (X \times X) \setminus \leftrightarrow$.*

We next recall the strong bisimulation game of Stirling [9], using notation from [1]. The intuition is Duplicator wants to show that two states are bisimilar, while Spoiler wants to show the difference in their behaviour (i.e., they are apart; see Definition 2). The game is a *turn-based two-player game on a graph* $(C = C_D \uplus C_S, \rightarrow_D, \rightarrow_S)$ where the players are called Duplicator (or just D) and Spoiler (or just S). We refer to the nodes C as *configurations*, to Duplicator moves by $\rightarrow_D \subseteq C_D \times C_S$, and Spoiler moves by $\rightarrow_S \subseteq C_S \times C_D$.

Definition 5 (Strong bisimulation game). *The set of configurations is given by* $C = C_S \uplus C_D$, *where* $C_S = X \times X$ *are Spoiler configurations, ranged over by tuples denoted by* $[x, y]$, *and* $C_D = X \times A \times X \times X$ *are Duplicator configurations, ranged over by tuples of the form* $\langle x, a, x', y \rangle$. *The moves are:*

- *Spoiler can move from a configuration* $[x, y]$ *as follows:*
  1. *to* $\langle x, a, x', y \rangle$ *if there is a transition of the form* $x \xrightarrow{a} x'$;
  2. *to* $\langle y, a, y', x \rangle$ *if there is a transition of the form* $y \xrightarrow{a} y'$;
- *Duplicator can move from a configuration* $\langle x, a, x', y \rangle$ *to* $[x', y']$ *if there exists a transition of the form* $y \xrightarrow{a} y'$.

Spoiler wins a play if and only if Duplicator cannot move. To formalise this, we need a few definitions regarding two-player games.

Definition 6. *We define plays and winning configurations as follows.*

- *A* play *from a Spoiler configuration* $[x, y]$ *is a finite or infinite sequence* $\sigma \in C^* \cup C^\omega$ *of configurations such that* $\sigma_0 = [x, y]$ *and for all* $i > 0$, σ_{i+1} *is a move from* σ_i.
- *A play is* maximal *if it is either infinite, or there is no move possible from the last configuration. A finite maximal play is winning for Spoiler if this last configuration is in* C_D *(that is, Duplicator is stuck); all other maximal plays are won by Duplicator.*
- *A* (positional) strategy *for player* $P \in \{D, S\}$ *is a map* π_P *from configurations* C_P *to moves for player* P *in that configuration. A play* σ *is* consistent *with a strategy* π_P *for player* P *if for all* i *such that* $\sigma_i \in C_P$, *we have that* $\pi_P(\sigma_i) = \sigma_{i+1}$.
- *A strategy* π_P *for player* P *is called* winning *from a configuration* $[x, y] \in C_S$ *if every maximal play that is consistent with* π_P *is winning for* P. *A Spoiler configuration* $[x, y] \in C_S$ *is* winning for player P *if there exists a winning strategy from* $[x, y]$ *for that player. The set of winning configurations of player* P *is called the* winning region *of* P, *and is denoted by* $\mathcal{W}_P$.

Bisimulation games are so-called reachability games for Spoiler, i.e., Spoiler wants to reach a particular set of configurations. Consequentially, they are so-called safety games for Duplicator, who wants to ensure that we do not reach those configurations. To make this explicit, we provide fixed-point characterisations of the winning regions of both players. Consider the usual definition of box $(\Box_D, \Box_S)$ and diamond $(\Diamond_D, \Diamond_S)$ modalities:

$$\Diamond_D W = \{c \in C_D \mid \exists c'.\ c \to_D c' \wedge c' \in W\}$$
$$\Box_D W = \{c \in C_D \mid \forall c'.\ c \to_D c' \text{ implies } c' \in W\},$$

where $W \subseteq C$. The modalities $\Diamond_S$, $\Box_S$ are defined analogously.

Proposition 7. *Let* $(C, \to_D, \to_S)$ *be an alternating two-player game.*

– *The winning region $\mathcal{W}_D$ of Duplicator is the largest set W such that*

$$W \subseteq \Box_S \Diamond_D W.$$

– *The winning region $\mathcal{W}_S$ of Spoiler is the least set W such that*

$$\Diamond_S \Box_D W \subseteq W.$$

Proof. The fixed-point characterisation for winning region of Duplicator is studied in, e.g., [6]. We consider the Spoiler case. We first show that $\mathcal{W}_S$ indeed satisfies $\Diamond_S \Box_D \mathcal{W}_S \subseteq \mathcal{W}_S$. To this end we introduce some notation: for a strategy π and configuration c, we denote by $\pi[c]$ the set of *reachable* configurations, i.e., the set of all $c' \in C_S$ such that c' occurs in a play from c consistent with π. Note that if π is a winning strategy from c, then it is winning from all $c' \in \pi[c]$.

Now suppose $c_0 \in \Diamond_S \Box_D \mathcal{W}_S$, so there exists $c_0 \to_S c'$ such that for all configurations c'' with $c' \to_D c''$, c'' is winning for Spoiler, i.e., $c'' \in \mathcal{W}_S$. We refer to these resulting Spoiler winning configurations as $c_1, \ldots, c_k$ (note that this is a finite set, since the LTS is image-finite; although this is not strictly needed for this side of the argument), and associate winning strategies $\pi_1, \ldots, \pi_k$ for each of these configurations. We construct a strategy π as follows:[1]

$$\pi(c) = \begin{cases} \pi_i(c) & \text{if } c \in \pi_i[c_i] \text{ and for all } j < i . c \notin \pi_j[c_j] \\ c' & \text{if } c = c_0 \text{ and } \forall j . c \notin \pi_j[c_j] \\ \text{any } c_* \in C_D & \text{otherwise} \end{cases}$$

This is a winning strategy from c_0: after one move from Spoiler, every move from Duplicator leads us to a configuration for which the strategy is winning.

Second, suppose W is an arbitrary set such that $\Diamond_S \Box_D W \subseteq W$. We have to show that $\mathcal{W}_S \subseteq W$. This is where we crucially rely on the assumption that our LTS is image-finite. Let $c \in \mathcal{W}_S$ and let π be a winning strategy from c (see Definition 5). Consider the tree of all possible plays from c consistent with π; the edges in this tree are given by Duplicator moves. Since the LTS is image-finite, Duplicator has finitely many choices at each point and therefore this tree is finitely branching. The crux is that it is also finite depth; for suppose it is not, then by König's lemma there would be an infinite play, which is winning for Duplicator and contradicts that π is winning for Spoiler from c. Thus, the length of maximal plays from c consistent with π is bounded by some $n \in \mathbb{N}$.[2]

Now, for each n, let $\mathcal{W}_S^n$ be the set of configurations $c \in \mathcal{W}_S$ for which there is a winning strategy π from c such that the length of the longest maximal play consistent with π is at most n. We prove $\mathcal{W}_S^n \subseteq W$ by induction on n. By the above argument, $\mathcal{W}_S = \bigcup_{n \in \mathbb{N}} \mathcal{W}_S^n$, so then we are done.

---

[1] The challenge in combining the strategies $\pi_1, \ldots, \pi_k$ is that there may be overlap between the reachable sets $\pi_1[c_1], \ldots, \pi_k[c_k]$. This issue arises specifically because we only consider *positional* strategies; if the first move (i.e., a choice of configuration c_i) is recorded, one can stick to the corresponding strategy π_i.

[2] Here the length of a play is the number of Spoiler configurations that appear in it.

1. In the base case, the longest maximal play consistent with a winning strategy is of the form $c \to_S c'$ and Duplicator is stuck. Then clearly $c \in \Diamond_S \Box_D W$. Since $\Diamond_S \Box_D W \subseteq W$ we are done.

2. For the inductive step, suppose $c \in \mathcal{W}_S^n$. Let π be a witnessing strategy where the longest maximal play consistent with π has length at most n, and suppose $\mathcal{W}_S^{n-1} \subseteq W$. Take any c' such that $\pi(c) \to_D c'$. Then π is winning in c' and every maximal play consistent with it has length at most $n-1$, so $c' \in \mathcal{W}_S^{n-1}$. By the induction hypothesis we get $c' \in W$. Since $c \to_S \pi(c) \to_D c'$ we get $c \in \Diamond_S \Box_D W$ and since $\Diamond_S \Box_D W \subseteq W$ we get $c \in W$.   □

Remark 8. The above proof seems non-constructive: we used König's lemma in a proof by contradiction, to argue that maximal plays consistent with a Spoiler winning strategy are bounded. This boundedness is based on the assumption that the LTS is image-finite. To avoid this use of König's lemma, one can assume the LTS to be *finite state*; then the length of maximal plays consistent with a winning strategy for Spoiler are bounded by the number of possible configurations.

The fixed-point characterisation of the winning region of Spoiler is very close to apartness. Indeed, instantiating the fixed-point characterisation for the winning region of Spoiler with the moves of the strong bisimulation game, we recover the definition of apartness. Thus we obtain:

Theorem 9. *Two states x, y of an LTS are apart (i.e. $x \# y$) iff Spoiler has a winning strategy from the configuration $[x, y]$ in the strong bisimulation game.*

The dual version, observed by Stirling, can be similarly obtained from the fixed-point characterisation of the winning region of Duplicator.

Theorem 10 ([9, Proposition 1]). *States $x, y \in X$ are bisimilar iff $[x, y]$ is winning for Duplicator in the strong bisimulation game.*

We recover the following from Theorem 10, Theorem 9 and Theorem 4.

Corollary 11 ([10]). *The bisimulation game is determined, i.e., every configuration is winning for exactly one of the two players.*

3   Branching Bisimilarity

In this section, we study LTSs with silent (internal) actions and consider branching bisimilarity [13] as the notion of behavioural equivalence. We recall the branching bisimulation game of [15] and connect Spoiler winning positions to the notion of branching apartness [4].

Throughout this section, we again fix an LTS $(X, A, \to)$ and assume that the set of labels contains A contains a distinguished silent action $\tau \in A$. We use α, β to range over A, and a, b for labels in $A \setminus \{\tau\}$. We write $x \Longrightarrow x'$ if there is a sequence of τ steps from x to x'. We use $x \xrightarrow{(\alpha)} x'$ to denote that either (1) $x \xrightarrow{\alpha} x'$ or (2) both $\alpha = \tau$ and $x = x'$. We assume that X is *finite*; as a consequence, $\Longrightarrow$ is finitely branching, that is, for each $x \in X$, the set $\{x' \mid x \Longrightarrow x'\}$ is finite.

Definition 12. *A symmetric relation $R \subseteq X \times X$ is a* branching bisimulation *if for all $(x, y) \in R$:*

- *if $x \xrightarrow{\alpha} x'$ then $\exists y', y''. y \Longrightarrow y' \xrightarrow{(\alpha)} y'' \wedge xRy' \wedge x'Ry''$.*

Two states $x, y \in X$ are branching bisimilar, *denoted $x \leftrightarrow_b y$, iff there exists a bisimulation R such that $x R y$.*

There is, accordingly, a natural notion of apartness [4].

Definition 13. *A* branching apartness relation *$R \subseteq X \times X$ is a symmetric relation satisfying the following rules.*

$$\frac{x \xrightarrow{\alpha} x' \quad \forall y', y''. y \Longrightarrow y' \xrightarrow{(\alpha)} y'' \text{ implies } (x \, R \, y' \vee x' \, R \, y'')}{x \, R \, y}$$

As usual we say x, y are branching apart, *denoted $x \mathrel{\#_b} y$, iff they are related by every branching apartness relation R.*

Example 14. Consider the LTS with silent action as given below.

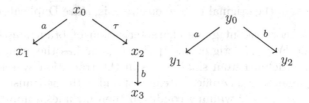

The states x_0 and y_0 are branching apart, as can be shown with the following proof tree.

$$\frac{x_0 \xrightarrow{\tau} x_2 \quad \forall y, y'. y_0 \text{ implies } y \xrightarrow{(\tau)} y' \text{ implies } \dfrac{\dfrac{\dfrac{y_0 \xrightarrow{a} y_1}{y_0 \mathrel{\#_b} x_2}}{(x_0 \mathrel{\#_b} y_0 \vee x_2 \mathrel{\#_b} y_0)}}{(x_0 \mathrel{\#_b} y \vee x_2 \mathrel{\#_b} y')}}{x_0 \mathrel{\#_b} y_0}$$

The following game for branching bisimilarity comes from [15], which is another turn-based two-player game. However, in the branching bisimulation game, Spoiler has two types of Spoiler moves or transition relations from its configurations. Contrary to the (strong) bisimulation game, the following game is no longer alternating, although it is "almost": after the first move from Spoiler, single moves of Duplicator are alternated with two consecutive moves from Spoiler.

Definition 15 (Branching bisimulation game). *The set of configurations is given by $C = C_S \uplus C_D$, where $C_S = X^2 \cup X^5$ are Spoiler configurations, ranged over by tuples denoted by $[x, y]$ and $[x, x', y, y', y'']$ respectively; and $C_D = X \times A \times X \times X$ are Duplicator configurations, ranged over by tuples of the form $\langle x, a, x', y \rangle$. The moves are:*

- *Spoiler can move as follows:*
    1. *from $[x, y]$ to $\langle x, \alpha, x', y \rangle$ if there is a transition of the form $x \xrightarrow{\alpha} x'$;*
    2. *from $[x, y]$ to $\langle y, \alpha, y', x \rangle$ if there is a transition of the form $y \xrightarrow{\alpha} y'$;*
    3. *from $[x, x', y, y', y'']$ to $[x, y']$ or to $[x', y'']$;*
- *Duplicator can move from a configuration $\langle x, \alpha, x', y \rangle$ to $[x, x', y, y', y'']$ if there exist transitions of the form $y \Longrightarrow y' \xrightarrow{(\alpha)} y'$.*

We model the Spoiler moves in Items 1 and 2 as a relation $\to_{S,1} \subseteq X^2 \times C_D$, and the Spoiler moves in Item 3 as a relation $\to_{S,2} \subseteq X^5 \times X^2$. Similarly, we write $c \to_D c'$ if there is a move from c to c' by Duplicator.

Plays and winning configurations are defined as in the strong bisimulation game.

In the branching bisimilarity game, Duplicator can answer with a sequence of τ-steps followed by an actual α-transition (or no transition at all, if $\alpha = \tau$). The key idea is to return the relevant information of this answer to Spoiler: the state just before and just after the α-transition. Spoiler can then choose which one to proceed with.

Remark 16. As we assume that X is finite, the set of Duplicator moves is also finite. Note that, contrary to the previous section, it does not suffice to assume image-finiteness of the original LTS to ensure this, since Duplicator can use $\Longrightarrow$.

Remark 17. A more recent game characterisation of branching bisimilarity is proposed by de Frutos-Escrig et al. [1]. That game has the advantage of being *local*: moves are defined from single steps in the transition system, as opposed to the above game, where Duplicator can respond with the transitive closure. To deal with divergence, the winning condition then includes a non-trivial liveness property. A correspondence between such games and apartness is left for future work; this could perhaps be based on an adapted "one-step" apartness rule.

Just like in the case of strong bisimulation, we provide a fixed-point characterisation of the winning regions of Duplicator and Spoiler. Notice the variety of box and diamond modalities, one for each type of Spoiler moves $\to_{S,i}$ (for $i \in \{1, 2\}$). We focus on Spoiler only.

Proposition 18. *In the branching bisimulation game, the winning region $\mathcal{W}_S$ of Spoiler is the least set W such that*

$$\Diamond_{S,1} \Box_D \Diamond_{S,2} W \subseteq W .$$

The proof is analogous to that of Proposition 7, with the key difference being that Spoiler has the "extra" move after every Duplicator move. When proving that $\Diamond_{S,1} \Box_D \Diamond_{S,2} \mathcal{W}_D \subseteq \mathcal{W}_D$, one extends the winning strategies with a first step as before; but this now includes two moves from Spoiler. In the proof that $\mathcal{W}_D \subseteq W$ whenever $\Diamond_{S,1} \Box_D \Diamond_{S,2} W \subseteq W$, the key is that Duplicator has finitely many possible moves, since the original LTS is finite; this means that the notion of longest maximal play consistent with a Spoiler strategy is once again well-defined.

The above characterisation helps in establishing a correspondence between Spoiler winning configurations and apartness proofs.

Theorem 19. *For any $x, y \in X$, we have that x and y are branching apart iff $[x, y]$ is winning for Spoiler in the branching bisimulation game.*

Proof. We prove the implication from left to right by induction on the proof tree of $x \#_b y$. Note that symmetry of the apartness relation corresponds to the Spoiler player having the choice which state to play from a pair. Suppose the rule is applied to conclude $x \#_b y$, with premise $x \xrightarrow{\alpha} x'$, so that for all $y \Longrightarrow y' \xrightarrow{\alpha} y''$ we have $x \#_b y'$ or $x' \#_b y''$. The induction hypothesis tells us that for each of these transitions, $[x, y']$ or $[x', y'']$ are winning for Spoiler (note that the base case is when there are no such transitions, Duplicator is stuck and Spoiler wins immediately). We have to show that $[x, y]$ is winning for Spoiler.

Indeed, Spoiler can move to $\langle x, \alpha, x', y \rangle$, and Duplicator has to answer $[x, x', y, y', y'']$ based on a transition $y \Longrightarrow y' \xrightarrow{\alpha} y''$. At this point Spoiler can move to $[x, y']$ or $[x', y'']$, one of which is a winning position.

For the converse, it suffices to show that the apartness relation $\#_b$ satisfies $\Diamond_{S,1} \Box_D \Diamond_{S,2} \#_b \subseteq \#_b$. By Proposition 18, we then get the desired implication. Indeed, suppose that $c \to_{S,1} c'$ and for any Duplicator move $c' \to_D c''$ there exists a Spoiler move $c'' \to_{S,2} c''' \in C_S$ such that $c''' \in \#_b$. Then by the definition of the moves, we can apply the apartness proof rule (possibly first with an application of symmetry) to obtain $c \in \#_b$. □

4 Future Work

The study of apartness as a dual to bisimilarity has been studied only recently [4] (although the notion of apartness for coalgebras is older, as explained in *op. cit.* which cites unpublished work from Jacobs written in 1995). In the current work we have connected apartness to games by relating Spoiler strategies with apartness proofs.

One notable limitation of our approach is that we assumed that the underlying LTS is image-finite, and in the case of branching bisimilarity even finite-state. In fact, in the proof of the characterisation of winning regions as a reachability game, we made use of a proof by contradiction and an appeal to König's lemma just to achieve a usable notion of size that allows us to carry out induction. Notice that in the general case, even apartness proofs will not be finite anymore. We note that one way around the problem of being finite-state might be to adopt the "one-step" games of de Frutos-Escrig et al., see Remark 17.

We have only analysed strong and branching bisimilarity for LTSs. One direction for future work is to develop similar results for other forms of bisimulation relations like weak bisimulation and branching bisimulation with explicit divergence. The former can already be handled by our results from Sect. 2 by working on the 'saturated' transition relation $\twoheadrightarrow \subseteq X \times A^* \times X$ instead of single step transition relation $\to$. In particular, $x \xrightarrow{w} x'$ iff x' is reachable from x under observation $w \in A^*$ with τ-steps interspersed between each observable step in A. As long as the state space is finite (the restriction required in the section on branching bisimilarity), Proposition 7 remains applicable.

Another direction for future is to try and extend these ideas to a more general coalgebraic framework. Thus, it would be interesting to relate apartness to existing work on coalgebraic games [2,6,7]; in particular, the work [7,14] that explicitly connect Spoiler strategies to distinguishing formulas. For instance, in [7], the authors give procedures to compute Spoiler strategies for a bisimulation game and construction of a distinguishing formula from a Spoiler strategy (both at the levels of coalgebras). Moreover, a Spoiler strategy is given by a pair of functions (instead of an apartness proof): one modelling the smallest index when two states are separated in the fixed-point computation of bisimilarity; while the other encodes the moves of Spoiler from a given pair of states in the bisimulation game. We leave the general coalgebraic study of the connection between apartness, distinguishing formulas and games for future work.

References

1. de Frutos-Escrig, D., Keiren, J.J.A., Willemse, T.A.C.: Games for bisimulations and abstraction. Log. Methods Comput. Sci. **13**(4) (2017)
2. Ford, C., Milius, S., Schröder, L., Beohar, H., König, B.: Graded monads and behavioural equivalence games. In: LICS, pp. 61:1–61:13. ACM (2022)
3. Geuvers, H.: Apartness and distinguishing formulas in Hennessy-Milner logic. In: Jansen, N., Stoelinga, M., van den Bos, P. (eds.) A Journey from Process Algebra via Timed Automata to Model Learning. Lecture Notes in Computer Science, vol. 13560, pp. 266–282. Springer, Cham (2022). https://doi.org/10.1007/978-3-031-15629-8_14
4. Geuvers, H., Jacobs, B.: Relating apartness and bisimulation. Log. Methods Comput. Sci. **17**(3) (2021)
5. Hennessy, M., Milner, R.: Algebraic laws for nondeterminism and concurrency. J. ACM **32**, 137–161 (1985)
6. Komorida, Y., Katsumata, S., Hu, N., Klin, B., Hasuo, I.: Codensity games for bisimilarity. In: LICS, pp. 1–13. IEEE (2019)
7. König, B., Mika-Michalski, C., Schröder, L.: Explaining non-bisimilarity in a coalgebraic approach: games and distinguishing formulas. In: Petrişan, D., Rot, J. (eds.) CMCS 2020. LNCS, vol. 12094, pp. 133–154. Springer, Cham (2020). https://doi.org/10.1007/978-3-030-57201-3_8
8. Park, D.: Concurrency and automata on infinite sequences. In: Deussen, P. (ed.) GI-TCS 1981. LNCS, vol. 104, pp. 167–183. Springer, Heidelberg (1981). https://doi.org/10.1007/BFb0017309
9. Stirling, C.: Modal and temporal logics for processes. LFCS ECS-LFCS-92-221, The University of Edinburgh (1992)
10. Stirling, C.: Bisimulation, modal logic and model checking games. Log. J. IGPL **7**(1), 103–124 (1999)
11. Turkenburg, R., Beohar, H., Kupke, C., Rot, J.: Forward and backward steps in a fibration. In: CALCO. LIPIcs, vol. 270, pp. 6:1–6:18. Schloss Dagstuhl - Leibniz-Zentrum für Informatik (2023)
12. Glabbeek, R.J.: The linear time—branching time spectrum II. In: Best, E. (ed.) CONCUR 1993. LNCS, vol. 715, pp. 66–81. Springer, Heidelberg (1993). https://doi.org/10.1007/3-540-57208-2_6
13. van Glabbeek, R.J., Weijland, W.P.: Branching time and abstraction in bisimulation semantics. J. ACM **43**(3), 555–600 (1996)

14. Wißmann, T., Milius, S., Schröder, L.: Quasilinear-time computation of generic modal witnesses for behavioural inequivalence. Log. Methods Comput. Sci. **18**(4) (2022)
15. Yin, Q., Fu, Y., He, C., Huang, M., Tao, X.: Branching bisimilarity checking for PRS. In: Esparza, J., Fraigniaud, P., Husfeldt, T., Koutsoupias, E. (eds.) ICALP 2014. LNCS, vol. 8573, pp. 363–374. Springer, Heidelberg (2014). https://doi.org/10.1007/978-3-662-43951-7_31

Fixed Point Theorems in Computability Theory

Sebastiaan A. Terwijn$^{(\boxtimes)}$

Department of Mathematics, Radboud University Nijmegen,
P.O. Box 9010, 6500 GL Nijmegen, The Netherlands
terwijn@math.ru.nl

Abstract. We give a quick survey of the various fixed point theorems in computability theory, partial combinatory algebra, and the theory of numberings, as well as generalizations based on those. We also point out several open problems connected to these.

1 Introduction

Let φ_e denote the e-th partial computable (p.c.) function. Then Kleene's recursion theorem simply states that for every computable function f there exists a number $e \in \omega$ such that

$$\varphi_{f(e)} = \varphi_e.$$

We can think of e as a fixed point of f, not literally in the sense that $f(e) = e$, but at the level of codes of p.c. functions.

The proof of the recursion theorem is very short, and it has an air of mystery. In Kleene's original paper [18], which is about ordinal notations, it is somewhat hidden at the end of Sect. 2, where it only occupies two cryptic lines, but even when written out it can be done in three or four lines (taking the S-m-n-theorem for granted). The fact that the proof is not very illuminating is perhaps due to the fact that it does not occur naturally in the context of p.c. functions. Kleene found the fixed point theorem in the lambda calculus, where it does occur in a natural way, and then translated it to computability theory to obtain the recursion theorem.[1] The following analogy between lambda calculus and computability theory may be helpful. To simplify matters we write $n \sim m$ for $\varphi_n = \varphi_m$, and we define (partial) application of numbers as $nm = \varphi_n(m)$. The left part of the following table follows Barendregt [6, 2.1.5].

---

[1] Explanations of the recursion theorem, elaborating on its short proof, are given in Owings [25] and Odifreddi [23].

V. Capretta et al. (Eds.): *Logics and Type Systems in Theory and Practice*, LNCS 14560, pp. 214–224, 2024.
https://doi.org/10.1007/978-3-031-61716-4_14

λ-calculus	computability
λ-terms	$n \in \omega$
FG	$nm = \varphi_n(m)$
fixed point theorem:	recursion theorem:
$\forall F\, \exists X\ FX = X$	$\forall f\, \exists x\ fx \sim x$
Proof: $W = \lambda x.F(xx)$	Proof: $bx \sim f(xx)$
$WW = F(WW)$ $\square$	$bb \sim f(bb)$ $\square$

That there exists b such that $bx \sim f(xx)$ (i.e. $\varphi_{\varphi_b(x)} = \varphi_{f(\varphi_x(x))}$) follows from the S-m-n-theorem.

The recursion theorem is a fundamental result of computability theory that has found many applications, even extending beyond pure computability theory, e.g. in set theory. The first application was to develop a theory of constructive ordinals, for which the recursion theorem is indispensable. The theorem and its many applications were excellently reviewed by Kleene's student Yiannis Moschovakis in [22]. The present quick survey is by no means intended to replace that much larger one, but we have a slightly different focus, especially with the view from combinatory algebra, and we will also discuss more recent results. We will not be including any proofs (apart from the proof above of the recursion theorem itself, and some short arguments), but mainly pointers to the literature. We will also not discuss applications, which is the main focus of [22].

Our notation for basic notions in computability theory is mostly standard. As already mentioned, φ_e denotes the e-th partial computable (p.c.) function, in some standard numbering of the p.c. functions. P.c. functions are denoted by lower case Greek letters, and (total) computable functions by lower case Roman letters. ω denotes the natural numbers. W_e denotes the e-th computably enumerable (c.e.) set, which is defined as the domain of φ_e. We write $\varphi_e(n)\downarrow$ if this computation is defined, and $\varphi_e(n)\uparrow$ otherwise. $\emptyset'$ denotes the halting set. For unexplained notions we refer to Odifreddi [23] or Soare [28]. Our presentation of partial combinatory algebra follows van Oosten [24].

2   The Second Recursion Theorem

The simple version of the recursion theorem stated at the beginning of this paper is fully effective, which means that we can compute the fixed point e effectively from a code of f. One way to state this is as follows: Let $h(x, n)$ be a computable binary function. By the recursion theorem, for every choice of n there exists e such that $\varphi_e = \varphi_{h(e,n)}$, and by Skolemization we can see e as a function $f(n)$ of n. Now the effectiveness means that we can choose f to be *computable*, so that

$$\varphi_{f(n)} = \varphi_{h(f(n),n)}$$

for every n. This is called the recursion theorem with parameters, or the *second recursion theorem*.[2] Another way to phrase this is as follows:[3]

Theorem 2.1 (The second recursion theorem, Kleene [18]). *There exists a computable function f such that for every n, if $\varphi_n(f(n))\downarrow$ then*

$$\varphi_{\varphi_n(f(n))} = \varphi_{f(n)}.$$

3   Partial Combinatory Algebra

Partial combinatory algebra was first introduced in the literature by Feferman [12] as an abstract axiomatic model of computation, though the concept had been known and discussed before. A *partial combinatory algebra* (pca) is a set $\mathcal{A}$ with a partial application operator $\cdot$ from $\mathcal{A} \times \mathcal{A}$ to $\mathcal{A}$. Instead of $a \cdot b$ we often simply write ab. We write $ab\downarrow$ if this is defined. By convention application associates to the left, so abc should be read as $(ab)c$. We call $f \in \mathcal{A}$ *total* if $fa\downarrow$ for every a. For terms (i.e. expressions built from elements of $\mathcal{A}$, variables, and application) t and s we write $t \simeq s$ if either both sides are undefined, or defined and equal. The defining property of a pca is that it should be *combinatory complete*, that is, for any term $t(x_1,\ldots,x_n,x)$, $n \geqslant 0$, there exists a $b\in\mathcal{A}$ such that for all $a_1,\ldots,a_n,a\in\mathcal{A}$,

(i) $ba_1\cdots a_n\downarrow$,
(ii) $ba_1\cdots a_n a \simeq t(a_1,\ldots,a_n,a)$.

Combinatory completeness is equivalent to the existence of the combinators s and k, familiar from combinatory algebra, cf. van Oosten [24].

Feferman proved the following version of the recursion theorem in pcas:

Theorem 3.1 (Feferman [12]). *Let $\mathcal{A}$ be a pca.*

(1) *There exists $f \in \mathcal{A}$ such that for all $g \in \mathcal{A}$, $g(fg) \simeq fg$.*
(2) *There exists a total $f \in \mathcal{A}$ such that $g(fg)a \simeq fga$ for every g and $a \in \mathcal{A}$.*

The prime example of a pca is ω, with application $nm = \varphi_n(m)$ as already defined above. From this we immediately recognize Theorem 3.1 (2) as a generalization of Theorem 2.1. However, there is a rich variety of other examples of pcas, drawing from lambda calculus, constructive mathematics, realizability, and computability theory. Examples and references may be found for example in van Oosten [24], Cockett and Hofstra [10], Longley and Normann [20], and Golov and Terwijn [14].

---

[2] The term second recursion theorem is from Rogers, and advocated by Moschovakis. Confusingly, Kleene called this form the first recursion theorem, a term which is now mostly used to refer to the simple version without parameters (also following Rogers).

[3] The two forms of the recursion theorem with parameters are equivalent because the numbering $n \mapsto \varphi_n$ of the p.c. functions is *precomplete*, cf. footnote 4.

4   The Theory of Numberings

A *numbering* of a set S is a surjection $\gamma \colon \omega \to S$. For every numbering we have an equivalence relation on ω defined by $n \sim_\gamma m$ if $\gamma(n) = \gamma(m)$. A numbering γ is *precomplete* if for every p.c. function ψ there exists a computable function f such that for every n

$$\psi(n)\downarrow \implies f(n) \sim_\gamma \psi(n). \tag{1}$$

Following Visser, we say that f *totalizes* ψ *modulo* $\sim_\gamma$.

Ershov [11] proved that the recursion theorem holds for every precomplete numbering: If γ is precomplete, then for every computable function f there exists $e \in \omega$ such that

$$f(e) \sim_\gamma e. \tag{2}$$

The recursion theorem is obtained from this by simply taking the numbering $n \mapsto \varphi_n$ of the p.c. functions. This numbering is easily seen to be precomplete by the S-m-n-theorem.

As for the recursion theorem, we have a version of Ershov's recursion theorem with parameters, which shows that the theorem is effective. The following formulation from Andrews, Badaev, and Sorbi [3] is completely analogous to Theorem 2.1 above.[4]

Theorem 4.1 (Ershov's recursion theorem [11]). *Let γ be a precomplete numbering. There exists a computable function f such that for every n, if $\varphi_n(f(n))\downarrow$ then*

$$\varphi_n(f(n)) \sim_\gamma f(n).$$

In order to combine the theorems of Feferman and Ershov, we consider *generalized numberings* $\gamma \colon \mathcal{A} \to S$, having as a base a pca $\mathcal{A}$ instead of ω. We call such numberings *precomplete*[5] if every $b \in \mathcal{A}$ can be totalized modulo $\sim_\gamma$, similarly to the definition of precompleteness (1) for ordinary numberings, namely if for every $b \in \mathcal{A}$ there exists a total element $f \in \mathcal{A}$ such that for all $a \in \mathcal{A}$,

$$ba\downarrow \implies fa \sim_\gamma ba.$$

---

[4] An alternative way to state Ershov's recursion theorem is: For every computable function $h(x, n)$ there is a computable function f such that for all n,

$$f(n) \sim_\gamma h(f(n), n).$$

The two forms are equivalent *for precomplete numberings*, see the discussion in Barendregt and Terwijn [7, Section 3]. Question 3.4 there asks if for arbitrary numberings γ the equivalence implies that γ is precomplete. This question is still open. A partial answer was obtained in [13], where it was shown that the answer for the relativized version of this question is negative..

[5] In [7] precompleteness was defined using terms instead of elements $b \in \mathcal{A}$, but the two definitions are equivalent by [7, Lemma 6.4].

Theorem 4.2 (Barendregt and Terwijn [7]). *Suppose $\mathcal{A}$ is a pca, and that $\gamma\colon \mathcal{A} \to S$ is a precomplete generalized numbering. Then there exists a total $f \in \mathcal{A}$ such that for all $g \in \mathcal{A}$, if $g(fg)\downarrow$ then*

$$g(fg) \sim_\gamma fg.$$

5   Overview

In this section we list the various forms of the recursion theorem discussed so far. To ease the comparison, we write them as succinctly as possible.

First consider the natural numbers ω as a pca, with application $nm = \varphi_n(m)$.

Kleene 1. $\forall n\, \exists m\, \forall a\; (nma \simeq ma).$

Kleene 2. $\exists f\; total\; \forall n\, \forall a\; (n(fn)a \simeq fna).$

Let $\gamma\colon \omega \to S$ be a precomplete numbering.

Ershov. $\exists f\; total\; \forall n\; (n(fn)\downarrow \Longrightarrow n(fn) \sim_\gamma fn).$
   Let $\mathcal{A}$ be a pca.

Feferman 1. $\exists f \forall g\; (g(fg) \simeq fg).$

Feferman 2. $\exists f\; total\; \forall g\, \forall a\; (g(fg)a \simeq fga).$

Let $\gamma\colon \mathcal{A} \to S$ be a precomplete generalized numbering.

BT. $\exists f\; total\; \forall g\; (g(fg)\downarrow \Longrightarrow g(fg) \sim_\gamma fg).$
   We have the following relations between these.

Kleene 2 $\Rightarrow$ Kleene 1: This is obvious, since fn provides the fixed point.

Ershov $\Rightarrow$ Kleene 2: The numbering $\gamma\colon n \mapsto \varphi_n$ is precomplete by the S-m-n-theorem, and $n(fn) \sim_\gamma fn$ iff $\forall a\; (n(fn)a \simeq fna).$

Feferman 2 $\Rightarrow$ Kleene 2: Immediate from the fact that ω is a pca.

BT $\Rightarrow$ Ershov: This is trivial.

BT $\Rightarrow$ Feferman 2: Let $a \sim_e b$ if $\forall x \in \mathcal{A}(ax \simeq bx)$. Then the natural map $\gamma\colon \mathcal{A} \to \mathcal{A}/\!\!\sim_e$ is a precomplete generalized numbering by [8, Proposition 4.2]. Applying BT to this numbering gives Feferman 2.

Feferman 1. by itself is very weak, and does not directly imply anything.
   The implications above are summarized in Fig. 1.

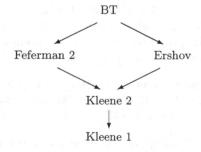

Fig. 1. The relation between various forms of the recursion theorem

6 Fixed Point Free Functions and Arslanov's Completeness Criterion

Recall that W_n is the n-th c.e. set. A function f is called *fixed point free*, or simply FPF, if $W_{f(n)} \neq W_n$ for every n. Note that by the recursion theorem no FPF function is computable. We will also consider *partial* functions without fixed points. Extending the above definition, we call a partial function δ FPF if for every n,

$$\delta(n)\downarrow \implies W_{\delta(n)} \neq W_n. \tag{3}$$

Below, by FPF function we will always mean a total function, unless explicitly stated otherwise.

The standard tool in computability theory to measure the complexity of sets is the notion of a Turing reduction. Informally, $B \leqslant_T A$ means that A can compute B. Here we are interested in the complexity of computing FPF functions. The following fact is well-known:

Proposition 6.1 (Jockusch et al. [15]). *The following are equivalent for any set A:*

(i) *A computes a FPF function,*
(ii) *A computes a function h such that $\varphi_{h(e)} \neq \varphi_e$ for every e.*

Proof (i)$\Rightarrow$(ii) is trivial. We give a direct proof of (ii)$\Rightarrow$(i), avoiding the detour via DNC functions as in Soare [28, p90].

Let ψ be p.c. such that $W_e \neq \emptyset \Rightarrow \psi(e) \in W_e$. Let p totalize ψ, i.e. p is computable such that $\varphi_{p(e)} = \varphi_{\psi(e)}$ for every e with $\psi(e)\downarrow$. (As mentioned before, every p.c. function can be totalized in this way.) Now suppose that h is as in (ii), and let f be A-computable such that $W_{f(e)} = \{h(p(e))\}$.

Suppose that $W_{f(e)} = W_e$. Then $\psi(e)\downarrow = h(p(e))$, hence

$$\varphi_{h(p(e))} = \varphi_{\psi(e)} = \varphi_{p(e)},$$

contradicting (ii). Hence f is a FPF function.   □

Arslanov (building on earlier work of Martin and Lachlan) extended the recursion theorem from computable functions to functions computable from an incomplete c.e. Turing degree. The Arslanov completeness criterion states that a c.e. set is Turing complete if and only if it computes a fixed point free function.

Theorem 6.2 (Arslanov completeness criterion [4]). *Suppose A is c.e. and A is Turing incomplete, i.e. $A <_T \emptyset'$. If f is an A-computable function, then f has a fixed point, i.e. an $e \in \omega$ such that $W_{f(e)} = W_e$.*

Note that Theorem 6.2 implies the recursion theorem by Proposition 6.1. Without the requirement that A is c.e. the theorem fails, as FPF functions can have low Turing degree (i.e. $A' \leqslant_T \emptyset'$) by the low basis theorem of Jockusch and Soare [16].

The Arslanov completeness criterion has been extended in various ways, by considering relaxations of the type of fixed point. For example, instead of requiring that $W_{f(e)} = W_e$, we can merely require $W_{f(e)}$ to be a finite variant of W_e (Arslanov), or for them to be Turing equivalent (Arslanov), or for the n-th jumps of these sets to be Turing equivalent (Jockusch). In this way the completeness criterion can be extended to all levels of the arithmetical hierarchy. For a discussion of these results we refer the reader to Soare [28, p270 ff] and Jockusch, Lerman, Soare, and Solovay [15]. The latter paper also contains an extension of Theorem 6.2 from c.e. degrees to d.c.e. degrees.

7   Further Generalizations

In this section we discuss several other generalizations of the recursion theorem.

Visser proved an extension called the ADN theorem (for "anti diagonal normalization theorem"), motivated by Rosser's extension of Gödel's incompleteness theorem.

Theorem 7.1 (ADN theorem, Visser [32]). *Suppose that δ is a partial computable fixed point free function. Then for every partial computable function ψ there exists a computable function f such that for every n,*

$$\psi(n)\downarrow \implies W_{f(n)} = W_{\psi(n)} \qquad (4)$$
$$\psi(n)\uparrow \implies \delta(f(n))\uparrow \qquad (5)$$

Note that (4) expresses that f totalizes ψ modulo the numbering $n \mapsto W_n$ of the c.e. sets. Also note that the ADN theorem implies the recursion theorem: The function δ cannot be total, for otherwise $f(n)$ could not exist when $\psi(n) \uparrow$. It follows that there can be no computable FPF function. By Proposition 6.1 this is equivalent to the statement of the recursion theorem.

For discussion about the motivation and applications of the ADN theorem we refer the reader to Visser [32] and Barendregt and Terwijn [7]. For example, it has interesting applications in the theory of ceers (c.e. equivalence relations), see Bernardi and Sorbi [9]. For recent results about diagonal functions for ceers see Badaev and Sorbi [1].

The following result simultaneously generalizes Arslanov's completeness criterion (Theorem 6.2) and the ADN theorem.

Theorem 7.2 (Joint generalization, Terwijn [29]). *Suppose A is a c.e. set such that $A <_T \emptyset'$, and suppose that δ is a partial A-computable fixed point free function. Then for every partial computable function ψ there exists a computable function f totalizing ψ avoiding δ, i.e. such that for every n,*

$$\psi(n)\downarrow \implies W_{f(n)} = W_{\psi(n)} \tag{6}$$
$$\psi(n)\uparrow \implies \delta(f(n))\uparrow \tag{7}$$

Note that the statement of Theorem 7.2 is identical to that of the ADN theorem, except that δ is now partial A-computable for A c.e. and incomplete, instead of just p.c. So Theorem 7.2 generalizes the ADN theorem in the same way that Arslanov's completeness criterion generalizes the recursion theorem. Theorem 7.2 implies Arslanov's completeness criterion, since in general f as in the theorem cannot satisfy (7) if δ is total. In particular, any total A-computable δ cannot be FPF, hence must have a fixed point (Fig. 2).

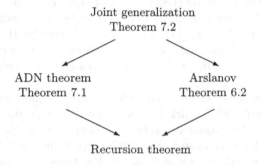

Joint generalization
Theorem 7.2

ADN theorem   Arslanov
Theorem 7.1   Theorem 6.2

Recursion theorem

Fig. 2. Generalizations of the recursion theorem

Visser actually proved the ADN theorem for arbitrary precomplete numberings, so that the ADN theorem also generalizes Ershov's recursion theorem.[6] Arslanov's completeness criterion also holds for precomplete numberings, as was proved by Selivanov [26]. So the obvious question at this point is whether this is also true for the joint generalization Theorem 7.2. This is currently open ([7, Question 5.2]). Since it is true for the two theorems that the joint generalization generalizes, the evidence seems to point in the positive direction. However, the proof of the joint generalization uses specific properties of c.e. sets that we do not have in general, so that the answer may still be negative.

---

[6] Not exactly the version with parameters Theorem 4.1, but the simpler statement (2) without parameters stated before it. That the ADN theorem with parameters fails was shown in Terwijn [30].

8   Effectiveness and Other Remarks

We discuss the extent to which the various fixed point theorems discussed above are effective. As mentioned in Sect. 2, the recursion theorem is effective, which is the content of Theorem 2.1. On the other hand, the proof of Arslanov's Theorem 6.2 does not effectively produce a fixed point, but rather an infinite c.e. set of numbers, at least one of which is a fixed point. That this is necessarily the case was discussed in Terwijn [30], where it was shown that Theorem 6.2 indeed is not effective, so that there is no version with parameters analogous to Theorem 2.1. This raises the question of exactly how noneffective Theorem 6.2 is. The matter of the complexity of the corresponding Skolem functions was discussed in Golov and Terwijn [13], and independently in Arslanov [5].

We already mentioned that the ADN theorem (Theorem 7.1) is not effective, although there is uniformity in some of its parameters. This was shown in [30]. Since neither the Arslanov completeness criterion nor the ADN theorem are effective, a fortiori the same holds for the joint generalization Theorem 7.2.

It is not known whether the Arslanov completeness criterion (appropriately formulated) holds for pcas in general. See Terwijn [31, Question 10.1] for a precise statement of this.

The role that the recursion theorem plays in the theory of pcas is interesting. For example it plays an important part in results about *embeddings* between pcas, see Shafer and Terwijn [27] and Golov and Terwijn [14]. (Note that [27] also contains results about another kind of fixed points, namely closure ordinals, but these are of a different kind than the ones that we have been discussing here).

We should also mention here the various forms of the recursion theorem in descriptive set theory, cf. Kechris [17, p289] and Moschovakis [21, p383]. These are formulated for the various pointclasses Γ occurring in descriptive set theory (effective or not) for which the Γ-computable functions on Polish spaces can be suitably parameterized, e.g. by reals in ω^ω. For these an analog of the S-m-n-theorem is available ([21, 7A.1]), which makes the proof of the recursion theorem work. Note that the idea of encoding continuous functions by reals is the same as the basic idea underlying Kleene's pca $\mathcal{K}_2$ (cf. [24]).

Finally we mention categorical approaches to the subject of diagonalization and fixed point theorems, starting with Lawvere [19]. Examples of Lawvere's basic scheme are discussed in Yanofsky [33] and Bauer [2], among others. Note, however, that these do not capture the more complex results such as Theorem 6.2 and Theorem 7.2, as these do not follow Lawvere's scheme.

Acknowledgements. We thank Dan Frumin and Anton Golov for discussions about the categorical view on the subject of fixed points.

References

1. Badaev, S.A., Sorbi, A.: Weakly precomplete computably enumerable equivalence relations. Math. Log. Q. **62**(1–2), 111–127 (2016)

2. Bauer, A.: On fixed-point theorems in synthetic computability. Tbilisi Math. J. **10**(3), 167–181 (2017)
3. Andrews, U., Badaev, S., Sorbi, A.: A survey on universal computably enumerable equivalence relations. In: Day, A., Fellows, M., Greenberg, N., Khoussainov, B., Melnikov, A., Rosamond, F. (eds.) Computability and Complexity. LNCS, vol. 10010, pp. 418–451. Springer, Cham (2017). https://doi.org/10.1007/978-3-319-50062-1_25
4. Arslanov, M.M.: On some generalizations of the fixed point theorem. Sov. Math. (Izvestiya VUZ. Matematika) **25**(5), 1–10 (1981). (English translation)
5. Arslanov, M.M.: Fixed-point selection functions. Lobachevskii J. Math. **42**, 685–692 (2021)
6. Barendregt, H.P.: The Lambda Calculus. Studies in Logic and the Foundations of Mathematics, vol. 103, 2nd edn. North-Holland, Amsterdam (1984)
7. Barendregt, H.P., Terwijn, S.A.: Fixed point theorems for precomplete numberings. Ann. Pure Appl. Logic **170**, 1151–1161 (2019)
8. Barendregt, H.P., Terwijn, S.A.: Partial combinatory algebra and generalized numberings. Theoret. Comput. Sci. **925**, 37–44 (2022)
9. Bernardi, C., Sorbi, A.: Classifying positive equivalence relations. J. Symb. Log. **48**(3), 529–538 (1983)
10. Cockett, J.R.B., Hofstra, P.J.W.: Introduction to Turing categories. Ann. Pure Appl. Logic **156**, 183–209 (2008)
11. Ershov, Y.L.: Theorie der Numerierungen II. Zeitschrift für Math. Logik Grundlagen Math. **21**, 473–584 (1975)
12. Feferman, S.: A language and axioms for explicit mathematics. In: Crossley, J.N. (ed.) Algebra and Logic. LNM, vol. 450, pp. 87–139. Springer, Heidelberg (1975). https://doi.org/10.1007/bfb0062852
13. Golov, A., Terwijn, S.A.: Fixpoints and relative precompleteness. Computability **11**(2), 135–146 (2022)
14. Golov, A., Terwijn, S.A.: Embeddings between partial combinatory algebras. Notre Dame J. Formal Logic **64**(1), 129–158 (2023)
15. Jockusch Jr., C.G., Lerman, M., Soare, R.I., Solovay, R.M.: Recursively enumerable sets modulo iterated jumps and extensions of Arslanov's completeness criterion. J. Symb. Logic **54**(4), 1288–1323 (1989)
16. Jockusch Jr., C.G., Soare, R.I.: Π_1^0 classes and degrees of theories. Trans. Am. Math. Soc. **173**, 33–56 (1972)
17. Kechris, A.S.: Classical Descriptive Set Theory. Springer, Heidelberg (1995). https://doi.org/10.1007/978-1-4612-4190-4
18. Kleene, S.C.: On notation for ordinal numbers. J. Symb. Log. **3**, 150–155 (1938)
19. Lawvere, F.W.: Diagonal arguments and cartesian closed categories. In: Category Theory, Homology Theory and their Applications II. LNM, vol. 92, pp. 134–145. Springer, Heidelberg (1969). https://doi.org/10.1007/BFb0080769. Republished in: Reprints in Theory and Applications of Categories, No. 15, 1–13 (2006)
20. Longley, J., Normann, D.: Higher-Order Computability. Springer, Heidelberg (2015). https://doi.org/10.1007/978-3-662-47992-6
21. Moschovakis, Y.N.: Descriptive Set Theory. North-Holland (1980)
22. Moschovakis, Y.N.: Kleene's amazing second recursion theorem. Bull. Symb. Logic **16**(2), 189–239 (2010)
23. Odifreddi, P.G.: Classical Recursion Theory, vol. 1. Studies in Logic and the Foundations of Mathematics, vol. 125. North-Holland (1989)
24. van Oosten, J.: Realizability: An Introduction to Its Categorical Side. Studies in Logic and the Foundations of Mathematics, vol. 152. Elsevier, Amsterdam (2008)

25. Owings, J.C.: Diagonalization and the recursion theorem. Notre Dame J. Formal Logic **14**, 95–99 (1973)
26. Selivanov, V.: Index sets of quotient objects of the Post numeration. Algebra Logika **27**(3), 343–358 (1988). (English translation 1989)
27. Shafer, P., Terwijn, S.A.: Ordinal analysis of partial combinatory algebras. J. Symb. Log. **86**(3), 1154–1188 (2021)
28. Soare, R.I.: Recursively Enumerable Sets and Degrees. Springer, Heidelberg (1987)
29. Terwijn, S.A.: Generalizations of the recursion theorem. J. Symb. Log. **83**(4), 1683–1690 (2018)
30. Terwijn, S.A.: The noneffectivity of Arslanov's completeness criterion and related theorems. Arch. Math. Logic **59**(5), 703–713 (2020)
31. Terwijn, S.A.: Computability in partial combinatory algebras. Bull. Symb. Logic **26**(3–4), 224–240 (2020)
32. Visser, A.: Numerations, λ-calculus, and arithmetic. In: Seldin, J.P., Hindley, J.R. (eds.) To H. B. Curry: Essays on Combinatory Logic, Lambda Calculus and Formalism, pp. 259–284. Academic Press (1980)
33. Yanofsky, N.S.: A universal approach to self-referential paradoxes, incompleteness and fixed points. Bull. Symb. Logic **9**(3), 362–386 (2003)

A New Perspective on Conformance Testing Based on Apartness

Frits Vaandrager$^{(\boxtimes)}$ (iD)

Institute for Computing and Information Sciences, Radboud University,
Nijmegen, The Netherlands
f.vaandrager@cs.ru.nl

Abstract. We revisit the classic problem of black box conformance testing and present a generalization of the k-completeness result of Vasilevskii and Chow, phrased entirely in terms of properties of the observation/prefix tree induced by a test suite, in particular in terms of apartness relations between states. The original result of Vasilevskii and Chow is then a corollary of our result. Also k-completeness results for other test methods that have been proposed in the literature, such as the Wp-method and the HSI-method, follow from our characterization. Based on the apartness relations in the observation tree, we may determine whether some test is redundant or can be shortened.

Keywords: conformance testing · finite state machines · Mealy machines · apartness · observation tree · k-complete test suites

1 Introduction

In this note, we revisit the classic problem of black box conformance testing. We consider the simple setting in which both the specification and the black box implementation can be described as (deterministic, complete) finite state machines (a.k.a. Mealy machines). In this setting, we refer to a sequence of inputs σ as a *test*. Given a specification S, we say that an implementation M *passes* test σ if σ triggers the same outputs in M and S. Otherwise, we say that M *fails* test σ. Implementation M *conforms* to specification S if it passes all tests. In our setting this means that M and S have equivalent behavior. The task of a tester of a black box implementation M is to find a test σ (if it exists) such that M fails σ. Figure 1 shows an example (taken from [11]). Here the test $\sigma = aba$ triggers outputs 010 in the specification, but outputs 011 in the implementation. Thus, M fails test σ and implementation M does not conform to specification S.

Ideally, given a specification S, a tester would like to compute a finite set of tests T, called a *test suite*, that is complete in the sense that, for any implementation M, M will pass all tests in T if and only if M conforms to S. Unfortunately,

Research supported by NWO TOP project 612.001.852 "Grey-box learning of Interfaces for Refactoring Legacy Software (GIRLS)".

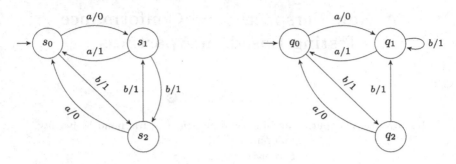

Fig. 1. A specification $\mathcal{S}$ (left) and a faulty implementation $\mathcal{M}$ (right).

such a test suite does not exist: for any finite test suite T the number of inputs in tests will be bounded by some number n. Thus, for any specification $\mathcal{S}$ and any n we may construct an implementation $\mathcal{M}$ that behaves exactly like $\mathcal{S}$ for the first n inputs, but behaves differently from that point onwards. Even though this $\mathcal{M}$ does not conform to $\mathcal{S}$, it will pass all tests in T.

In two classic papers, Vasilevskii [20] and Chow [3] independently showed that, for any specification $\mathcal{S}$ and any natural number k, a finite test suite T can be constructed that is k-*complete* in the sense that, for any implementation $\mathcal{M}$ with at most k more states than $\mathcal{S}$, $\mathcal{M}$ passes T if and only if $\mathcal{M}$ conforms to $\mathcal{S}$. The so-called W-*method* proposed by Vasilevskii [20] and Chow [3] for constructing a k-complete test suite is actually quite simple:

- First, construct a *state cover* for $\mathcal{S}$: a finite, prefix closed set of input sequences A such that every state of $\mathcal{S}$ is reached by some sequence in A. For example, the set $A = \{\epsilon, a, b\}$ is a state cover for the specification $\mathcal{S}$ from Fig. 1, since s_0 is reached by the empty sequence ϵ, s_1 is reached by a, and s_2 is reached by b.
- Next, construct a *characterization set* for $\mathcal{S}$: a nonempty, finite set of input sequences W, such that for each pair of inequivalent states s and t, W contains a separating sequence for s and t. Here a *separating sequence* for s and t is a sequence σ such that when σ is applied in s the resulting outputs are different from those when σ is applied in t. Two states s and t of $\mathcal{S}$ are *equivalent* if there exists no separating sequence for them. For example, aa is a separating sequence for all pairs of distinct states of $\mathcal{S}$ since in s_0 it triggers outputs 01, in s_1 it triggers outputs 10, and in s_2 it triggers outputs 00. Thus $W = \{aa\}$ is a characterization set for $\mathcal{S}$.
- Finally, a k-complete test suite T is defined by $T = A \cdot I^{\leq k+1} \cdot W$, where $I^{\leq k+1}$ is the set of all input sequences with length less than or equal to $k+1$, and "$\cdot$" denotes concatenation of sequences, extended to sets of sequences by pointwise extension.

Intuitively, in the simple case with $k = 0$, input sequences $A \cdot W$ test whether all states of the specification are present in the implementation, and input sequences $A \cdot I \cdot W$ test whether all transitions produce the correct output and lead to the correct target state. If we apply the W-method to construct a 0-complete test suite for the specification of Fig. 1, we obtain $T = \{aa, aaa, baa, aaaa, abaa, baaa, bbaa\}$. We can always omit tests that are a prefix of another test: if an implementation fails on a prefix of a test then it will also fail on the full test. Thus we can omit tests aa, aaa and baa, and obtain a 0-complete test suite with 4 tests. Indeed, the implementation from Fig. 1 fails the test $abaa$.

Numerous variations and improvements of the W-method have been proposed in the literature; we refer to [4,9,11] for overviews and further references. Still, for all these methods it is unclear how to (1) establish whether certain tests are redundant or can be shortened (for instance, in the example test suite T, $abaa$ can be replaced by the shorter aba without compromising 0-completeness), (2) compute which tests should be selected first in order to maximize the likelihood that bugs will be discovered, and related (3) quantify how performing some test will reduce our uncertainty whether an implementation conforms to a specification.

In order to be able address these fundamental questions, we present a new perspective on conformance testing, drawing inspiration from the recently developed $L^\#$ learning algorithm [19]. Actually, given the natural duality between learning and testing, first observed by Weyuker [21], it is not surprising that ideas regarding active learning of finite state machines can be applied in the setting of conformance testing. A first idea from the $L^\#$ algorithm that we will use is to store the outcomes of all experiments/tests in a single data structure, the *observation tree* (a.k.a. prefix tree). This allows us, for instance, to base the choice of the next test on the set of tests executed thus far. A second idea that we adapt from $L^\#$ is the concept of *apartness*, a constructive form of inequality [18]. The notion of apartness is standard in constructive real analysis and goes back to Brouwer, with Heyting giving an axiomatic treatment in [7]. The importance of apartness for automata theory and concurrency theory was first observed by Herman Geuvers and Bart Jacobs [6]. This insight was an inspiration for the work of [19] and for the present note.

The main result that we present here is a generalization of the k-completeness result of [3,20], phrased entirely in terms of properties of the observation tree, in particular in terms of apartness relations between states. The original result of [3,20] is then a corollary of our result. Also k-completeness results for other test methods that have been proposed in the literature follow from our characterization. Based on the apartness relations in the observation tree, we may determine whether a certain test is redundant or can be shortened.

The rest of this note is structured as follows. First we recall the formal definitions of (partial) Mealy machines, observation trees and apartness in Sect. 2. Next, we present our main result in Sect. 3. The connections with existing k-completeness results are discussed in Sect. 4. In Sect. 5, we discuss implications of our results and directions for future research.

2   Partial Mealy Machines and Apartness

In this preliminary section, which is largely copied from [19] (but with the word "learning" replaced by "testing"), we formalize some of the key concepts that play a role in our result: Mealy machines, observation trees and apartness. Since observation trees are (a specific type of) partial Mealy machines, we need to fix some notation for partial maps.

We write $f\colon X \rightharpoonup Y$ to denote that f is a partial function from X to Y and write $f(x)\!\downarrow$ to mean that f is defined on x, that is, $\exists y \in Y\colon f(x) = y$, and conversely write $f(x)\!\uparrow$ if f is undefined for x. Often, we identify a partial function $f\colon X \rightharpoonup Y$ with the set $\{(x, y) \in X \times Y \mid f(x) = y\}$. There is a partial order on $X \rightharpoonup Y$ defined by $f \sqsubseteq g$ for $f, g\colon X \rightharpoonup Y$ iff for all $x \in X$, $f(x)\!\downarrow$ implies $g(x)\!\downarrow$ and $f(x) = g(x)$.

Throughout this paper, we fix a nonempty, finite set I of *inputs* and a set O of *outputs*.

Definition 2.1. *A **Mealy machine** is a tuple $\mathcal{M} = (Q, q_0, \delta, \lambda)$, where*

- *Q is a finite set of **states** and $q_0 \in Q$ is the **initial state**,*
- *$\langle \lambda, \delta \rangle\colon Q \times I \rightharpoonup O \times Q$ is a partial map whose components are an **output function** $\lambda\colon Q \times I \rightharpoonup O$ and a **transition function** $\delta\colon Q \times I \rightharpoonup Q$.*

*We use superscript $\mathcal{M}$ to disambiguate to which Mealy machine we refer, e.g. $Q^{\mathcal{M}}$, $q_0^{\mathcal{M}}$, $\delta^{\mathcal{M}}$ and $\lambda^{\mathcal{M}}$. We write $q \xrightarrow{i/o} q'$, for $q, q' \in Q$, $i \in I$, $o \in O$ to denote $\lambda(q, i) = o$ and $\delta(q, i) = q'$. We call $\mathcal{M}$ **complete** iff δ is total, i.e., $\delta(q, i)$ is defined for all states q and inputs i. We generalize the transition and output functions to input words of length $n \in \mathbb{N}$ by composing $\langle \lambda, \delta \rangle$ n times with itself: we define maps $\langle \lambda_n, \delta_n \rangle\colon Q \times I^n \rightharpoonup O^n \times Q$ by $\langle \lambda_0, \delta_0 \rangle = \mathrm{id}_Q$ and*

$$\langle \lambda_{n+1}, \delta_{n+1} \rangle\colon \; Q \times I^{n+1} \xrightarrow{\langle \lambda_n, \delta_n \rangle \times \mathrm{id}_I} O^n \times Q \times I \xrightarrow{\mathrm{id}_{O^n} \times \langle \lambda, \delta \rangle} O^{n+1} \times Q$$

Whenever it is clear from the context, we use λ and δ also for words.

Definition 2.2 (Equivalence and minimality). *The semantics of a state q is a map $[\![q]\!]\colon I^* \rightharpoonup O^*$ defined by $[\![q]\!](\sigma) = \lambda(q, \sigma)$. States q, q' in possibly different Mealy machines are **equivalent**, written $q \approx q'$, iff $[\![q]\!] = [\![q']\!]$. Mealy machines $\mathcal{M}$ and $\mathcal{N}$ are **equivalent** iff their respective initial states are equivalent: $q_0^{\mathcal{M}} \approx q_0^{\mathcal{N}}$. Mealy machine $\mathcal{M}$ is **minimal** iff, for all pairs of states q, q', $q \approx q'$ iff $q = q'$.*

In our testing setting, an *undefined* value in the partial transition map represents lack of knowledge. We consider maps between Mealy machines that preserve existing transitions, but possibly extend the knowledge of transitions:

Definition 2.3 (Simulation). *For Mealy machines $\mathcal{M}$ and $\mathcal{N}$, a **functional simulation** $f\colon \mathcal{M} \to \mathcal{N}$ is a map $f\colon Q^{\mathcal{M}} \to Q^{\mathcal{N}}$ with*

$$f(q_0^{\mathcal{M}}) = q_0^{\mathcal{N}} \qquad \text{and} \qquad q \xrightarrow{i/o} q' \text{ implies } f(q) \xrightarrow{i/o} f(q').$$

Intuitively, a functional simulation preserves transitions and the initial state.

Lemma 2.4. *For a functional simulation* $f: \mathcal{M} \to \mathcal{N}$ *and* $q \in Q^{\mathcal{M}}$*, we have* $[\![q]\!] \sqsubseteq [\![f(q)]\!]$.

For a given machine $\mathcal{M}$, an observation tree is simply a Mealy machine itself which represents the inputs and outputs we have observed so far during testing. Using functional simulations, we define it formally as follows.

Definition 2.5 (Observation tree). *A Mealy machine* T *is a* **tree** *iff for each* $q \in Q^T$ *there is a unique sequence* $\sigma \in I^*$ *s.t.* $\delta^T(q_0^T, \sigma) = q$. *We write* access($q$) *for the sequence of inputs leading to* q. *A tree* T *is an* **observation tree** *for a Mealy machine* $\mathcal{M}$ *iff there is a functional simulation* $f: T \to \mathcal{M}$.

Figure 2 shows an observation tree for the Mealy machine displayed on the right. The functional simulation f is indicated via state colors. By performing tests, a tester may construct an observation tree T for implementation $\mathcal{M}$. The observation tree of Fig. 2, for instance, may be constructed by performing tests a, ba, and bba. There is a one-to-one correspondence between observation trees (up to isomorphism) and finite test suites T that do not contain redundant prefixes (i.e., if $\sigma, \rho \in T$ with σ a prefix of ρ then we require $\sigma = \rho$.)

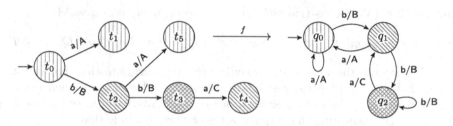

Fig. 2. An observation tree (left) for a Mealy machine (right). (Color figure online)

Suppose T is an observation tree for implementation $\mathcal{M}$. If all tests, as recorded in the tree, have passed then T is also an observation tree for S. However, as soon as a test fails then T is no longer an observation tree for S: there exists no functional simulation from T to S, since the observation tree contains a state q such that the output in response to input sequence access(q) is different for S and $\mathcal{M}$.

Even though the tester constructs the observation tree T, they do not know the functional simulation. However, by analysis of the observation tree, a tester may infer that certain states in the tree cannot have the same color, that is, they cannot be mapped to same states of $\mathcal{M}$ by a functional simulation. In this analysis, the concept of *apartness*, a constructive form of inequality, plays a crucial role [6,18].

Definition 2.6 (Apartness). *For a Mealy machine* $\mathcal{M}$*, we say that states* $q, p \in Q^{\mathcal{M}}$ *are* **apart** *(written* $q \mathbin{\#} p$*) iff there is some* $\sigma \in I^*$ *such that* $[\![q]\!](\sigma)\!\downarrow$, $[\![p]\!](\sigma)\!\downarrow$, *and* $[\![q]\!](\sigma) \neq [\![p]\!](\sigma)$. *We say that* σ *is the* **witness** *of* $q \mathbin{\#} p$ *and write* $\sigma \vdash q \mathbin{\#} p$.

Note that the apartness relation $\# \subseteq Q \times Q$ is irreflexive and symmetric. Within conformance testing theory, a witness is commonly called *separating sequence* [17]. For the observation tree of Fig. 2 we may derive the following apartness pairs and corresponding witnesses:

$$a \vdash t_0 \# t_3 \qquad a \vdash t_2 \# t_3 \qquad b\,a \vdash t_0 \# t_2$$

The apartness of states $q \# p$ expresses that there is a conflict in their semantics, and consequently, apart states can never be identified by a functional simulation:

Lemma 2.7. *For a functional simulation* $f : T \to M,$

$$q \# p \text{ in } T \qquad \Longrightarrow \qquad f(q) \not\approx f(p) \text{ in } M \qquad \text{for all } q, p \in Q^T.$$

Thus, whenever states are apart in the observation tree T, the learner knows that these are distinct states in the hidden Mealy machine M.

The apartness relation satisfies a weaker version of *co-transitivity*, stating that if $\sigma \vdash r \# r'$ and q has the transitions for σ, then q must be apart from at least one of r and r', or maybe even both:

Lemma 2.8 (Weak co-transitivity). *In every Mealy machine* $M,$

$$\sigma \vdash r \# r' \wedge \delta(q, \sigma)\!\downarrow \quad \Longrightarrow \quad r \# q \vee r' \# q \qquad \text{for all } r, r', q \in Q^M, \sigma \in I^*.$$

A tester may use the weak co-transitivity property during testing. For instance in Fig. 2, by performing the test aba, consisting of the access sequence for t_1 concatenated with the witness ba for $t_0 \# t_2$, co-transitivity ensures that $t_0 \# t_1$ or $t_2 \# t_1$. By inspecting the outputs, a tester may conclude that $t_2 \# t_1$.

3   Main Result

In this section, we describe a sufficient condition for a test suite to be k-complete, phrased in terms of the corresponding observation tree. This tree should contain access sequences for each state in the specification, successors for these states for all possible inputs should be present up to depth $k + 1$, and certain apartness relations between states of the tree should hold.

In order to present our condition and the proof of its correctness, we first need to introduce some auxiliary termininology.

Definition 3.1 (Stratification). *Let* $A \subseteq I^*$ *be a nonempty, finite, prefix closed set of input sequences, and let* T *be an observation tree. Then* A *induces a* **stratification** *of* Q^T *as follows:*

1. *A state* q *of* T *is called a* **basis state** *iff* $\text{access}(q) \in A$. *We write* B *to denote the set of basis states:* $B = \{q \in Q^T \mid \text{access}(q) \in A\}$. *Note that, since* A *is nonempty and prefix closed, initial state* q_0^T *is in the basis, and all states on the path leading to a basis state are basis states as well.*

2. We write F^0 for the set of immediate successors of basis states that are not basis states themselves: $F^0 := \{q' \in Q^T \setminus B \mid \exists q \in B, i \in I : q' = \delta^T(q, i)\}$. We refer to F^0 as the 0-**level frontier**.

3. For $k > 0$, the k-**level frontier** F^k is the set of immediate successors of $k-1$-level frontier states: $F^k := \{q' \in Q^T \mid \exists q \in F^{k-1}, i \in I : q' = \delta^T(q, i)\}$.

We say that basis B is **complete** if for each $\sigma \in A$ there is a state $q \in B$ with $\delta^T(q_0^T, \sigma) = q$. The 0-level frontier is **complete** if for each $q \in B$ and for each $i \in I$, $\delta^T(q, i) \downarrow$. For $k > 0$, the k-level frontier is **complete** if for each $q \in F^{k-1}$ and for each $i \in I$, $\delta^T(q, i) \downarrow$.

For each state q of an observation tree, we define the **candidate set** $C(q)$ as the set of basis states that are not apart from q: $C(q) = \{q' \in B \mid \neg(q \# q')\}$. A state q of an observation tree is **identified** if its candidate set is a singleton.

Example 3.2. Figure 3 shows the stratification for an observation tree for specification S from Fig. 1 induced by $A = \{\epsilon, a, b\}$. States from sets B, F^0, F^1 and F^2 are marked with different colors. In Fig. 3, B and F^0 are complete, but F^1 and F^2 are incomplete (since states of F^0 and F^1 have no outgoing b-transitions). Witness aa shows that the three basis states are pairwise apart, and therefore identified. The four F^0 states are also identified since $C(2) = \{0\}$, $C(5) = \{8\}$, $C(9) = \{0\}$, and $C(12) = \{1\}$. Two states in F^1 are identified since $C(3) = C(10) = \{1\}$, whereas the other two are not since $C(6) = C(13) = \{0, 8\}$. Since states in F^2 have no outgoing transitions, they are not apart from any other state, and thus $C(4) = C(7) = C(11) = C(14) = \{0, 1, 8\}$.

In the proof of our main result, we use a bisimulation relation and the well-known fact that for (deterministic) Mealy machines bisimulation equivalence coincides with the standard behavioral equivalence for Mealy machines from Definition 2.2.

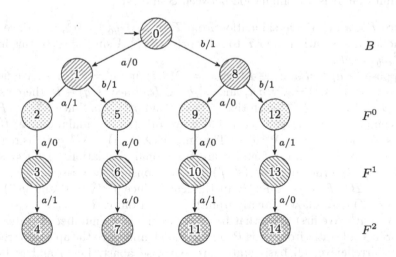

Fig. 3. Stratification of an observation tree induced by $A = \{\epsilon, a, b\}$.

Definition 3.3 (Bisimulation). *A **bisimulation** between Mealy machines $\mathcal{M}$ and $\mathcal{N}$ is a relation $R \subseteq Q^{\mathcal{M}} \times Q^{\mathcal{N}}$ satisfying, for all $q \in Q^{\mathcal{M}}$, $r \in Q^{\mathcal{N}}$, $i \in I$, $o \in O$,*

$$q_0^{\mathcal{M}} \, R \, q_0^{\mathcal{N}} \quad and \quad q \, R \, r \wedge q \xrightarrow{i/o} q' \Rightarrow \exists r' : r \xrightarrow{i/o} r' \wedge q' \, R \, r'$$

We write $\mathcal{M} \simeq \mathcal{N}$ if there exists a bisimulation relation between $\mathcal{M}$ and $\mathcal{N}$.

The next lemma, a variation of the classical result of [13], is easy to prove.

Lemma 3.4. *Let $\mathcal{M}$ and $\mathcal{N}$ be complete Mealy machines. Then $\mathcal{M} \simeq \mathcal{N}$ iff $\mathcal{M} \approx \mathcal{N}$.*

We are now prepared to state and prove our main result.

Theorem 3.5. *Let $\mathcal{M}$ and $\mathcal{S}$ be complete Mealy machines and let $\mathcal{T}$ be an observation tree for both $\mathcal{M}$ and $\mathcal{S}$. Let $A \subseteq I^*$ be a state cover for $\mathcal{S}$, and let $B, F^0, F^1, \ldots$ be the stratification of $Q^{\mathcal{T}}$ induced by A. Let $k \geq 0$. Suppose that B and $F^0, \ldots, F^k$ are all complete, all states in B and $F^0, \ldots, F^k$ are identified, and the following co-transitivity property holds*

$$\forall r \in B \; \forall t' \in F^k \; \forall t'' \in F^0 \cup \cdots \cup F^{k-1} : r \# t' \implies r \# t'' \vee t' \# t'' \quad (1)$$

Suppose $\mathcal{M}$ has at most k more states than $\mathcal{S}$. Then $\mathcal{S} \approx \mathcal{M}$.

Proof. Let f be a functional simulation from $\mathcal{T}$ to $\mathcal{S}$ and let g be a functional simulation from $\mathcal{T}$ to $\mathcal{M}$. Define relation $R \subseteq Q^{\mathcal{S}} \times Q^{\mathcal{M}}$ by

$$(s, q) \in R \Leftrightarrow \exists t \in B \cup F^0 \cup \cdots \cup F^{k-1} : f(t) = s \wedge g(t) = q.$$

We claim that R is a bisimulation between $\mathcal{S}$ and $\mathcal{M}$.

1. Since f is a functional simulation from $\mathcal{T}$ to $\mathcal{S}$, $f(q_0^{\mathcal{T}}) = q_0^{\mathcal{S}}$, and since g is a functional simulation from $\mathcal{T}$ to $\mathcal{M}$, $g(q_0^{\mathcal{T}}) = q_0^{\mathcal{M}}$. Using $q_0^{\mathcal{T}} \in B$, this implies $(q_0^{\mathcal{S}}, q_0^{\mathcal{M}}) \in R$.
2. Suppose $(s, q) \in R$ and $i \in I$. Let $s' = \delta^{\mathcal{S}}(s, i)$ and $q' = \delta^{\mathcal{M}}(q, i)$. We need to show that $\lambda^{\mathcal{S}}(s, i) = \lambda^{\mathcal{M}}(q, i)$ and $(s', q') \in R$. Since $(s, q) \in R$, there exists a $t \in B \cup F^0 \cup \cdots \cup F^{k-1}$ such that $f(t) = s$ and $g(t) = q$. Since $F^0, \ldots, F^k$ are all complete, $\delta^{\mathcal{T}}(t, i) \downarrow$. Since f and g are functional simulations, $\lambda^{\mathcal{T}}(t, i) = \lambda^{\mathcal{S}}(s, i)$ and $\lambda^{\mathcal{T}}(t, i) = \lambda^{\mathcal{M}}(q, i)$. This implies $\lambda^{\mathcal{S}}(s, i) = \lambda^{\mathcal{M}}(q, i)$, as required. Let $t' = \delta^{\mathcal{T}}(t, i)$. Since f and g are functional simulations, $f(t') = s'$ and $g(t') = q'$. In order to prove $(s', q') \in R$, we consider two cases:
   (a) $t' \in B \cup F^0 \cup \cdots \cup F^{k-1}$. In this case, since $f(t') = s'$ and $g(t') = q'$, $(s', q') \in R$ follows directly from the definition of R.
   (b) $t' \in F^k$. We first show that basis B has the same number of states as $\mathcal{S}$. Since all states from basis B are identified, and since the apartness relation is irreflexive, all basis states are pairwise apart. Let q and q' be two distinct states in B. Since $q \# q'$, we may conclude by Lemma 2.7 that

$f(q) \not\approx f(q')$. Thus in particular $f(q) \neq f(q')$ and so f restricted to B is injective.

Let u be a state of $\mathcal{S}$. Since A is a state cover for $\mathcal{S}$, there exists a $\sigma \in A$ with $\delta^{\mathcal{S}}(q_0^{\mathcal{S}}, \sigma) = u$. Since basis B is complete, there exists a state $v \in B$ with $\delta^{\mathcal{T}}(q_0^{\mathcal{T}}, \sigma) = v$. Using that f is a functional simulation, we can show by induction on the length of σ that $f(v) = u$. Thus f restricted to B is surjective.

This implies that f restricted to B is a bijection between B and $Q^{\mathcal{S}}$, which proves B has the same number of states as $\mathcal{S}$.

With the same argument that we used to prove that f restricted to B is injective, we may also show that g restricted to B is injective. Thus $g(B)$ contains exactly $|Q^{\mathcal{S}}|$ states. Now either $g(B \cup F^0)$ contains the same number of states as $g(B)$, which implies that $\mathcal{M}$ contains $|Q^{\mathcal{S}}|$ states, or $g(B \cup F^0)$ contains at least one extra state. In the latter case we may continue: either $g(B \cup F^0 \cup F^1)$ contains the same number of states as $g(B \cup F^0)$, which implies that we have seen all states of $\mathcal{M}$, or $g(B \cup F^0 \cup F^1)$ contains at least one extra state. Etc. Since $\mathcal{M}$ has at most k more states than $\mathcal{S}$, we may conclude that $g(B \cup F^0 \cup \cdots \cup F^{k-1})$ contains all states of $\mathcal{M}$.

This means that $B \cup F^0 \cup \cdots \cup F^{k-1}$ contains a state t'' such that $g(t'') = q'$. We claim that t' and t'' have the same candidate set. The proof is by contradiction. Assume $C(t') \neq C(t'')$. We consider two cases:

i $t'' \in B$. Then $C(t'') = \{t''\}$ and $t'' \notin C(t')$. Hence $t' \# t''$. But now Lemma 2.7 gives $g(t') \neq g(t'')$. This is a contradiction, and therefore $C(t') = C(t'')$.

ii $t'' \in F^0 \cup \cdots \cup F^{k-1}$. Let $C(t'') = \{r\}$. Then $r \# t'$ and $\neg(r \# t'')$. Therefore, by the co-transitivity assumption (1), $t' \# t''$. But now Lemma 2.7 gives $g(t') \neq g(t'')$. This is a contradiction, and therefore $C(t') = C(t'')$.

We claim that for any state $u \in B \cup F^0 \cup \cdots \cup F^{k-1}$, $C(u) = \{r\}$ implies $f(u) = f(r)$. The proof of the claim is by contradiction. Assume $f(u) = v' \neq v = f(r)$. Since f restricted to B is a bijection, there exists an $r' \in B$ with $r \neq r'$ and $f(r') = v'$. Then $u \# r'$. Let σ be a witness. By definition of apartness $\lambda^{\mathcal{T}}(u, \sigma) \neq \lambda^{\mathcal{T}}(r', \sigma)$. But since f is a functional simulation and $f(r') = v' = f(u')$, $\lambda^{\mathcal{T}}(u, \sigma) = \lambda^{\mathcal{S}}(v', \sigma) = \lambda^{\mathcal{T}}(r', \sigma)$. Contradiction and therefore the claim follows.

The above claim implies that $f(t'') = s'$. This in turn implies that $(s', q') \in R$, which completes the proof that R is bisimulation.

The theorem now follows by application of Lemma 3.4.

Remark 3.6. The assumption that A is a state cover for $\mathcal{S}$ implies that $\mathcal{S}$ is connected, that is, all states are reachable from the initial state. The assumptions that B is complete and all states in B are identified, in combination with Lemma 2.7, imply that $\mathcal{S}$ is minimal. Connectedness and minimality of specifications are common assumptions in conformance testing.

Example 3.7. A simple example of the application of Theorem 3.5, is provided by the observation tree from Fig. 3 for the specification S from Fig. 1. This observation tree corresponds to a test suite T of 4 tests that is generated by the W-method for $k = 0$, as described in the introduction. Note that the co-transitivity property vacuously holds when $k = 0$. All other conditions of the theorem are also met: both B and F^0 are complete and all states in B and F^0 are identified. Therefore, according to our theorem, test suite T is 0-complete. We may slightly optimize the test suite by removing state 14 from the observation tree (i.e., replacing test *bbaa* by test *bba*), since all conditions of the theorem are still met for the reduced tree.

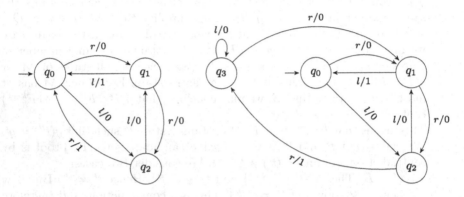

Fig. 4. A specification S (left) and a faulty implementation M (right).

Example 3.8. Probably the most interesting condition in Theorem 3.5 is the co-transitivity requirement. The Mealy machines S and M of Fig. 4 illustrate why we need it. Note that these machines are not equivalent: input sequence *rrrlll* distinguishes them. Mealy machine M has one more state than S, so $k = 1$. The extra state q_3 of M behaves similar as state q_0 of S, but is not equivalent. Figure 5 shows an observation tree T for both S and M. One way to think of T is that M cherry picks distinguishing sequences from S to ensure that the F^1 states are identified by a sequence for which S and M agree. Note that B, F^0 and F^1 are all complete, and all states in B, F^0 and F^1 are identified. However, the tree does not satisfy the co-transitivity condition as $t_{13} \# t_0$, but neither $t_0 \# t_6$ nor $t_{13} \# t_6$.

Example 3.9. In 2012, Arjan Blom, then a student at Radboud University, performed a security analysis of the E.dentifier2 system of the ABN AMRO bank, in which customers use a USB-connected device - a smartcard reader with a display and numeric keyboard - to authorise transactions with their bank card and PIN code. He found a security vulnerability in the E.dentifier2 that was so serious that he even made it to the evening news on Dutch national TV. He did

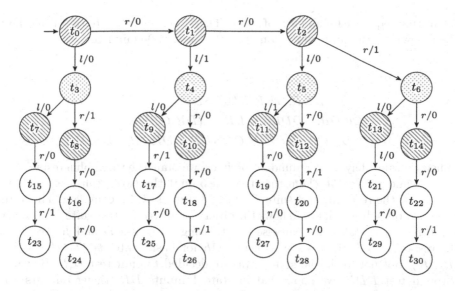

Fig. 5. Observation tree for FSMs S and $\mathcal{M}$ from Fig. 4.

not use systematic testing techniques to find the vulnerability, but Chalupar et al. [1] used model learning to reverse engineer models of the E.dentifier2, demonstrating that this technique could have easily revealed the security problem.

Figure 6 shows the FSM S that specifies the required behavior of the E.dentifier2. There are three states $\{q_0, q_1, q_2\}$, five inputs $\{C, D, G, R, S\}$ and four outputs $\{C, L, T, OK\}$. We use commas to indicate multiple transitions. For instance, in S there is both a transition $q_1 \xrightarrow{C/OK} q_1$ and a transition $q_1 \xrightarrow{R/T} q_1$. We refer to Chalupar et al. [1] for a detailled explanation of the model. Note that $A = \{\epsilon, C, CD\}$ is an access sequence set for S, and sequence DR is a

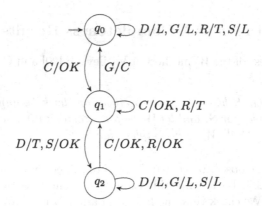

Fig. 6. FSM S that specifies the required behavior of the E.dentifier2.

separating sequence for all pairs of states. Therefore, according to the W-method, the following test suite, which comprises 18 tests, is 0-complete:[1]

$$T = \{DR, CDR, CDDR,$$
$$CDR, DDR, GDR, RDR, SDR,$$
$$CCDR, CDDR, CGDR, CRDR, CSDR,$$
$$CDCDR, CDDDR, CDGDR, CDRDR, CDSDR\}$$

Since we can safely omit redundant prefixes, we omit the three blue tests CDR, $CDDR$ and CDR. We claim that if we also omit the two green tests DR and $CDDDR$, the resulting test suite T' with 13 tests is still 0-complete. The argument is a bit subtle. Figure 7 shows the observation tree for test suite T'. Clearly, the three basis states are pairwise apart, using witnesses D and R. We know the response to distinguishing sequence DR for basis states t_1 (T OK) and t_2 (L OK), but not for initial state t_0 (as we decided to omit test DR). However, through test DDR, we know that in state 1 inputs DR trigger outputs LT. Under the assumption that $\mathcal{M}$ has three states, state 1 must be equivalent to either state t_0, t_1 or t_2. Therefore, the outcome of test DR in the initial state must be LT! Once we extend the observation tree with the inferred outcome of test DR, all F^0 states are easily identified. Since the basis and F^0 are complete, we may apply Theorem 3.5 to conclude that T' is 0-complete. Clearly, we cannot omit any other test, since then at least one F^0 state would no longer be visited, and $\mathcal{M}$ would no longer be uniquely determined.

Remark 3.10. Suppose observation tree $\mathcal{T}$ has N states. Then we can check in $\mathcal{O}(N^2)$ time for each pair of states of $\mathcal{T}$ whether they are apart or not. Using this information, we can check in $\mathcal{O}(N^2)$ time whether the conditions of Theorem 3.5 hold. This means that we can also check in $\mathcal{O}(N^2)$ time whether a test can be removed without compromising k-completeness. Note, however, that the size N of $\mathcal{T}$ grows exponentially in k.

4   Deriving Previous k-Completeness Results

The k-completeness of the W-method of Vasilevskii [20] and Chow [3] is a corollary of Theorem 3.5.

Corollary 4.1. *Let S be a minimal specification, let k be any natural number, let A be a state cover for S, and let W be a characterization set for S. Then the test suite $T = A \cdot I^{\leq k+1} \cdot W$ is k-complete for S.*

Proof. Consider the observation tree $\mathcal{T}$ for S obtained by running all tests from T. Let $B, F^0, F^1, \ldots$ be the stratification of $Q^{\mathcal{T}}$ induced by A. Let ρ be an element from W. We check that the following assumptions of Theorem 3.5 hold:

---

[1] Test $CDGDR$ reveals the problem with the faulty implementation of the E.dentifier2 that was discovered by Arjan Blom.

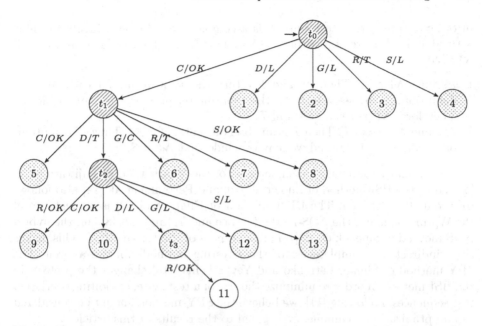

Fig. 7. Observation tree for test suite T' (for readability, we omitted the distinguishing sequence DR from the leaves in F^0)

1. Basis B is complete: Assume $\sigma \in A$. Then $\sigma\rho \in T$. Since $\mathcal{T}$ is constructed by running all tests from T, this implies that $\mathcal{T}$ contains a state q that is reached by input sequence σ. By definition, $q \in B$.

2. Frontier F^0 is complete: Suppose $q \in B$ and $i \in I$. Since q is a basis state, q is reached by a sequence $\sigma \in A$. Since $\sigma\, i\, \rho \in T$, $\delta^{\mathcal{T}}(q, i) \downarrow$.

3. For each $0 < j \le k$, frontier F^j is complete: Suppose $q \in F^{j-1}$ and $i \in I$. By induction, we may show that there exists a basis state r and a sequence $\tau \in I^j$ such that q is reached from r via input sequence τ. Let σ be the access sequence for state r. Then $\sigma \cdot \tau \cdot i \cdot \rho \in T$. This implies $\delta^{\mathcal{T}}(q, i) \downarrow$, as required.

4. All states in B and $F^0, \ldots, F^k$ are identified: Suppose $q \in B \cup F^0 \cup \cdots \cup F^k$. Then by construction of T, for each $\rho \in W$, $\delta^{\mathcal{T}}(q, \rho) \downarrow$. Now suppose r and r' are two distinct states in B. Then, since W is a characterization set for $\mathcal{S}$, $\mathcal{T}$ is an observation tree for $\mathcal{S}$, A is a state cover for $\mathcal{S}$, and $B, F^0, F^1, \ldots$ is the stratification induced by A, there exists a $\rho \in W$ such that $\rho \vdash r \mathbin{\#} r'$. Therefore, by the weak co-transitivity Lemma 2.8, either $q \mathbin{\#} r$ or $q \mathbin{\#} r'$. Since r and r' were chosen arbitrarily, this implies that state q is identified.

5. Co-transitivity: Suppose $r \in B$, $t' \in F^k$, $t'' \in F^0 \cup \cdots \cup F^{k-1}$ and $r \mathbin{\#} t'$. By the previous item, state t' is identified, that is, there exists a r' such that $C(t') = r'$. Since $r \mathbin{\#} t'$, we know that $r \ne r'$. Therefore, repeating the argument from the previous item, there exists a $\rho \in W$ such that $\rho \vdash r \mathbin{\#} r'$. By construction of T, for each $\tau \in W$, $\delta^{\mathcal{T}}(t', \tau) \downarrow$ and $\delta^{\mathcal{T}}(t'', \tau) \downarrow$. Therefore, by weak co-transitivity and because $C(t') = r'$, $\rho \vdash r \mathbin{\#} t'.$, Another application of weak co-transitivity now gives $\rho \vdash r \mathbin{\#} t''$ or $\rho \vdash t' \mathbin{\#} t''$.

In order to prove that test suite T is k-complete, let $\mathcal{M}$ be a Mealy machine with at most k more states than $\mathcal{S}$. We show that $\mathcal{M}$ passes T if and only if $\mathcal{M} \approx \mathcal{S}$:

1. Assume $\mathcal{M} \approx \mathcal{S}$. Then $\mathcal{M}$ and $\mathcal{S}$ return the same result for each test. By definition, $\mathcal{M}$ passes a test σ if the resulting outputs are the same as for $\mathcal{S}$. Therefore, $\mathcal{M}$ passes all tests in T.
2. Assume $\mathcal{M}$ passes T. Then T is an observation tree for $\mathcal{M}$. Thus all conditions of Theorem 3.5 hold, and we may conclude that $\mathcal{M} \approx \mathcal{S}$.

Via similar arguments, k-completeness of the Wp-method of Fujiwara et al. [5] and of the HSI-method of Luo et al. [10] and Petrenko et al. [14] also follows from our Theorem 3.5. The UIOv-method of Chan et al. [2] is an instance of the Wp-method, and the ADS-method of Lee and Yannakakis [8] and the hybrid ADS method of Smeenk et al. [16] are instances of the HSI-method. This means that, indirectly, k-completeness of these testing methods follows as well. The SPY-method of Simao, Petrenko and Yevtushenko [15] changes the prefixes in the HSI-method in order to minimize the size of a test suite, exploiting overlap in test sequences. Following [11], we believe the SPY-method should be considered as an optimization technique, orthogonal to the results of this article.

5   Conclusions and Future Work

We provided a sufficient condition for k-completeness in terms of apartness of states of the observation/prefix tree induced by a test suite. Our condition can be checked efficiently (in terms of the size of the test suite) and can be used to prove k-completeness of several methods for test suite generation that have been proposed in the literature.

Our characterization of k-completeness in terms of apartness triggers several questions. For instance:

1. Closest to our work is probably the result of Moerman [11, Proposition 2, Chapter 2], which provides sufficient conditions for a test suite of a certain shape to be k-complete. It would be interesting to see if this result of [11] follows from our result.
2. An intriguing question is whether it is possible to strengthen our result, that is, if weaker conditions exist that are still sufficient for k-completeness. An exciting perspective would be to come up with conditions that are not only sufficient but also necessary for k-completeness.
3. Our result suggest simple progress measures for performing a k-complete test suite, namely the sum of the number of elements of $B \cup F^0 \cup \cdots \cup F^k$ and the number of established apartness pairs required for state identification and co-transitivity. This progress measure in a way quantifies our uncertainty about the correctness of the implementation. A natural question then is to search for test queries that lead to a maximal increase of the progress measure. In order to reduce our uncertainty as fast as possible and/or to find bugs as quickly as possible, it makes sense to give priority to these tests.

4. It will be interesting to explore if our characterization can be used to develop efficient test suite generation algorithms, or efficient algorithms for pruning test suites that have been generated by other methods. Of course, scalability may become an issue: for large specifications, large input alphabets and large values of k it may no longer be feasible to store the full observation tree in main memory. However, for many practical benchmarks (see e.g. [12]) it should not be a problem to handle the observation trees for k-complete test suites for $k = 2$ or $k = 3$.

Acknowledgements. Many thanks to Joshua Moerman, Jurriaan Rot and the anonymous reviewers for their feedback on an earlier version of this article.

References

1. Chalupar, G., Peherstorfer, S., Poll, E., de Ruiter, J.: Automated reverse engineering using Lego. In: Proceedings 8th USENIX Workshop on Offensive Technologies (WOOT 2014), San Diego, California. IEEE Computer Society, Los Alamitos (2014)
2. Chan, W.Y.L., Vuong, C.T., Otp, M.R.: An improved protocol test generation procedure based on UIOs. In: SIGCOMM 1989, pp. 283–294. Association for Computing Machinery, New York (1989). https://doi.org/10.1145/75246.75274
3. Chow, T.: Testing software design modeled by finite-state machines. IEEE Trans. Softw. Eng. **4**(3), 178–187 (1978)
4. Dorofeeva, R., El-Fakih, K., Maag, S., Cavalli, A.R., Yevtushenko, N.: FSM-based conformance testing methods: a survey annotated with experimental evaluation. Inf. Softw. Technol. **52**(12), 1286–1297 (2010). https://doi.org/10.1016/j.infsof.2010.07.001
5. Fujiwara, S., Bochmann, G.V., Khendek, F., Amalou, M., Ghedamsi, A.: Test selection based on finite state models. IEEE Trans. Softw. Eng. **17**(6), 591–603 (1991)
6. Geuvers, H., Jacobs, B.: Relating apartness and bisimulation. Logical Methods Comput. Sci. **17**(3) (2021). https://doi.org/10.46298/lmcs-17(3:15)2021
7. Heyting, A.: Zur intuitionistischen Axiomatik der projektiven Geometrie. Math. Ann. **98**, 491–538 (1927)
8. Lee, D., Yannakakis, M.: Testing finite-state machines: state identification and verification. IEEE Trans. Comput. **43**(3), 306–320 (1994)
9. Lee, D., Yannakakis, M.: Principles and methods of testing finite state machines – a survey. Proc. IEEE **84**(8), 1090–1123 (1996)
10. Luo, G., Petrenko, A., v. Bochmann, G.: Selecting test sequences for partially-specified nondeterministic finite state machines. In: Mizuno, T., Higashino, T., Shiratori, N. (eds.) Protocol Test Systems. ITIFIP, pp. 95–110. Springer, Boston, MA (1995). https://doi.org/10.1007/978-0-387-34883-4_6
11. Moerman, J.: Nominal techniques and black box testing for automata learning. Ph.D. thesis, Radboud University Nijmegen (2019)
12. Neider, D., Smetsers, R., Vaandrager, F., Kuppens, H.: Benchmarks for automata learning and conformance testing. In: Margaria, T., Graf, S., Larsen, K.G. (eds.) Models, Mindsets, Meta: The What, the How, and the Why Not? LNCS, vol. 11200, pp. 390–416. Springer, Cham (2019). https://doi.org/10.1007/978-3-030-22348-9_23

13. Park, D.: Concurrency and automata on infinite sequences. In: Deussen, P. (ed.) GI-TCS 1981. LNCS, vol. 104, pp. 167–183. Springer, Heidelberg (1981). https://doi.org/10.1007/BFb0017309
14. Petrenko, A., Yevtushenko, N., Lebedev, A., Das, A.: Nondeterministic state machines in protocol conformance testing. In: Rafiq, O. (ed.) Protocol Test Systems, VI, Proceedings of the IFIP TC6/WG6.1 Sixth International Workshop on Protocol Test systems, Pau, France, 28–30 September 1993. IFIP Transactions, vol. C-19, pp. 363–378. North-Holland (1993)
15. da Silva Simão, A., Petrenko, A., Yevtushenko, N.: On reducing test length for FSMs with extra states. Softw. Test. Verification Reliab. **22**(6), 435–454 (2012). https://doi.org/10.1002/STVR.452
16. Smeenk, W., Moerman, J., Vaandrager, F., Jansen, D.N.: Applying automata learning to embedded control software. In: Butler, M., Conchon, S., Zaïdi, F. (eds.) ICFEM 2015. LNCS, vol. 9407, pp. 67–83. Springer, Cham (2015). https://doi.org/10.1007/978-3-319-25423-4_5
17. Smetsers, R., Moerman, J., Jansen, D.N.: Minimal separating sequences for all pairs of states. In: Dediu, A.-H., Janoušek, J., Martín-Vide, C., Truthe, B. (eds.) LATA 2016. LNCS, vol. 9618, pp. 181–193. Springer, Cham (2016). https://doi.org/10.1007/978-3-319-30000-9_14
18. Troelstra, A.S., Schwichtenberg, H.: Basic Proof Theory. Cambridge Tracts in Theoretical Computer Science, 2 edn. Cambridge University Press (2000). https://doi.org/10.1017/CBO9781139168717
19. Vaandrager, F., Garhewal, B., Rot, J., Wißmann, T.: A new approach for active automata learning based on apartness. In: Fisman, D., Rosu, G. (eds.) TACAS 2022. LNCS, vol. 13243, pp. 223–243. Springer, Cham (2022). https://doi.org/10.1007/978-3-030-99524-9_12
20. Vasilevskii, M.: Failure diagnosis of automata. Cybern. Syst. Anal. **9**(4), 653–665 (1973). https://doi.org/10.1007/BF01068590. (Translated from Kibernetika, No. 4, pp. 98–108, July–August 1973)
21. Weyuker, E.J.: Assessing test data adequacy through program inference. ACM Trans. Program. Lang. Syst. **5**(4), 641–655 (1983). https://doi.org/10.1145/69575.357231

The Interval Domain in Homotopy Type Theory

Niels van der Weide[1]([⊠]) and Dan Frumin[2]

[1] Radboud University Nijmegen, Nijmegen, The Netherlands
nweide@cs.ru.nl
[2] University of Groningen, Groningen, The Netherlands
d.frumin@rug.nl

Abstract. Even though the real numbers are the cornerstone of many fields in mathematics, it is challenging to formalize them in a constructive setting, and in particular, homotopy type theory. Several approaches have been established to define the real numbers, and the most prominent of them are based on Dedekind cuts and on Cauchy sequences. In this paper, we study a different approach towards defining the real numbers. Our approach is based on domain theory, and in particular, the interval domain, and we build forth on recent work on domain theory in univalent foundations. All the results in this paper have been formalized in Coq as part of the UniMath library.

1 Introduction

The real numbers are one of the basic objects in mathematics, with a wide variety of applications ranging from geometry to probablistic programming. Classically, there are many established approaches for defining the reals. However, formalizing real numbers in a constructive setting remains a challenging task.

There are two causes for this difficulty. Firstly, there are many different representations of real numbers, such as Dedekind cuts and Cauchy sequences (see the work of Geuvers et al. [23] for discussion of some representations). These two representations are equivalent to each other in a classical setting, and then it does not matter which one is used. However, this equivalence depends on the axiom of countable choice, meaning that these two representations are not equivalent in a constructive setting.

Depending on the precise foundations, Dedekind cuts and Cauchy sequences have their own drawbacks. Since Dedekind cuts are defined using power sets, they raise the universe level in a predicative setting. If one assumes impredicativity (e.g., propositional resizing), then one can avoid going to a larger universe. For Cauchy reals, on the other hand, one needs to use either quotients or setoids to guarantee that the equality relation is correct. As a consequence, one either has

---

This paper is dedicated to our supervisor and teacher, Herman Geuvers, on the occasion of his birthday. Herman, you teachings and research in domain theory, type theory, and mechanized proofs have greatly inspired us, and we hope that you enjoy this little expedition in the world of constructive real numbers, combining all the topics above.

© The Author(s), under exclusive license to Springer Nature Switzerland AG 2024
V. Capretta et al. (Eds.): *Logics and Type Systems in Theory and Practice*, LNCS 14560, pp. 241–256, 2024.
https://doi.org/10.1007/978-3-031-61716-4_16

to assume the axiom of countable choice to tame the quotients, or one has to manually deal with setoids. Note that if one assumes quotient inductive-inductive types, then one can avoid both of these [25].

Another challenge comes from defining the division operation on the real numbers. Since we cannot divide by zero, division is a partial operation: its input consists of a numerator, a denominator, and a proof that the denominator is distinct from 0. To define this operation constructively, the proof that the denominator is non-zero should be positive, because this allows us to extract a suitable approximation. For this reason, constructive versions of the notion of fields make use of *apartness relations*.

In this paper we study a constructive formalization of the (Dedekind) real numbers based on the ideas from domain theory. This approach, inspired by recent work of De Jong [17,18], allows us to formulate real numbers without setoids, quotients, or higher inductive-inductive types, and gives us an easier treatment of apartness, by appealing to generic domain theoretic principles. For example, the fact that arithmetic operations are strongly extensional (i.e. they reflect apartness) follow from general facts of domain theory. The main result is a construction of an ordered field of real numbers in univalent foundations. While this theorem is not new, the method that we used to prove it, is new.

More specifically, we construct a domain $\mathbb{IR}$ of *interval reals* (sometimes referred to as "partial real numbers"), and define real numbers to be the maximal elements of this domain. Classically, the domain $\mathbb{IR}$ contains intervals $[x, y] \subseteq \mathbb{R}$ of real numbers ordered by reverse inclusion. From the computational point of view, an interval $[x, y]$ represents a computation of a real number that only results in an approximate information, that the real number lies in the interval $[x, y]$. The higher we go in the domain, the more information we get, and the smaller the interval becomes. The maximal elements of this domain are then the singleton intervals $[x, x]$ representing exact real numbers.

Since our goal is to construct the real numbers from the interval domain, there are slight differences in our approach. If we would use the definition of $\mathbb{IR}$ as above, then we would get a circular definition. Instead, we construct the domain $\mathbb{IR}$ out of rational numbers using general techniques from domain theory. More specifically, we define $\mathbb{IR}$ as the rounded ideal completion of rational intervals ordered by reverse strict inclusion. Secondly, the notion of maximality is not suitable in a constructive setting, and instead, the correct notion of maximality in a constructive setting is *strong maximality* [16,17]. For instance, if we stick with regular maximality, then proving that the maximal elements of $\mathbb{IR}$ are the Dedekind reals requires weak excluded middle.

Just constructing the type of real numbers is, of course, not enough. We show that the real numbers form a constructive ordered Archimedean field. To do so, we define a number of arithmetical operations on the real numbers. Domain theory helps us with that: we can extend operations on rational intervals to the real numbers. The main challenge then lies in proving that these operations preserve strongly maximal elements (i.e. they give rise to operations on the real numbers). To help us with that task, we follow the approach of Bauer and Taylor [5], and we identify a number of *locatedness* properties for interval reals. In addition, a constructive field comes with an apartness relation that should satisfy several properties. Here another application of domain theory arises, because every DCPO comes with an intrinsic apartness relation [17], which is well-behaved on strongly maximal elements.

Foundations. The results in this paper are formalized in Coq using the Uni-Math library [36]. Note that even though UniMath is based on univalent foundations, this paper is written in the language of set theory, for ease of understanding. In this paper, we also assume propositional resizing, which says that every proposition in universe level is equivalent to one in the lowest universe.

The formalization can be found online in a "frozen" state at https://zenodo.org/doi/10.5281/zenodo.10664690. The entry point to our results is the module `UniMath.OrderTheory.DCPOs.Examples.Reals`. Our formalization is mostly complete, with several admitted results about interval arithmetic, and we are in the proceess of merging the formalization into the upstream UniMath repository.

Synopsis. The rest of the paper is organized as follows. In Sect. 2 we recall the preliminaries on domain theory. After that, we construct the DCPO of interval reals in Sect. 3, and we define the real numbers to be the strongly maximal interval reals. In Sect. 4, we characterize strong maximality of interval reals via a notion of *locatedness*. In Sect. 5 we define arithmetic operations on $\mathbb{IR}$, and we show that they restrict to operations on $\mathbb{R}$. We show that the real numbers form a constructive ordered field in Sect. 6. Finally, in Sect. 7 we conclude and discuss related work.

2 Preliminaries on Domain Theory

In this section we briefly recall the notions from domain theory needed in the remainder of the paper. Most the material here is standard (we refer an interested reader to a classical text on domain theory [1]), or, when mentioned, to the recent work of De Jong on constructive domain theory [17].

2.1 (Continuous) DCPOs

A *directed-complete partial order* (DCPO) is a set D together with a partial order $\sqsubseteq$ such that every directed subset X of D (i.e. a subset that contains upper bounds for each pair of elements) has the least upper bound (lub) $\bigsqcup X$. Morphisms between DCPOs are *Scott-continuous* functions, i.e. monotone functions that preserve the least upper bounds.

Remark 1. Type theoretically, a directed set in a partial order $(D, \sqsubseteq)$ is given by a type I and a function $f : I \to D$ such that I is inhabited (written $\|I\|$), and that $\{f(x) \mid x \in I\}$ is directed in the usual sense. Notice that here we follow De Jong and Escardó [27] and take directed sets to be inhabited. In particular, this means that DCPOs do not have to be pointed.

Given two elements $x, y \in D$ of a DCPO, we say that x is *way below* y, denoted as $x \ll y$, if for any directed set $X \subseteq D$, if $y \sqsubseteq \bigsqcup X$, then there exists some element $b \in X$ for which $x \sqsubseteq b$.

We are particularly interested in DCPOs that are "generated" by the way below relation. We say that a DCPO D is *continuous* if for every $x \in D$ the set $\{y \in D \mid y \ll x\}$ is directed and its supremum is x itself. There is a systematic way of constructing continuous DCPO out of what is called an *abstract basis*.

Definition 2. *An **abstract basis** is a set B together with a transitive relation $\prec$ that satisfies the following interpolation property:*

- *for each $a \in B$ there is a element $b \in B$ such that $b \prec a$;*
- *for each $a_1, a_2, b \in B$ such that $a_1, a_2 \prec b$, there is an interpolant $a \in B$ such that $a_1, a_2 \prec a \prec b$.*

Given an abstract basis $(B, \prec)$, we construct a DCPO $\mathsf{RIdl}(B, \prec)$ in which the way below relation is induced by the $\prec$ relation of the basis.

Definition 3. *A **rounded ideal** is a set $X \subseteq B$ of basis elements that is inhabited, downwards closed, and contains upper bounds: if $a_1, a_2 \in X$ then there exists some $b \in X$ such that $a_1, a_2 \prec b$.*

*The **rounded ideal completion** of B is defined to be the DCPO whose underlying set consists of all rounded ideals over the basis $(B, \prec)$, and whose order is given by the subset relation.*

There is a monotone map from B to $\mathsf{RIdl}(B, \prec)$ sending a basis element b to the *principal rounded ideal* $\downarrow(b) = \{a \in B \mid a \prec b\}$. There are some important facts about rounded ideal completions that we use.

Lemma 4. *Every element $X \in \mathsf{RIdl}(B, \prec)$ is the least upper bound of the basis elements included in it: $X = \bigcup\{\downarrow b \mid b \in X\}$.*

Lemma 5. *The way below relation on $\mathsf{RIdl}(B, \prec)$ is "induced" by $\prec$. In the sense that for any elements a, b of the basis we have:*

- *If $a \prec b$ then $\downarrow a \ll \downarrow b$;*
- *If $a \in X$ then $\downarrow a \ll X$;*
- *If $X \ll Y$ then there exists $b \in Y$ such that $X \subseteq \downarrow b$.*

The rounded ideal completion satisfies the universal property which allows us to lift monotone functions from the basis to the ideal completion.

Definition 6. *Let $f : B \to Y$ be a function from an abstract basis B to the DCPO Y such that for all $a, b \in B$ with $\downarrow a \subseteq \downarrow b$, we have $f(a) \sqsubseteq f(b)$. Then we extend f to a Scott-continuous function $f^* : \mathsf{RIdl}(B, \prec) \to Y$ defined as*

$$f^*(X) = \bigsqcup\{f(b) \mid b \in X\}.$$

The extension commutes with the principal ideals up to inequality, which means that $f^*(\downarrow b) \sqsubseteq f(b)$. It is the greatest function with such property. More specifically, for every Scott-continuous function g that satisfies $g(\downarrow b) \sqsubseteq f(b)$ we have $g(X) \sqsubseteq f^*(X)$ for any ideal X.

2.2   Scott Topology and Apartness

Every DCPO comes with an intrinsic notion of topology, in which the open sets are *Scott-open*. A set $X \subseteq D$ is Scott-open if it is upwards closed and whenever $\bigsqcup Y \in X$ then there is already some $y \in Y$ for which $y \in X$. This forms a topology in the usual sense, and if the DCPO D is a rounded ideal completion, then the topology is generated by the principal open sets over the principal ideals. Using the Scott topology, we define the *intrinsic apartness* relation. We say that x is apart from y, written $x \# y$, if there exists a Scott open set containing x and not containing y, or the other way around.

Intrinsic apartness is irreflexive and symmetric, but in general it is not tight or cotransitive (see [17, Theorem 58]). A stronger notion, also stemming from topology is *Hausdorff-separatedness*. The elements x and y are Hausdorff-separated if there are disjoint Scott-open sets S_1, S_2 such that $x \in S_1$ and $y \in S_2$. We have the following characterization of Hausdorff-separatedness in continuous DCPOs.

Proposition 7 ([18, Lemma 76])**.** *Two elements $x, y \in D$ of a continuous DCPO with a basis B are Hausdorff-separated iff there are elements $a, b \in B$ such that (i) $a \ll x$, (ii) $b \ll y$, (iii) there is no $z \in D$ such that $a \ll z$ and $b \ll z$.*

Proposition 8. *Scott continuous maps reflect apartness. More specifically, given a Scott continuous map $f : D_1 \to D_2$ and elements $x, y : D_1$ such that $f(x) \# f(y)$, we have $x \# y$.*

2.3   Strongly Maximal Elements

The right notion of maximality in a constructive setting is *strong maximality*. In the context of impredicative set theory it was studied by De Jong [17], based on a notion of constructively maximal elements, studied in classical meta-theory by Smyth [34]. The work by De Jong carries over to homotopy type theory with resizing without any essential modification.

Definition 9 ([17, Definition 70])**.** *An element $x \in D$ of a DCPO is **strongly maximal** if for any $u, v \in D$ with $u \ll v$ either $u \ll x$ or v and x are Hausdorff-separated. If D is a continuous DCPO, then it suffices to consider the elements u, v from a basis of D.*

Proposition 10 ([17, Proposition 80])**.** *Strongly maximal elements in a continuous DCPO are maximal. If $x \in D$ is strongly maximal and $x \sqsubseteq y$, then $x = y$.*

Proposition 11 ([17, Proposition 85])**.** *The relative Scott topology on the strongly maximal elements in a continuous DCPO is Hausdorff, and the intrinsic apartness on strongly maximal elements coincides with Hausdorff-separatedness.*

Proposition 12. *The intrinsic apartness relation is tight and cotransitive on strongly maximal elements. That is to say, $\neg(x \# y)$ implies $x = y$, and $x \# y$ implies $x \# z \lor z \# y$.*

In [17], Proposition 12 is proven by showing that every strongly maximal element is sharp (see [17, Definition 59] for the precise definition), and that the corresponding statement holds for sharp elements.

3   The Interval Domain

In this section we construct the set $\mathbb{R}$ of real numbers, using the domain-theoretic techniques described in Sect. 2. To do so, first define the set of *open rational intervals*, and we show that it gives rise to an abstract basis in Proposition 13. We also define the arithmetical operations on rational intervals following [31], which we use in Sect. 5 to define arithmetic operations on real numbers. The set of *interval reals* is defined to be the rounded ideal completion of the rational intervals (Definition 14), and the *real numbers* are defined to be the strongly maximal interval reals (Definition 15).

An (open, non-empty) rational interval is a pair $I = (d, u) \in \mathbb{Q} \times \mathbb{Q}$ such that $d < u$. We write $\underline{I}$ and $\overline{I}$ for left and right endpoints of $I = (d, u)$, i.e. d and u respectively. By $\mathbf{IQ}$ we denote the set of rational intervals. We write $|I|$ for the size $\overline{I} - \underline{I}$ of the interval I.

Proposition 13. *We have an abstract basis whose underlying set consists of rational intervals. The order relation of this abstract basis is given by reverse strict inclusion $\supsetneq$, which is defined by*

$$I \supsetneq J \iff \underline{I} < \underline{J} \wedge \overline{J} < \overline{I}.$$

To prove Proposition 13, we need to show that $(\mathbf{IQ}, \supsetneq)$ satisfies the two interpolation properties in Definition 2. If we have an interval I, we must find a J such that $J \supsetneq I$. We can take J to be $(\underline{I} - 1, \overline{I} + 1)$. Furthermore, if we have intervals I_1, I_2, K such that $I_1 \supsetneq K$ and $I_2 \supsetneq K$, we need to find an interval J such that $I_1 \supsetneq J$, $I_2 \supsetneq J$, and $K \supsetneq J$. We define J as follows:

$$J = (\frac{\max\{\underline{I_1}, \underline{I_2}\} + \underline{K}}{2}, \frac{\min\{\overline{I_1}, \overline{I_2}\} + \overline{K}}{2}).$$

In Sect. 5, we define operations on real numbers, and for that, we need several arithmetical operations on real intervals [31]. Chiefly among them are various algebraic operations that are lifted from $\mathbb{Q}$ to $\mathbf{IQ}$. Some of the operations are given in Fig. 1, and the full implementations are found in the Coq source code.

Not all arithmetical laws hold for the operations in Fig. 1. For example, addition and multiplication on intervals is associative and commutative. However, distributivity and neutrality for addition and multiplication do not hold.

Next we define the *interval domain*. Usually, one first defines the real numbers, and then the interval domain is defined to be the collection of real intervals. One can then show that this is a continuous DCPO, and that the set of rational intervals forms a basis for this DCPO. Our definition is reversed compared to this approach: using the fact that the rational intervals form an abstract basis, we can take the rounded ideal completion to acquire a continuous DCPO with that basis.

$$I + J = (\underline{I} + \underline{J}, \overline{I} + \overline{J}) \qquad\qquad -I = (-\overline{I}, -\underline{I})$$

$$I * J = (\min P, \max P) \qquad\qquad \text{where } P = \{\underline{I} \cdot \underline{J}, \underline{I} \cdot \overline{J}, \overline{I} \cdot \underline{J}, \overline{I} \cdot \overline{J}\}$$

$$I \vee J = (\max\{\underline{I}, \underline{J}\}, \max\{\overline{I}, \overline{J}\}) \qquad I \wedge J = (\min\{\underline{I}, \underline{J}\}, \min\{\overline{I}, \overline{J}\})$$

$$I <_{\mathrm{IQ}} J \text{ iff } \overline{I} < \underline{J}$$

Fig. 1. Selected operations on the rational intervals.

Definition 14. *The **interval domain** is defined to be the rounded ideal completion* $\mathsf{RIdl}(\mathrm{I}\mathbb{Q}, \supsetneq)$ *of open rational intervals ordered by reverse strict inclusion. We denote the interval domain by* $(\mathrm{I}\mathbb{R}, \subseteq)$, *and call elements of* $\mathrm{I}\mathbb{R}$ ***interval reals***.

Note that we can recover the real numbers from $\mathrm{I}\mathbb{R}$. This is because the real numbers can be identified with intervals consisting of only one element. Those intervals are the largest with respect to reverse inclusion. As such, we recover the real numbers by looking at the strongly maximal elements.

Definition 15. *The set* $\mathbb{R}$ *of **real numbers** is defined to be the collection of strongly maximal elements of* $\mathrm{I}\mathbb{R}$.

Note that from Definition 15 we directly obtain the apartness relation on the real numbers. This is because every DCPO has an apartness relation on its strongly maximal elements that is irreflexive, symmetric, cotransitive, and tight by Proposition 12.

4 Strong Maximality and Locatedness

To construct a real number using Definition 15, we need to do two things. We must describe an interval real (i.e. an element of $\mathrm{I}\mathbb{R}$), which describes all approximations to the real number, and then we must show this interval real is strongly maximal. Giving a direct proof for such facts is rather complicated, and for that reason, we give a characterization for strong maximality of interval reals in this section.

The characterization of strong maximality that we use, is based on *locatedness*. Our notion of locatedness is similar to the locatedness condition of Dedekind cuts, but phrased using intervals.

Definition 16. *A interval real* $x \in \mathrm{I}\mathbb{R}$ *is **order located** (or simply **located**) if for any rational interval I there exists a rational interval $J \in x$ such that either* $\underline{I} < \underline{J}$ *or* $\overline{J} < \overline{I}$.

Before we show that order locatedness coincides with strong maximality, we will need the following characterization of Hausdorff separatedness.

Lemma 17. *Interval reals x and y are Hausdorff separated iff there are nonintersecting rational intervals I, J such that $I \in x$ and $J \in y$.*

Theorem 18. *An interval real x is order located if and only if x is strongly maximal.*

Proof. Suppose that x is order located and let I, J be intervals such that $\downarrow I \ll \downarrow J$ (which, by Lemma 22 is equivalent to $I \supsetneq J$), and we are to decide if $\downarrow I \ll x$ (i.e. $I \in x$) or $\downarrow J$ and x are Hausdorff-separated.

Since $I \supsetneq J$, we can consider an interval $L = (\underline{I}, \underline{J})$. By order locatedness, there exists an interval $K \in x$ such that either $\underline{L} = \underline{I} < \underline{K}$ or $\overline{K} < \overline{L} = \underline{J}$. In the latter case we know that K and J are completely disjoint: by Lemma 17 x and J are Hausdorff-separated. In the former case, we consider an interval $R = (\overline{J}, \overline{I})$, and locate it within x. We get an interval $K' \in x$ such that either $\underline{R} = \overline{J} < \underline{K'}$ or $\overline{K'} < \overline{R} = \overline{I}$; we again consider two cases. In the latter case we have $\overline{K'} < \overline{I}$ and $\underline{I} < \underline{K}$. Since x is a rounded ideal, there exists an interval $N \in x$ which refines both K and K': $K, K' \supsetneq N$. It follows that $I \supsetneq N$, and, therefore $I \in x$.

In the former case we have $\overline{J} < \underline{K'}$. Then J and K' are disjoint, and it follows that x and $\downarrow J$ are Hausdorff-separated by Lemma 17.

For the converse, we suppose that I is an interval, and we need to locate it within x. We can always find a smaller interval $I \supsetneq J$; so $\downarrow I \ll \downarrow J$. By strong maximality, either $\downarrow I \ll x$ (equivalently, $I \in x$), or x and $\downarrow J$ are Hausdorff separated.

In the former case, by roundedness of x, there exists some interval $K \in x$ such that $I \supsetneq K$, and it follows that $\underline{I} < \underline{K}$.

In the latter case, by Lemma 17 there are $J_1 \subsetneq J$ and $J_2 \in x$ such that J_1 and J_2 do not intersect. If J_1 lies to the left of J_2, then $\underline{I} < \overline{J} < \overline{J_1} \leq \underline{J_2}$. If J_1 lies to the right of J_2, then $\overline{J_2} \leq \underline{J_1} < \underline{J} < \overline{I}$.

From Theorem 18, we directly obtain that our real numbers are equivalent to the Dedekind real numbers, which was originally proven by De Jong in [16].

Corollary 19 ([16, Theorem 102]). *The set of real numbers from Definition 15 is equivalent to the set of Dedekind real numbers.*

As an application of Theorem 18, we construct the inclusion from the rational numbers to the real numbers. First, we show that every rational number gives rise to an interval real.

Definition 20. *Let q be a rational number. We define an interval real $\lceil q \rceil$ to be the supremum:* $\bigcup \{ \downarrow I \mid q \in I \}$.

Note that the supremum exists in Definition 20, because the set $\{ \downarrow I \mid q \in I \}$ is directed. Intuitively, this definition says that approximations of $\lceil q \rceil$ are given by intervals that contain q.

Before we show that $\lceil q \rceil$ actually gives rise to a real number, we characterize the way below relation for $\lceil q \rceil$.

Lemma 21. *For all $q \in \mathbb{Q}$ and rational intervals I, we have $\downarrow I \ll \lceil q \rceil$ iff $q \in I$.*

Proof. Follows from Lemma 5 and the definition of $\lceil - \rceil$.

If $x \in \mathbb{IR}$, then from Lemma 5 we know that $I \in x$ implies $\downarrow I \ll x$, as in any continuous DCPO. However, for $\mathbb{IR}$ we also have the converse, giving us a characterization of the approximation of interval reals by rational intervals.

Lemma 22. *For any $I \in \mathbb{IQ}$ and any $x \in \mathbb{IR}$, we have $\downarrow I \ll x$ iff $I \in x$.*

Proof. Follows from the density of rationals; see also [17, Lemma 100].

Now we show that $\lceil q \rceil$ actually is strongly maximal.

Proposition 23. *For any $q \in \mathbb{Q}$, the interval real $\lceil q \rceil$ is strongly maximal.*

Proof. We use Theorem 18 and show that $\lceil q \rceil$ is order located. Given an interval $I \in \mathbb{IR}$, we know that either $\underline{I} < q$ or $q < \overline{I}$, by cotransitivity of rational numbers.

Without loss of generality, suppose that $\underline{I} < q$. Then we consider an interval $J = (\frac{\underline{I}+q}{2}, q + 1)$. By Lemmas 21 and 22, we have that $J \in \lceil x \rceil$. Furthermore, $\underline{I} < \underline{J} = \frac{\underline{I}+q}{2}$, concluding the proof.

5   Real Arithmetic

So far we have defined the set $\mathbb{R}$ of real numbers and the inclusion $\mathbb{Q} \to \mathbb{R}$. In this section we define the arithmetic operations (addition, multiplication, subtraction, and division) on $\mathbb{R}$, the order $<$, and the lattice operations. More specifically, we show that $\mathbb{R}$ forms an Archimedean field.

Definition 24 ([35, Definition 11.2.7]). *An **ordered field** consists of a set F together with elements $0 \in F$ and $1 \in F$, binary operations $+, *, \min, \max : F \to F \to F$, binary relations $\leq, <, \#$, and operations $- : F \to F$ and $(-)^{-1} : \{x \in F \mid x \# 0\} \to F$. The operations need to satisfy a number of standard laws [35, Definition 11.2.7].*
*A field F is called **Archimedean** if for all $x, y \in F$ such that $x < y$, there merely exists a $q \in \mathbb{Q}$ such that $x < q < y$.*

To define these operations on $\mathbb{R}$, we first lift a corresponding operation on intervals from Fig. 1 to the level of interval reals. After that, we show that the lifted operation maps real numbers to real numbers, i.e. they preserve strongly maximal elements. To prove that these operations preserve strong maximality, we use ideas from Bauer and Taylor [5]. More specifically, we define alternative notions of *locatedness* of interval reals, namely *arithmetic locatedness* and *multiplicative locatedness*, that are used to prove locatedness. We define all the operations in the same three steps: 1. identify the corresponding operation rational intervals; 2. lift the operation to interval reals; 3. prove that the lifted operation preserves strong maximality. To lift the functions to real numbers, we use the extension from Definition 6. We define a monotone function $f : \mathbb{IQ}^n \to \mathbb{IQ}$, and we define its extension $f_* : \mathbb{IR}^n \to \mathbb{IR}$ as follows

$$f_*(x_1, \ldots, x_n) = \bigcup \{\downarrow (f(I_1, \ldots, I_n)) \mid I_1, \ldots, I_n \in x_1, \ldots x_n\}.$$

We use the following properties of the lifting of operations.

Lemma 25. *The following statements hold.*

- $f_*(\downarrow I_1, \ldots, \downarrow I_n) \subseteq \downarrow f(I_1, \ldots, I_n)$;
- *If* $I_1 \in x_1, \ldots, I_n \in x_n$ *then* $f(I_1, \ldots, I_n) \in f_*(x_1, \ldots, x_n)$;
- *If* $K \in f_*(x_1, \ldots, x_n)$ *then there are* $I_1 \in x_1, \ldots, I_n \in x_n$ *such that* $f(I_1, \ldots, I_n) \subsetneq K$.

We now show how to use this lifting to define operations on interval reals, and show that they preserve strong maximality.

Additive Inverse. We start by defining the additive inverse function $-x$. It is a lifted extension of the corresponding function on the rational intervals:

$$-x = \bigcup\{\downarrow -I \mid I \in X\}.$$

Lemma 26. *If* $x \in \mathbb{IR}$ *is order located, then so is* $-x$.

Proof. Suppose that $I \in \mathbb{IQ}$. We are to locate I within $-x$. First, we use order-locatedness of x w.r.t. $-I$: there exists $J \in x$ such that $\underline{-I} < \underline{J} \vee \overline{J} < \overline{-I}$. Then, by Lemma 25 $-J \in -x$, and, furthermore, $\underline{I} < \underline{-J} = -\overline{J}$ or, similarly, $-\underline{J} < \overline{I}$.

Addition. Addition on reals is defined as follows

$$x + y = \bigcup\{\downarrow (I + J) \mid I \in x, J \in y\}.$$

In order to show that addition preserves strong maximality, we use the auxiliary notion of *arithmetic locatedness*.

Definition 27. *An interval real* $x \in \mathbb{IR}$ *is* **arithmetically located** *if for any rational number* $q > 0$ *there is a rational interval* $I \in x$ *such that the interval size* $|I|$ *satisfies* $|I| < q$.

Proposition 28. *Let* x *be arithmetically located, and* y *be order located. Then* $x + y$ *is order located.*

Proof. Let I be an interval that we are to locate in $x + y$. By arithmetic located-ness, there exists a $J_0 \in x$ with $|J_0| < |I|$. Writing it out, we have $\overline{J_0} - \underline{J_0} < \overline{I} - \underline{I}$, or, equivalently, $\underline{I} - \underline{J_0} < \overline{I} - \overline{J_0}$.

We then use order locatedness of y with respect to the interval $(\underline{I} - \underline{J_0}, \overline{I} - \overline{J_0})$. We get an interval $J_1 \in y$ such that either $\underline{I} - \underline{J_0} < \underline{J_1}$ or $\overline{J_1} < \overline{I} - \overline{J_0}$. Equivalently, $\underline{I} < \underline{J_0} + \underline{J_1}$ or $\overline{J_0} + \overline{J_1} < \overline{I}$. And, furthermore, by Lemma 25 we have $J_0 + J_1 \in x + y$, thus locating I in $x + y$.

If we show that every order located interval real is arithmetically located, that would conclude the construction of the addition operation on reals. In order to show this, we use the following auxiliary lemma.

Lemma 29. *Let $x \in \mathbb{IR}$ be order located and let $I \in x$. Suppose that J_0, J_1 are overlapping intervals that cover I. Then there exists $J' \in x$ such that either $J_0 \supsetneq J'$ or $J_1 \supsetneq J'$.*

Proposition 30. *If $x \in \mathbb{IR}$ is order located, then it is arithmetically located.*

Proof. Suppose that x is order located and $q > 0$ is a rational number. By roundedness, x contains some interval $I \in x$. Then we can cover the whole interval I with n overlapping intervals of size q, for some natural number n. We then use induction on n and Lemma 29 to find some interval in x that is strictly included in one of the covering intervals. By construction, the size of that interval will be strictly smaller than q.

Multiplication. For defining multiplication and proving that it preserves strong maximality, we follow the same approach as for addition. As the operation itself we take the lifting

$$x * y = \bigcup \{ \downarrow (I * J) \mid I \in x, J \in y \}.$$

In order to show that this operation preserves strong maximality we use an intermediate notion of multiplicative locatedness:

Definition 31. *An interval real x is **multiplicatively located** if for any rational interval J that lies to the right of 0 (i.e. $0 < \underline{J}$), there exists an interval $K \in x$ such that K lies to the right of 0, and $\underline{J} \cdot \overline{K} < \overline{J} \cdot \underline{K}$. Equivalently, we can say $\overline{K}/\underline{K} < \overline{J}/\underline{J}$.*

Lemma 32. *Suppose that x is an order located interval real that is positive (i.e. there is some $I \in x$ that lies to the right of 0). Then x is multiplicatively located.*

While the lemma above is stated only for positive reals, by playing around with signs and using multiplicative locatedness we can show the following.

Lemma 33. *Suppose that x and y are strongly maximal interval reals. Then $x * y$ is strongly maximal as well.*

Multiplicative Inverse. The multiplicative inverse requires special treatment, because it is defined only for interval reals apart from zero. For this reason, we do not define the multiplicative inverse by lifting an operation, but instead, we define it as a particular supremum.

We define a reciprocal of rational intervals as $I^{-1} = (\overline{I}^{-1}, \underline{I}^{-1})$, and we define the reciprocal of interval reals as the following supremum:

$$x^{-1} = \bigcup \{ \downarrow (I^{-1}) \mid I \in x \wedge I \# 0 \}.$$

This is the supremum over the set $\{ \downarrow (I^{-1}) \mid I \in x \wedge I \# 0 \}$ ranging over the intervals in x that do not contain 0 (denoted, abusing the notation slightly, as $I \# 0$). This set is always semidirected; furthermore it is inhabited, and therefore directed, whenever x is apart from zero.

Remark 34. The operation I^{-1} is only defined for intervals that do not contain zero – otherwise the operation does not map intervals to intervals. In type theory this is represented by the signature

$$(-,-)^{-1} : (\sum_{I:\mathbb{IQ}} I\#0) \to \mathbb{IQ}.$$

The reciprocal not only takes an interval as its argument, but also a proof that it is apart from 0. That means that the directed family in the definition of x^{-1} is represented in type theory as a function with the signature $(\sum_{I \in x} I\#0) \to \mathbb{IR}$, given by $(I, H) \mapsto \downarrow ((I, H)^{-1})$.

Similarly, the inverse operation for interval reals takes a proof of apartness from 0 as one of the arguments.

To show that the multiplicative inverse maps real numbers to real numbers, we prove the following:

Lemma 35. *If x is order located and apart from 0, then x^{-1} is also order located.*

Lattice Structure. Finally, we define the lattice operation on the real numbers (minimum, maximum), and the strict order. The strict order is defined as follows: given interval reals x, y, we say that $x < y$ if there are intervals $I \in x, J \in y$ such that $I <_{\mathbb{IQ}} J$. For the minimum and maximum, we use the same approach as for addition. We show that these these operations preserve strongly maximal elements, using only order locatedness.

6   Arithmetic Laws

Finally, we prove that the real numbers $\mathbb{R}$, together with the operations defined in the previous section, form a constructive ordered field. Some of the constructive field laws follow automatically: intrinsic apartness is reflected by Scott-continuous functions by Proposition 8. Thus, addition and multiplication automatically reflect apartness, as they are defined as Scott continuous extensions of operations on rational intervals. For the remaining constructive field laws, we need to put in a bit more work. Due to space reasons, we only sketch the proof for associativity of addition, and we note that that similar ideas are used to prove the other laws.

Proposition 36. *For any $x, y, z \in \mathbb{R}$ we have $(x + y) + z = x + (y + z)$.*

Proof. First of all, we notice that since we are working with strongly maximal elements, in order to show an equality it suffices to find a common upper bound h such that $(x + y) + z \subseteq h$ and $x + (y + z) \subseteq h$. Secondly, since we are working with an expression with three variables x, y, z, we introduce an intermediate *ternary* version of addition: we write $h(x, y, z)$ for an "unbiased" addition that sums the three numbers together directly

$$h(x, y, z) = \bigcup \{\downarrow (I_1 + I_2 + I_3) \mid I_1 \in x, I_2 \in y, I_3 \in z\}.$$

This approach of using "unbiased" operations is inspired by the "unbiased" monoidal products in monoidal categories [29, Section 3.1]; a similar trick was used to show associativity of smash products in the context of homotopy type theory [30].

Let us then look at how to show $(x + y) + z \subseteq h(x, y, z)$ for all x, y, z. Since h is defined as an extension of a monotone function on the basis, by the universal property it suffices to show $(\downarrow I_1 + \downarrow I_2) + \downarrow I_3 \subseteq \downarrow (I_1 + I_2 + I_3)$. Note that the addition symbol on the right hand side represents addition of rational intervals, not addition of reals. This inclusion then holds by Lemma 25.

Theorem 37. $\mathbb{R}$ *is an Archimedean constructive ordered field.*

7   Conclusions and Related Work

In this paper we presented a formalization of real numbers in the setting of univalent mathematics. We constructed the set of real numbers and we showed that it is an ordered archimedean field. In our formalization we took a novel approach to formulating Dedekind reals using domain theory, and several results were proven more generally. In the future work we would like to show completeness of $\mathbb{R}$ as well.

The topic of constructive real numbers and its formalization has received a lot of attention and has been studied broadly, going back all the way to the pioneering work of Bishop [7,8]. For an overview, see, for example, the surveys [23] (with a focus on constructivity, type theory and domain theory) and [9] (with a focus on formalization). We finish our paper by discussing selected related work on interval reals and domain theory in univalent foundations, and on constructive formalizations of real numbers in type theory.

Domain Theory and Interval Reals. The domain $\mathbb{IR}$ of interval real numbers was used to study computations with real numbers in the setting of domain theory, for example in semantics of RealPCF [19–21]. The interval domain has also been studied in the context of realizability [4], for the purposes of extracting programs for computing with exact real numbers.

In terms of formalizations of domain theory in type theory, we build upon the recent work of De Jong and Escardo on domain theory in the context of univalent mathematics [15,16,18,27].

Finally, while not directly related to interval reals, we would like to mention the work of Bauer and Taylor on constructing Dedekind reals in the context of Abstract Stone Duality [5]. Their work has also served as an inspiration to ours, especially with regard to different notions of locatedness.

Formalization of Reals in Type Theory. Specifically in the context of homotopy type theory/univalent foundations, real numbers have already been considered (with both Dedekind and Cauchy flavors) in the HoTT book [35, Chapter 11], but were not formalized at the time. Dedekind reals were later formalized as part of the UniMath library by Catherine Lelay. The formalization of Cauchy reals usually requires use of quotients and/or countable choice. However, the approach in the HoTT book sidesteps those issues by using a higher inductive-inductive types. This approach has been extended to generic metric space completion and formalized in [25], and it was further used for formalizing synthetic topology [6]. Another approach to the real numbers in univalent foundations is given by the Escardó-Simpson reals [20], which are equivalent to the Cauchy reals [11]. Those have been formalized by Ghica and Ambridge [2,24]. Booij also studied *locators* in univalent foundations [10]. Locators are an additional structure on top of a constructive field, such as the Dedekind reals, and it allows one to assign a decimal expansion to a Dedekind real. Booij also showed that a Dedekind real is a Cauchy real if and only if there is a locator for that number.

It is also worth mentioning the formalization projects from Nijmegen related to real numbers. Initially, as part of the FTA project [13] (formalization of the fundamental theorem of algebra), the real numbers were defined axiomatically, as an abstract interface, to facilitate proof modularity. Later, Niqui and Geuvers [22] developed an implementation of that interface, based on Cauchy reals. This implementation became the basis for FTA, and later became a part of C-CoRN [14], but the implementation was not suitable for extraction and evaluation. O'Conor then proposed another formalization of Cauchy reals [32,33] aimed at extracting and running programs. This approach was further refined by Krebbers and Spitters [28], utilizing the type class approach for formalizing algebraic hierarchies.

Among other formalizations, there is the ALEA Coq library [3], which builds up monadic semantics for a probabilistic programming language based on the real interval. However, the real interval is axiomatized as an abstract type and is not implemented. Another formalization of reals in Coq [12] defines real numbers through (coinductive) infinite streams. The formalization of real numbers in LEGO [26] is similar to ours, as it is based on (converging) nested rational intervals, but the general setting is quite different, and they forgo domain theory.

Acknowledgements. We would like to thank the anonymous reviewers for their comments and suggestions. The authors also thank Andrej Bauer and Tom de Jong for useful pointers to the literature.

References

1. Abramsky, S., Jung, A.: Domain theory (corrected and expanded version). In: Handbook of Logic in Computer Science, pp. 1–168. Oxford University Press (1994)
2. Ambridge, T.W.: Exact Real Search: Formalised Optimisation and Regression in Constructive Univalent Mathematics (2024). https://doi.org/10.48550/arXiv.2401.09270
3. Audebaud, P., Paulin-Mohring, C.: Proofs of randomized algorithms in Coq. Sci. Comput. Program. **74**(8), 568–589 (2009). https://doi.org/10.1016/j.scico.2007.09.002

4. Bauer, A., Kavkler, I.: A constructive theory of continuous domains suitable for implementation. Ann. Pure Appl. Logic **159**(3), 251–267 (2009). https://doi.org/10.1016/j.apal.2008.09.025
5. Bauer, A., Taylor, P.: The Dedekind reals in abstract Stone duality. Math. Struct. Comput. Sci. **19**(4), 757–838 (2009). https://doi.org/10.1017/S0960129509007695
6. Bidlingmaier, M.E., Faissole, F., Spitters, B.: Synthetic topology in homotopy type theory for probabilistic programming. Math. Struct. Comput. Sci. **31**(10), 1301–1329 (2021). https://doi.org/10.1017/S0960129521000165
7. Bishop, E.: Foundations of Constructive Analysis. McGraw-Hill (1967)
8. Bishop, E., Bridges, D.: Constructive Analysis. Springer, Berlin (1985)
9. Boldo, S., Lelay, C., Melquiond, G.: Formalization of real analysis: a survey of proof assistants and libraries. Math. Struct. Comput. Sci. **26**(7), 1196–1233 (2016). https://doi.org/10.1017/S0960129514000437
10. Booij, A.B.: Extensional constructive real analysis via locators. Math. Struct. Comput. Sci. **31**(1), 64–88 (2021). https://doi.org/10.1017/S0960129520000171
11. Booij, A.B.: The HoTT reals coincide with the Escardó-Simpson reals (2017). http://arxiv.org/abs/1706.05956
12. Ciaffaglione, A., Di Gianantonio, P.: A co-inductive approach to real numbers. In: Coquand, T., Dybjer, P., Nordström, B., Smith, J. (eds.) TYPES 1999. LNCS, vol. 1956, pp. 114–130. Springer, Heidelberg (2000). https://doi.org/10.1007/3-540-44557-9_7
13. Cruz-Filipe, L.: A constructive formalization of the fundamental theorem of calculus. In: Geuvers, H., Wiedijk, F. (eds.) TYPES 2002. LNCS, vol. 2646, pp. 108–126. Springer, Heidelberg (2003). https://doi.org/10.1007/3-540-39185-1_7
14. Cruz-Filipe, L., Geuvers, H., Wiedijk, F.: C-CoRN, the constructive Coq repository at Nijmegen. In: Asperti, A., Bancerek, G., Trybulec, A. (eds.) MKM 2004. LNCS, vol. 3119, pp. 88–103. Springer, Heidelberg (2004). https://doi.org/10.1007/978-3-540-27818-4_7
15. de Jong, T.: The Scott model of PCF in univalent type theory. Math. Struct. Comput. Sci. **31**(10), 1270–1300 (2021). https://doi.org/10.1017/S0960129521000153
16. de Jong, T.: Sharp elements and apartness in domains. Electron. Proc. Theor. Comput. Sci. **351**, 134–151 (2021). https://doi.org/10.4204/EPTCS.351.9
17. de Jong, T.: Apartness, sharp elements, and the Scott topology of domains. Math. Struct. Comput. Sci. 1–32 (2023). https://doi.org/10.1017/S0960129523000282
18. de Jong, T.: Domain theory in constructive and predicative univalent foundations. Ph.D. thesis, University of Birmingham (2023)
19. Escardó, M., Hofmann, M., Streicher, T.: On the non-sequential nature of the interval-domain model of real-number computation. Math. Struct. Comput. Sci. **14**(6), 803–814 (2004). https://doi.org/10.1017/S0960129504004360
20. Escardó, M.H.: PCF extended with real numbers. Theor. Comput. Sci. **162**(1), 79–115 (1996). https://doi.org/10.1016/0304-3975(95)00250-2
21. Escardó, M.H., Streicher, T.: Induction and recursion on the partial real line with applications to Real PCF. Theor. Comput. Sci. **210**(1), 121–157 (1999). https://doi.org/10.1016/S0304-3975(98)00099-1
22. Geuvers, H., Niqui, M.: Constructive reals in Coq: axioms and categoricity. In: Callaghan, P., Luo, Z., McKinna, J., Pollack, R., Pollack, R. (eds.) TYPES 2000. LNCS, vol. 2277, pp. 79–95. Springer, Heidelberg (2002). https://doi.org/10.1007/3-540-45842-5_6
23. Geuvers, H., Niqui, M., Spitters, B., Wiedijk, F.: Constructive analysis, types and exact real numbers. Math. Struct. Comput. Sci. **17**(1), 3–36 (2007). https://doi.org/10.1017/S0960129506005834

24. Ghica, D.R., Ambridge, T.W.: Global optimisation with constructive reals. In: 36th Annual ACM/IEEE Symposium on Logic in Computer Science, LICS 2021, Rome, Italy, 29 June–2 July 2021, pp. 1–13. IEEE (2021). https://doi.org/10.1109/LICS52264.2021.9470549

25. Gilbert, G.: Formalising real numbers in homotopy type theory. In: Proceedings of the 6th ACM SIGPLAN Conference on Certified Programs and Proofs, CPP 2017, pp. 112–124. Association for Computing Machinery, New York (2017). https://doi.org/10.1145/3018610.3018614

26. Jones, C.: Completing the rationals and metric spaces in LEGO. Logical Environ. 297–316 (1993)

27. de Jong, T., Escardó, M.H.: Domain theory in constructive and predicative univalent foundations. In: 29th EACSL Annual Conference on Computer Science Logic (CSL 2021). Leibniz International Proceedings in Informatics (LIPIcs), vol. 183, pp. 28:1–28:18. Schloss Dagstuhl–Leibniz-Zentrum für Informatik, Dagstuhl, Germany (2021). https://doi.org/10.4230/LIPIcs.CSL.2021.28

28. Krebbers, R., Spitters, B.: Type classes for efficient exact real arithmetic in Coq. Logical Methods Comput. Sci. **9**(1) (2013). https://doi.org/10.2168/LMCS-9(1:1)2013

29. Leinster, T.: Higher Operads, Higher Categories (2003). https://doi.org/10.48550/arXiv.math/0305049

30. Ljungstrom, A.: Symmetric Monoidal Smash Products in Homotopy Type Theory (2024). https://arxiv.org/abs/2402.03523

31. Moore, R.E.: Interval Analysis, vol. 4. Prentice-Hall Englewood Cliffs (1966)

32. O'Connor, R.: A monadic, functional implementation of real numbers. Math. Struct. Comput. Sci. **17**(1), 129–159 (2007). https://doi.org/10.1017/S0960129506005871

33. O'Connor, R.: Certified exact transcendental real number computation in Coq. In: Mohamed, O.A., Muñoz, C., Tahar, S. (eds.) TPHOLs 2008. LNCS, vol. 5170, pp. 246–261. Springer, Heidelberg (2008). https://doi.org/10.1007/978-3-540-71067-7_21

34. Smyth, M.B.: The constructive maximal point space and partial metrizability. Ann. Pure Appl. Logic **137**(1), 360–379 (2006). https://doi.org/10.1016/j.apal.2005.05.032

35. Univalent Foundations Program, T.: Homotopy Type Theory: Univalent Foundations of Mathematics (2013). https://homotopytypetheory.org/book, Institute for Advanced Study

36. Voevodsky, V., Ahrens, B., Grayson, D., et al.: UniMath—a computer-checked library of univalent mathematics. http://unimath.org. https://doi.org/10.5281/zenodo.7848572. https://github.com/UniMath/UniMath

Characterizing Morphic Sequences

Hans Zantema[✉]

Department of Computer Science, TU Eindhoven, P.O. Box 513,
5600 MB Eindhoven, The Netherlands
h.zantema@tue.nl

Abstract. Morphic sequences form a natural class of infinite sequences, extending the well-studied class of automatic sequences. Where automatic sequences are known to have several equivalent characterizations and the class of automatic sequences is known to have several closure properties, for the class of morphic sequences similar closure properties are known, but only limited equivalent characterizations. In this paper we extend the latter. We discuss a known characterization of morphic sequences based on automata and we give a characterization of morphic sequences by finiteness of a particular class of subsequences. Moreover, we relate morphic sequences to rationality of infinite terms and describe them by infinitary rewriting.

1 Introduction

The simplest class of infinite sequences over a finite alphabet are *periodic*: sequences of the shape u^∞ for some finite non-empty word u. The one-but-simplest are *ultimately periodic*: sequences of the shape vu^∞ for some finite non-empty words u, v. But these are still boring in some sense. More interesting are well-structured sequences that are simple to define, but not being ultimately periodic. One class of such sequences are *morphic sequences*, being the topic of this paper. As a basic example consider the morphism f replacing 0 by the word 01 and replacing 1 by the symbol 0. Then we obtain the following sequence of words:

$$f(0) = 01,$$

$$f^2(0) = f(f(0)) = f(01) = 010,$$

$$f^3(0) = f(f^2(0)) = f(010) = 01001,$$

$$f^4(0) = f(f^3(0)) = f(01001) = 01001010,$$

and so on. We observe that in this sequence of words for every word its predecessor is a prefix, so we can take the limit of the sequence of words, being the binary *Fibonacci sequence* fib. This is a typical example of a (pure) morphic sequence.

From 2007 until his retirement in 2022 Zantema had an exchange position with Herman Geuvers, in which Zantema was employed for one day a week at Radboud University in Nijmegen and Geuvers for one day a week at TU in Eindhoven.

V. Capretta et al. (Eds.): *Logics and Type Systems in Theory and Practice*, LNCS 14560, pp. 257–272, 2024.
https://doi.org/10.1007/978-3-031-61716-4_17

In general, a *pure morphic sequence* over a finite alphabet Γ is of the shape $f^\infty(a)$ for a finite alphabet Γ and $f : \Gamma \to \Gamma^+$, $a \in \Gamma$ and $f(a) = au$, $u \in \Gamma^+$. Here $f^\infty(a)$ is defined as the limit of $f^n(a)$, which is well-defined as for every n the word $f^n(a) = auf(u) \cdots f^{n-1}(u)$ is a prefix of $f^{n+1}(a) = auf(u) \cdots f^n(u)$. So

$$f^\infty(a) = auf(u)f^2(u)f^3(u) \cdots .$$

If we have a (typically smaller) finite alphabet Σ, and a *coding* $\tau : \Gamma \to \Sigma$, then a *morphic sequence* over the alphabet Σ is defined to be of the shape $\tau(\sigma)$ for a pure morphic sequence σ. Clearly any pure morphic sequence is morphic by choosing $\Sigma = \Gamma$ and τ to be the identity. Basic properties of morphic sequences are extensively described in the books [1,4]. A characterization of morphic sequences based on automata is given in [5]. Papers relating morphic sequences to rewriting and checking properties automatically include [8,13,14].

A typical example of a morphic sequence that is not pure morphic is

$$\mathsf{spir} = 1101001000100001 \cdots ,$$

consisting of infinitely many ones and for which the numbers of zeros in between two successive ones is $0, 1, 2, 3, \ldots$, respectively. Choosing $\Gamma = \{0, 1, 2\}$ and $f(2) = 21, f(1) = 01, f(0) = 0$, we obtain $f^2(2) = 2101$, $f^3(2) = 2101001$ and so on, yielding $f^\infty(2) = 2101001000100001 \cdots$, for which indeed by choosing $\Sigma = \{0, 1\}$, $\tau(0) = 0, \tau(1) = \tau(2) = 1$ we obtain $\mathsf{spir} = \tau(f^\infty(2))$, showing that spir is morphic. But spir is not pure morphic. If it was then we have $\mathsf{spir} = f^\infty(1)$ for f satisfying $f(1) = 1u$, $f(0) = v$, yielding contradictions for all cases: u should contain a 0, but no 1 (otherwise $f^\infty(1)$ would contain a pattern 10^k1 infinitely often for a fixed k), and v should contain a 1 (otherwise $f^\infty(1)$ would contain only a single 1), but as $f^\infty(1)$ purely consists of $f(0) = v$ and $f(1) = 1u$ it cannot contain unbounded groups of zeros.

If you see this definition of morphic sequences for the first time, it looks quite ad hoc. However, there are several reasons to consider this class of morphic sequences as a natural class of sequences. One of them is the observation that the class of morphic sequences is closed under several kinds of operations. For instance, if the result of any morphism $g : \Sigma \to \Sigma^*$ applied on any morphic sequence over Σ is an infinite sequence, then it is always again morphic. A more general result states that if applying a *finite state transducer* on a morphic sequence yields an infinite sequence, then it is always morphic too. Results like these are not easy, and can be found in [1], Corollary 7.7.5 and Theorem 7.9.1.

The class of morphic sequences is a generalization of the class of *k-automatic sequences*, being the main topic of [1]. These k-automatic sequences can be defined in several ways, but one way corresponds to our definition of morphic sequences with the extra requirement that the morphism $f : \Gamma \to \Gamma^+$ is k-*uniform*, that is, the length $|f(b)|$ of $f(b)$ is equal to k for every $b \in \Gamma$.

As can be found in [1], this class of k-automatic sequences can be defined in several equivalent ways. One way is by means of an automaton (justifying the name 'automatic'), another one is by finiteness of the k-kernel of the sequence, being a particular set of subsequences of the sequence. This has the flavor of a main well-known result from formal language theory, namely that the class of *regular languages* can be defined in several equivalent ways: DFAs, NFAs, regular expressions, right-linear grammars, and so on.

Similar characterizations for the class of morphic sequences are less known. A main goal of this paper is to give an overview of such characterizations. Apart from the closure properties, having several equivalent characterizations indicate that the class of morphic sequences is a natural class of sequences to consider, just like the class of regular languages is a natural class of languages to consider.

In [5], page 76, Theorem 2.24, an automaton-based characterization for morphic sequences is given, generalizing the automaton-based characterization of k-automatic sequences. For k-automatic sequences the corresponding type of automaton is a DFAO, a DFA with output, over the alphabet $\{0, 1, \ldots, k-1\}$, in which the transition function $\delta : Q \times \Sigma \to Q$ is total, where Q is the set of states and Σ is the alphabet. In the generalization this transition function is partial. More precisely, when numbering the symbols from Σ from 0 to $n-1$, then every state $q \in Q$ has an arity $0 < \text{ar}(q) \leq n$, and $\delta(q, i)$ is defined if and only if $i < \text{ar}(q)$. Surprisingly, this is exactly the same notion of a *mix-DFAO* as it is used in [2] to define and investigate *mix-automatic sequences*, being an extension of automatic sequences closed under zip operations. But the way in which words are entered to the automaton is different. One of the main results of [2] is that morphic sequences and mix-automatic sequences are incomparable.

Another characterization more in the flavor of finiteness of the kernel that we give is that we show how an infinite tree structure on the natural numbers defines for every natural number a subsequence of a given sequence, and we show that a sequence is morphic if and only if a tree structure exists such that the set of corresponding subsequences is finite.

Both sequences and these infinite trees can be seen as infinite terms: for sequences it is only over unary symbols, and for trees also symbols of arity >1 appear. Infinite terms are called *rational* if they only contain finitely many distinct subterms. We reformulate our result on finitely many distinct subsequences in terms of rationality of terms, and we also describe morphic sequences as infinitary normal forms using infinitary rewriting.

Another more esthetic argument to consider morphic sequences is that they give rise to interesting *turtle graphics*, [9]. Having a sequence σ over a finite alphabet Σ, one chooses an angle $\phi(b)$ for every $b \in \Sigma$. Next a picture is drawn in the following way: choose an actual angle that is initialized in some way, and next proceed as follows: for $i = 0, 1, 2, 3, \ldots$ the actual angle is turned by $\phi(\sigma(i))$ and after every turn a unit segment is drawn in the direction of the actual angle. For every segment its end point is used a the starting point for the next segment. In this way infinitely many segments are drawn. The resulting figure is called a *turtle figure*.

In [9] turtle figures are investigated for several morphic sequences, sometimes yielding finite pictures, that is, after a finite but typically very big number of steps only segments will be drawn that have been drawn before by which the picture will not change any more. In other cases the turtle figure will be *fractal*.

To give the flavor of these figures, we show the turtle figure for the very simple pure morphic (and even 2-automatic) sequence $f^\infty(0) = 010001010100\cdots$ for $f(0) = 01$, $f(1) = 00$, sometimes called the *period doubling sequence*, and angles $\phi(0) = 140^o$ and $\phi(1) = -80^o$, so for every 0 the angle turns $140^\circ = \frac{7\pi}{9}$ to the left and for every 1 the angle turns $80^\circ = \frac{4\pi}{9}$ to the right, while after every turn a unit segment is drawn. The result is as follows:

A simple program of only a few lines that draws this particular turtle figure shows that after 6000 steps not yet the full figure is drawn, but in [11,12] it is shown that after $9216 = 2^{10} \times 3^2$ steps, only segments will be drawn that were drawn before, so the full picture will be equal to the picture drawn by only this finite part, and that is the picture we show here. Other morphic sequences (many of which not being automatic) and other angles give rise to a wide range of remarkable turtle figures, many of which are shown in [9,11,12]. Turtle figures can be made for all kinds of sequences, but morphic sequences show up a nice balance between the boring and very regular figures for ultimately periodic sequences on the one hand, and the complete chaos of random sequences on the other hand. The notation for the sequence spir we saw is motivated by the fact that it gives rise to spiral shaped turtle figures.

This paper is organized as follows. In Sect. 2 we give some notation and preliminaries. In Sect. 3 we show how the definition of a (pure) morphic sequence gives rise to a tree structure on the natural numbers. This tree structure is fully defined by a weakly monotone and surjective function $P : \mathbb{N}_{>0} \to \mathbb{N}$ that plays an important role in the rest of the paper. Then we give the various equivalent characterizations, starting in Sect. 4 where we characterize morphic sequences by means of mix-DFAOs, a particular kind of automata. Next in Sect. 5 we show that morphic sequences are characterized by the property that a particular class of subsequences consist of only finitely many distinct sequences. In Sect. 6 the same result is reformulated in terms of infinite terms. In Sect. 7 we present how morphic sequences can be characterized by infinitary rewriting. We conclude in Sect. 8. For some proofs we refer to the extended version [10] of this paper.

2   Notation and Preliminaries

We write $\mathbb{N}$ for the set of natural numbers (including 0) and $\mathbb{N}_{>0}$ for the set of natural numbers >0. A function $f : \mathbb{N}_{>0} \to \mathbb{N}$ is called *weakly monotone* if $f(n+1) \geq f(n)$ for all $n \in \mathbb{N}_{>0}$.

For a *word* $w \in \Sigma^*$ over an alphabet Σ we write $|w|$ for the length of w. The set of non-empty words over Σ is denoted by Σ^+. The empty word of length 0 is denoted by ϵ.

A *sequence* σ over an alphabet Σ can be seen as a mapping $\sigma : \mathbb{N} \to \Sigma$, so

$$\sigma = \sigma(0)\sigma(1)\sigma(2)\cdots.$$

So by sequence we always mean an infinite sequence. In some texts they are called *streams* or *infinite words*. The set of sequences over Σ is denoted by Σ^∞.

Morphisms $f : \Sigma \to \Sigma^+$ can also be applied on words and sequences, for instance, $f(\sigma)$ is the concatenation of the following words:

$$f(\sigma) = f(\sigma(0))f(\sigma(1))f(\sigma(2))\cdots.$$

Apart from applying morphisms another simple operation on sequences is the tail tail that simply removes the first element, so $\text{tail}(\sigma) = \sigma(1)\sigma(2)\sigma(3)\cdots$.

From the introduction we recall:

Definition 1. *A pure morphic sequence over a finite alphabet Γ is of the shape $f^\infty(a)$ for a finite alphabet Γ and $f : \Gamma \to \Gamma^+$, $a \in \Gamma$ and $f(a) = au$, $u \in \Gamma^+$.*

A morphic sequence over a finite alphabet Σ is of the shape $\tau(\sigma)$ for some coding $\tau : \Gamma \to \Sigma$ and some pure morphic sequence σ over some finite alphabet Γ.

Here $f^\infty(a)$ is defined as the limit of $f^n(a)$, which is well-defined as for every n the word $f^n(a) = auf(u) \cdots f^{n-1}(u)$ is a prefix of $f^{n+1}(a) = auf(u) \cdots f^n(u)$. So

$$f^\infty(a) = auf(u)f^2(u)f^3(u) \cdots .$$

It is easy to see that it is the unique *fixed point* of f starting in a, that is, $f(f^\infty(a)) = f^\infty(a)$.

In the above notation the coding $\tau : \Gamma \to \Sigma$ is lifted to $\tau : \Gamma^\infty \to \Sigma^\infty$ in which τ is applied on alle separate elements.

A slightly more relaxed definition of morphic sequence also allows $f(b) = \epsilon$ for some elements $b \in \Gamma$. In [1], Theorem 7.5.1, it is proved that this is equivalent to our definition, so we may and will always assume that $f(b) \in \Gamma^+$ for all $b \in \Gamma$.

A sequence over Σ can also be seen as an *infinite term* over Σ in which every symbol $b \in \Sigma$ has *arity* 1, notation $\mathsf{ar}(b) = 1$. More general, if every symbol $b \in \Sigma$ has arity $\mathsf{ar}(b) \geq 1$, an infinite term over Σ is defined by saying which symbol is on which position. Here a *position* $p \in \mathbb{N}^*$ is a sequence of natural numbers. In order to be a proper term, some requirements have to be satisfied as indicated in the following definition. Here we write $\bot$ for undefined, use f for symbols in Σ and every symbol f has an arity $\mathsf{ar}(f) \in \mathbb{N}_{>0}$.

Definition 2. *An infinite* term *over a signature Σ is defined to be a map $t :$ $\mathbb{N}^* \to \Sigma \cup \{\bot\}$ such that*

- *the root $t(\epsilon)$ of the term t is defined, so $t(\epsilon) \in \Sigma$, and*
- *for all $p \in \mathbb{N}^*$ and all $i \in \mathbb{N}$ we have*

$$t(pi) \in \Sigma \iff t(p) \in \Sigma \land 0 \leq i < \mathsf{ar}(t(p)).$$

A word $p \in \mathbb{N}^*$ such that $t(p) \in \Sigma$ is called a position *of t.*

The subterm *$t_p : \mathbb{N}^* \to \Sigma \cup \{\bot\}$ at position $p \in \mathbb{N}^*$ satisfying $t(p) \in \Sigma$ is defined by $t_p(q) = t(pq)$ for all $q \in \mathbb{N}^*$.*

An infinite term is called rational *if it contains only finitely many distinct subterms.*

When allowing *constants*, that is, symbols f with $\mathsf{ar}(f) = 0$, this definition would also cover usual finite terms. As we only consider infinite terms with $\mathsf{ar}(f) > 0$ for all $f \in \Sigma$, we do not allow constants and all our terms are defined on infinitely many positions including 0^n for all $n \in \mathbb{N}$. In many presentations positions are words over numbers from 1 to $\mathsf{ar}(f)$. Due to the relationship with number representations we prefer to number from 0 to $\mathsf{ar}(f) - 1$, just like digits in decimal numbers are numbered from 0 to 9 and not from 1 to 10.

A sequence $\sigma : \mathbb{N} \to \Sigma$ now can be seen as an infinite term t_σ over Σ where every symbol has arity 1, and any number $n \in \mathbb{N}$ is identified with the position 0^n, so $t_\sigma(n)(0^n) = \sigma(n)$ for all $n \in \mathbb{N}$ and $t_\sigma(p) = \bot$ for all $p \in \mathbb{N}^*$ not being of the shape 0^n.

3   Tree Structures of Morphic Sequences

In this section we show how a (pure) morphic sequence $f^\infty(a)$ gives rise to a tree structure on the natural numbers. Mapping every natural number to its position in the tree may be seen as an *enumeration system*, being the main topic of [4,5].

A pure morphic sequence $f^\infty(a)$ is the unique fixed point of f starting in a, and is of the shape

$$auf(u)f^2(u)f^3(u)\cdots.$$

For every $i > 0$ the element in $f^\infty(a)(i)$ is obtained as an element of $f(b)$ for some $b = f^\infty(a)(j)$ for $j < i$, so occurring earlier in the sequence. If we draw an arrow between every element in the sequence and the element from which it is obtained, we get a tree structure, as is illustrated in the following example, where $\Gamma = \{0,1\}$, $f(0) = 01$, $f(1) = 0$. So the sequence $f^\infty(0)$ is the binary Fibonacci sequence fib as we saw in the introduction. The first few levels of the tree look as follows:

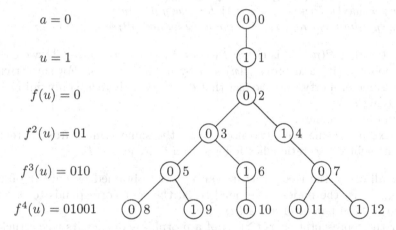

$a = 0$ $\quad$ ⓪ 0

$u = 1$ $\quad$ ① 1

$f(u) = 0$ $\quad$ ⓪ 2

$f^2(u) = 01$ $\quad$ ⓪ 3   ① 4

$f^3(u) = 010$ $\quad$ ⓪ 5   ① 6   ⓪ 7

$f^4(u) = 01001$ ⓪ 8   ① 9   ⓪ 10   ⓪ 11   ① 12

On the first level we have a, in this example being 0, denoted inside the node, numbered by 0. On the second level we have u, in this example being 1, again inside the node, numbered by 1. If $|u| > 1$ this level does not consist of a single node but of $|u|$ nodes. On the next levels we have $f(u)$, $f^2(u)$, and so on. So the full pure morphic sequence $f^\infty(a) = auf(u)f^2(u)\cdots$ is obtained by concatenating all node labels in the order of the node numbers. This applies similarly for any morphic sequence, not only for this example.

The tree can be seen as an infinite term over the signature $\Gamma \cup \{a_0\}$, where the arity of a symbol $b \in \Gamma$ is $|f(b)|$ and the arity of a_0 is $|f(a)| - 1$. Here a_0 only occurs at the root.

Fixing f and a, we write $P : \mathbb{N}_{>0} \to \mathbb{N}$ for the function mapping node number i to its parent in the tree. So $P(i) = j$ if $f^\infty(a)(i)$ is obtained from $f(b)$ for $b = f^\infty(a)(j)$. We write $R : \mathbb{N}_{>0} \to \mathbb{N}$ for the function mapping i to the corresponding position in $f(b)$, being 0 for the first position and $|f(b)| - 1$ for the last. For instance, the top right 1 numbered by 12 is $f^\infty(0)(12)$, obtained as the second element of $f(0)$ for 0 being $f^\infty(0)(7)$, so $P(12) = 7$ and $R(12) = 1$.

These functions P, R can be defined inductively as follows:

- $P(1) = 0$, $R(1) = 1$,
- $P(n + 1) = P(n)$ and $R(n + 1) = R(n) + 1$ if $R(n) + 1 < |f(b)|$ for $b = f^\infty(a)(P(n))$,
- $P(n+1) = P(n)+1$ and $R(n+1) = 0$ if $R(n)+1 = |f(b)|$ for $b = f^\infty(a)(P(n))$.

For f being k-uniform, that is, $|f(b)| = k > 1$ for all $b \in \Gamma$, we have $P(i) = i \div k$ and $R(i) = i \mod k$ for all $i > 0$.

We collect some basic properties:

Theorem 1. *For $f : \Gamma \to \Gamma^+$, $a \in \Gamma$ and $f(a) = au$, $u \in \Gamma^+$ and P, R as defined above, we have*

1. *$P : \mathbb{N}_{>0} \to \mathbb{N}$ is weakly monotone and surjective, and $P(n) < n$ for every $n > 0$,*
2. *$f^\infty(a)(n) = f(b)(R(n))$ for $b = f^\infty(a)(P(n))$, for every $n > 0$,*
3. *$R(n) = \max\{k | P(n - k) = P(n)\}$ for every $n > |u|$.*
4. *The infinite term representing the corresponding tree is rational.*

Proof. 1: Using $P(n + 1)$ is either $P(n)$ or $P(n) + 1$ for all $n \in \mathbb{N}$, we conclude weak monotonicity and prove $P(n) < n$ by induction on n. For surjectivity we use the same property, and the fact that for every n there are only finitely many k satisfying $P(k) = n$.

2 and 3: immediate.

4: Except for the root, two nodes with the same symbol from Γ represent the same subtree, so rationality follows from finiteness of Γ.   □

For all cases, the tree structure can be fully obtained by only the function $P : \mathbb{N}_{>0} \to \mathbb{N}$: the nodes correspond to $\mathbb{N}$, the root corresponds to 0, and for every $n > 0$ the parent of n is $P(n)$.

For the representation $\tau(f^\infty(a))$ of a morphic sequence its tree structure is defined to be the tree defined by the function $P : \mathbb{N}_{>0} \to \mathbb{N}$ for the pure morphic sequence $f^\infty(a)$.

4 Morphic Sequences Characterized by Automata

A full characterization theorem of morphic sequences by automata is given as Theorem 2.24 in [5]; for a precise formulation we also refer to [10]. Here we focus on a variant of this theorem that we independently found ourselves. The key concept is a mix-DFAO, being an extension of a standard DFAO in which the number of outgoing arrows may depend on the state.

Definition 3. *A mix deterministic finite automaton with output (mix-DFAO) is a sixtupel $(Q, \mathsf{ar}, \delta, q_0, \Sigma, \lambda)$, where*

- *Q is a finite set of states,*
- *$\mathsf{ar} : Q \to \mathbb{N}_{>0}$ is the arity function,*

- $\delta : Q \times \mathbb{N} \to Q$ is a *partial* transition function *for which* $\delta(q, a)$ *is defined if and only if* $0 \le a < \mathsf{ar}(q)$,
- $q_0 \in Q$ *is the* initial state,
- Σ *is a finite* alphabet, *and*
- $\lambda : Q \to \Sigma$ *is the* output function.

In [2] mix-DFAO's were used to define and investigate *mix-automatic sequences*, being an extension of automatic sequences closed under zip operations. Here the ith element of the sequence is obtained by feeding the interpretation of i in some dynamic radix enumeration system to the automaton and taking the output of the resulting state. So the key idea is to represent i by a word w, and feed w to the automaton.

Surprisingly, we will use exactly the same notion of mix-DFAO to characterize morphic sequences. Given such a mix-DFAO M we will define a sequence σ_M over Σ that will be morphic. We do this by defining $\sigma_M(i) \in \Sigma$ for every $i \in \mathbb{N}$. As such an element in Σ may be defined as $\lambda(q)$ for some $q \in Q$, we do this by mapping natural numbers to states, using the transition function. This idea is the same as for mix-automatic sequences. The difference is that we use a different representation of the number i by a word. While for mix-automatic sequences the dynamic radix interpretation is taken from right to left, starting at the least significant digit, we will do it the other way around, so starting at the most significant digit. For k-automatic sequences these correspond to the two ways to define such a sequence by a DFAO: a natural number i is either mapped to $\delta(q_0, w_i)$ or to $\delta(q_0, w_i^R)$, where w_i is the k-ary notation of i, and every state has arity k, and R stands for reversing the word. In our morphic setting the representation is slightly more complicated as different states may have different arities.

Let $n_M = \max\{\mathsf{ar}(q) \mid q \in Q\}$ and $\Delta = \{0, 1, \ldots, n_M - 1\}$. As $\delta(q, a)$ is undefined for every $a \notin \Delta$, we may consider δ as a partial function $\delta : Q \times \Delta \to Q$. As usual in automata we extend δ allowing words in the second argument rather than single symbols. Due to the partial character of δ, the result of δ on a word may be undefined. We define L_M to be the set of words w over Δ not starting in 0 for which $\delta(q_0, w) \in Q$ is defined. So $0^* L_M$ is the smallest prefix closed language over Δ satisfying $\epsilon \in L_M$ and $wa \in L_M$ for every $w \in L_M$ and $0 \le a < \mathsf{ar}(\delta(q_0, w))$.

We define the enumeration function $\phi_M : \mathbb{N} \to L_M$ inductively as follows: $\phi_M(0) = \epsilon$ and $\phi_M(1) = 1$. For the inductive part we have the invariant that if $\phi_M(n) = wa$ for $w \in L_M$ and $a \in \Gamma$, then there exists exactly one $n' < n$ satisfying $\phi_M(n') = w$. Indeed for $n = 1, n' = 0$ this holds for $w = \epsilon$. For defining $\phi_M(n + 1)$ for $n > 0$ write $\phi_M(n) = wa$. In case $a + 1 < \mathsf{ar}(\delta(q_0, w))$ then we define $\phi_M(n + 1) = w(a + 1)$, otherwise we define $\phi_M(n + 1) = \phi_M(n' + 1)0$ for $n' < n$ satisfying $\phi_M(n') = w$.

Note that for the k-automatic case where $\mathsf{ar}(q) = k$ for all $q \in Q$, $\phi_M(n)$ coincides with the k-ary notation of n.

For a mix-DFAO $M = (Q, \text{ar}, \delta, q_0, \Sigma, \lambda)$ the sequence σ_M over Σ is defined by

$$\sigma_M(i) = \lambda(\delta(q_0, \phi_M(i)))$$

for all $i \in \mathbb{N}$. Now we are ready to formulate our characterization of morphic sequences by mix-DFAOs, for the proof of this theorem we refer to [10].

Theorem 2. *A sequence σ over an alphabet Σ is morphic if and only if a mix-DFAO $M = (Q, \text{ar}, \delta, q_0, \Sigma, \lambda)$ exists such that $\sigma = \sigma_M$.*

Note that L_M is closely related to the set of words $p \in \mathbb{N}^*$ for which $t(p) \neq \bot$ in the definition of infinite terms; the only difference is in the first symbol: non-empty words in L_M always start in $b > 0$, where in the position notation this is replaced by $b - 1$.

Using the standard automaton notation where $\delta(q, i) = r$ is denoted by an arrow from node q to node r labeled by i, the start state q_0 is denoted by an incoming arrow, and the output $\lambda(q)$ of a state q is denoted inside the node q, a DFAO for the binary Fibonacci sequence $f^\infty(0)$ for $f(0) = 01, f(1) = 0$ reads as follows:

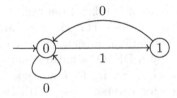

The tree we earlier gave for this sequence can be obtained by unwinding this automaton, starting from q_0 and ignoring the self-loop of q_0 at the first step. This observation holds for morphic sequences in general.

5   Morphic Sequences Characterized by Subsequences

As is presented in [1], k-automatic sequences can be characterized by finiteness of the k-kernel of the sequence. For $k = 2$ this means that by applying the operations even and odd on the sequence any number of times in any order, only finitely many distinct sequences are obtained. In this section we show that the much more general class of morphic sequences can also be characterized by finiteness of some class of subsequences.

In Sect. 3 we saw that the tree structure of a morphic sequence $\tau(f^\infty(a))$ is fully described by a surjective weakly monotone function $P : \mathbb{N}_{>0} \to \mathbb{N}$. Now we will use such a function P to define subsequences corresponding to subtrees.

As a first observation we state the following lemma:

Lemma 1. *Let $P : \mathbb{N}_{>0} \to \mathbb{N}$ be weakly monotone and surjective. Then for every $n \in \mathbb{N}_{>0}$:*

1. $P(n) < n$, and
2. *there exists $k > 0$ such that $P^i(n) > 0$ for all $i < k$ and $P^k(n) = 0$.*

Proof. We apply induction on n, first we do (1). For $n = 1$ it holds since $P(1) = 0$, which holds since $P(1) > 0$ would contradict surjectivity. If $P(n+1) > P(n)+1$ then $P(n) + 1$ would not be in the image of P, contradicting surjectivity. So $P(n + 1) \leq P(n) + 1$. Using the induction hypothesis $P(n) < n$ this yields $P(n+1) < n + 1$, proving (1).

Also for (2) we apply induction: for $n = 1$ it holds since $P(1) = 0$, and for $n > 1$ it follows from the induction hypothesis applied to $P(n) < n$. □

A weakly monotone and surjective function $P : \mathbb{N}_{>0} \to \mathbb{N}$ we shortly call a *tree function*, as it defines a tree structure on $\mathbb{N}$ in which 0 is the root and i is the parent of j in the tree if and only if $P(j) = i$. For our purpose we need an extra requirement, namely that the tree is *rational*, that is, contains only finitely many distinct subtrees. A *rational tree function* is defined to be a tree function $P : \mathbb{N}_{>0} \to \mathbb{N}$ for which the corresponding tree is rational.

Lemma 2. *Let $P : \mathbb{N}_{>0} \to \mathbb{N}$ be the tree function corresponding to a pure morphic function $f^\infty(a)$ over a finite alphabet Σ as described in Sect. 3. Then P is a rational tree function.*

Proof. Let p, q be two nodes of the corresponding tree, both distinct from the root. Then if p and q are labeled by he same symbol from Σ, then by construction the subtrees having p and q as their roots are equal. Now the lemma follows from finiteness of Σ. □

For instance, for the tree we presented for the binary Fibonacci sequence fib in Sect. 3 there are three distinct subtrees: the full tree having node 0 as its root, and the two subtrees having nodes 1 and 2 as its root, respectively. Every other subtree of which the root is labeled by 0 is equal to subtree having 2 as its root, and every other subtree of which the root is labeled by 1 is equal to subtree having 1 as its root.

Let $P : \mathbb{N}_{>0} \to \mathbb{N}$ be a tree function. For $n \in \mathbb{N}$ we define

$$S_P(n) = \{m \in \mathbb{N} \mid \exists k \in \mathbb{N} : P^k(m) = n\}.$$

Note that for every $n \in \mathbb{N}$ the set $S_P(n) \subseteq \mathbb{N}$ is infinite.

For any infinite set $S \subseteq \mathbb{N}$ we can uniquely write $S = \{s_0, s_1, s_2, \ldots\}$ with $s_0 < s_1 < s_2 < \cdots$. For a sequence σ and an infinite set $S \subseteq \mathbb{N}$ we define σ^S being the subsequence

$$\sigma^S = \sigma_{s_0} \sigma_{s_1} \sigma_{s_2} \cdots,$$

so $\sigma^S(i) = \sigma(s_i)$ for all $i \in \mathbb{N}$.

Now we arrive at our next characterization of morphic sequences; for the proof we refer to [10].

Theorem 3. *A sequence σ over Σ is morphic if and only if a rational tree function $P : \mathbb{N}_{>0} \to \mathbb{N}$ exists such that the set*

$$\{\sigma^{S_P(n)} \mid n \in \mathbb{N}\}$$

of subsequences of σ is finite.

For instance, for the binary Fibonacci sequence fib the set $\{\sigma^{S_P(n)} \mid n \in \mathbb{N}\}$ consists of three sequences: $\sigma^{S_P(0)} = $ fib, $\sigma^{S_P(1)} = $ tail(fib) and $\sigma^{S_P(2)} = $ tail(tail(fib)), corresponding to the three subtrees of the the rational tree we already observed. For more complicated examples of morphic sequences σ typically $\{\sigma^{S_P(n)} \mid n \in \mathbb{N}\}$ also contains sequences that cannot be obtained from σ by only applying the tail function tail.

It is a natural question whether the requirement for the tree function P being rational is essential in Theorem 3. It is, as is shown by the following construction. For any sequence natural numbers $n_1, n_2, n_3, \ldots$ of natural numbers satisfying $1 \leq n_1 \leq n_2 \leq n_3 \leq \cdots$ we can make a tree T for which the number of nodes with distance i to the root is exactly n_i for all $i \geq 1$. Next we consider the following tree:

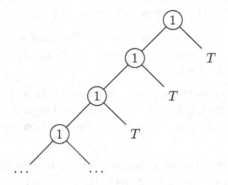

Here the indicated nodes are labeled by 1, and all nodes in all copies of T are labeled by 0. Let σ be the sequence and P the tree function corresponding to this tree. Then $\{\sigma^{S_P(n)} \mid n \in \mathbb{N}\}$ consists only of two sequences: the sequence σ described by any of all equal subtrees having any node labeled by 1 at its root, and the sequence purely consisting of zeros having any node labeled by 0 at its root. The sequence σ contains infinitely many ones, and the numbers of zeros between them are $0, 1, 1 + n_1, 1 + n_1 + n_2, 1 + n_1 + n_2 + n_3, \ldots$, respectively. But there are only countably many morphic sequences, and the uncountably many choices for $1 \leq n_1 \leq n_2 \leq n_3 \leq \cdots$ yield uncountably many distinct sequences σ, so not all of these are morphic, showing that Theorem 3 does not hold if the requirement for P being a rational tree function is weakened to only being any tree function.

We started this section by comparing our new construction of subsequences by the k-kernel of which finiteness is equivalent to being k-automatic. We want to stress that in case of the rational tree function $P : \mathbb{N}_{>0} \to \mathbb{N}$ corresponds to k-automatic sequences, so $P(n) = n \div k$, then the set $\{\sigma^{S_P(n)} \mid n \in \mathbb{N}\}$ does not coincide with the k-kernel. More precisely, the k-kernel consists of the subsequences σ^S of σ where S runs over all numbers of which the k-ary notation ends with some fixed word w, while our subsequences $\sigma^{S_P(n)}$ run over the subsequences σ^S of σ where S runs over all numbers of which the k-ary notation starts with some fixed word w.

6   Morphic Sequences Characterized by Rational Infinite Terms

In Theorem 3 the subsequences $\sigma^{S_P(n)}$ are described by subtrees. In this section we reformulate this theorem by describing these subtrees as subterms of infinite terms. The idea is to represent the tree represented by the rational tree function $P : \mathbb{N}_{>0} \to \mathbb{N}$ by a rational term over the signature consisting of all finitely many subtrees, in which the arity corresponds to the number of children of the tree. Next the nodes of this tree are numbered in a breadth-first way, and for a sequence σ every node with number i is labeled by $\sigma(i)$. Now Theorem 3 essentially states that σ is morphic if and only if this labeled tree is rational.

More precisely, we start by a rational tree function $P : \mathbb{N}_{>0} \to \mathbb{N}$. Let Δ consist of the finitely many subtrees, and the arity of $t \in \Delta$ is its number of children. Now the tree coincides with an infinite term t_P over Δ, described by a function $t_P : \mathbb{N}^* \to \Delta \cup \{\bot\}$. We define a mix-DFAO M in which the set of states is Δ, the initial state is the full tree described by the root, the arity of every state is the number of children of the root of the corresponding subtree, and $\delta(q, i)$ is the $(i+1)$th child of the root of the corresponding subtree. The output function λ does not play a role here. Next we define the function $\phi_M : \mathbb{N} \to L_M$ as in Sect. 4, where L_M is the language of words w not starting in 0 for which $\delta(q_0, w)$ is defined. Note that $0^* L_M$ coincides exactly with the set of positions of the infinite term t_P. As $\phi_M : \mathbb{N} \to L_M$ is a bijection, we can also consider its inverse $\phi_M^{-1} : L_M \to \mathbb{N}$, mapping every position to its number.

For any sequence σ over a finite alphabet Σ, we extend the signature Δ to $\Delta \times \Sigma$, where for every $(t, a) \in \Delta \times \Sigma$ the arity is $\mathsf{ar}(t)$. We consider the infinite term over this extended signature by labeling every node in the term t_P by $\sigma(i)$, where i is the number of the node. More precisely, the term $t_P : \mathbb{N}^* \to \Delta \cup \{\bot\}$ is extended to $t_{P,\sigma} : \mathbb{N}^* \to \Delta \times \Sigma \cup \{\bot\}$ defined by $t_{P,\sigma}(p) = (t_P(p), \sigma(\phi_M^{-1}(p)))$ for every $p \in \mathbb{N}^*$.

Now the reformulation of Theorem 3 reads as follows:

Theorem 4. *A sequence σ over the alphabet Σ is morphic if and only if a rational tree function $P : \mathbb{N}_{>0} \to \mathbb{N}$ exists such that the infinite term $t_{P,\sigma}$ is rational.*

7 Morphic Sequences Characterized by Infinitary Rewriting

In this section we describe how any morphic sequence can be obtained by infinitary rewriting. For basics on rewriting we refer to [6]. Infinitary rewriting extends rewriting to infinite terms and infinite reductions. Infinite terms were already introduced in our preliminaries. An infinite reduction is said to *converge* if for every n it holds that after a finite number of steps no reductions take place any more on level $\leq n$. In that case the limit of the reduction is well-defined as an infinite term, and that is defined to be the *infinitary normal form* of the reduction. For more basics on infinitary rewriting and such reductions to infinitary normal forms we refer to the chapter [3] on infinitary rewriting in [6] and [7].

So the goal of this section is to describe any morphic sequence to be the infinitary normal form of a particular start term with respect to a particular rewrite system. The rewrite system is designed in such a way that it follows the definition of morphic sequence quite directly.

For a word $w = w_0 w_1 \cdots w_{k-1}$ of unary symbols $w_0, w_1, \ldots, w_{k-1}$ and a term t we use $w(t)$ as an abbreviation for $w_0(w_1(\cdots(w_{k-1}(t)\cdots)))$.

Theorem 5. *For a morphic sequence $\sigma = \tau(f^\infty(a))$ over Σ for $\tau : \Gamma \to \Sigma$, $f : \Gamma \to \Gamma^+$, $f(a) = au$, $a \in \Gamma$, $u \in \Gamma^+$, we define the signature $\Delta = \Sigma \cup \Gamma \cup \overline{\Gamma} \cup \{S, E\}$, where $\overline{\Gamma} = \{\overline{b} \mid b \in \Gamma\}$, the symbol E is a constant marking the end, and all other symbols are unary. The rewrite system R_σ consists of the rules*

$$
\begin{aligned}
S(b(x)) &\to \tau(b)(S(S(\overline{b}(x)))) && \text{for all } b \in \Gamma, \\
\overline{b}(c(x)) &\to c(\overline{b}(x)) && \text{for all } b, c \in \Gamma, \\
\overline{b}(E) &\to f(b)(E) && \text{for all } b \in \Gamma
\end{aligned}
$$

Then σ is the unique infinitary normal form of $\tau(a)(S(u(E)))$.

Proof. For any word $w \in \Gamma^+$ and $b \in \Gamma$ we obtain $\overline{b}(w(E)) \to_{R_\sigma}^+ w(\overline{b}(E))$ by using the second type of rules. For any word $w \in \Gamma^+$ and $b \in \Gamma$ combining this by using the other types of rules yields

$$S(b(w(E))) \to_{R_\sigma} \tau(b)(S(\overline{b}w(E))) \to_{R_\sigma}^+ \tau(b)(S(w(\overline{b}(E)))) \to_{R_\sigma} \tau(b)(S(wf(b)(E))).$$

Repeating this $|w|$ times yields $Sw(E) \to_{R_\sigma}^+ \tau(w)Sf(w)(E)$ for any word $w \in \Gamma^+$. Hence

$$\tau(a)(S(u(E))) \to_{R_\sigma}^+ \tau(au)Sf(u)(E) \to_{R_\sigma}^+ \tau(auf(u))Sf^2(u)(E) \to_{R_\sigma}^+ \cdots$$

$$\to_{R_\sigma}^+ \tau(auf(u)\cdots f^{n-1}(u))Sf^n(u)(E) = \tau(f^n(a))Sf^n(u)(E)$$

for all $n > 0$. Due to the shape of the rules, in further rewriting $\tau(f^n(a))Sf^n(u)(E)$ the initial part $\tau(f^n(a))$ will not change any more, hence by repeating this for increasing n converges to $\tau(f^\infty(a)) = \sigma$.

As R_σ is orthogonal, no other infinitary normal form of $\tau(a)(S(u(E)))$ exists. $\square$

8   Conclusions

The main result of this paper is the series of alternative characterizations of morphic sequences: Theorems 2, 3 and 4. They all give if-and-only-if characterizations of morphic sequences, just like in any text book on automata theory you find a range of if-and-only-if characterizations of regular languages. The most remarkable theorem is Theorem 2, characterizing morphic sequences by mix-DFAOs, exactly the same notion of automata defining mix-automatic sequences in a slightly different way. In this paper we focus on the characterizations themselves. Applying them, like applying finiteness of DFAs gives rise to the pumping lemma as a standard technique to prove that a particular language is not regular, is a topic of further research.

The tree structure of a representation of a morphic sequence as described in Sect. 3 can be seen as an unwinding of the mix-DFAO as used in Theorem 2. Many distinct morphic sequences, like all 2-automatic sequences, may have representations with the same tree structure. This follows from the fact that for every k a sequence is k-automatic if and only if it can be represented by the tree structure in which every node has degree k. Now Cobham's Theorem ([1], Theorem 11.2.2) implies that every not ultimately periodic 2-automatic sequence does not have a representation with the tree structure in which every node has degree 3.

A topic of further research is to investigate which tree structures may occur for some given morphic sequence. Indeed, a single morphic sequence may be represented by distinct tree structures. For instance, the binary Fibonacci sequence fib is not only represented by fib $= f^\infty(0)$ for $f(0) = 01$, $f(1) = 0$, but also by fib $= \tau(g^\infty(0))$ for $g(0) = 02$, $g(1) = 021$, $g(2) = 102$, $\tau(0) = \tau(1) = 0$, $\tau(2) = 1$, having a completely different tree structure. The equality $f^\infty(0) = \tau(g^\infty(0))$ may be proven by simultaneously proving that $f^{2n}(0) = \tau(g^{n-1}(021))$ and $f^{2n}(1) = \tau(g^{n-1}(02))$ for all $n \geq 1$, by induction on n. This proof is given fully automatically by a very first prototype of a tool, that also succeeds in proving equality of two representations of even(fib) which was mentioned as an open problem in the conclusion of [10].

References

1. Allouche, J.-P., Shallit, J.: Automatic sequences: theory, applications, generalizations (2003)
2. Endrullis, J., Grabmayer, C., Hendriks, D.: Mix-automatic sequences. In: Dediu, A.-H., Martín-Vide, C., Truthe, B. (eds.) LATA 2013. LNCS, vol. 7810, pp. 262–274. Springer, Heidelberg (2013). https://doi.org/10.1007/978-3-642-37064-9_24
3. Kennaway, R., de Vries, F.-J.: Infinitary rewriting. In: Term Rewriting Systems, by Terese, pp. 668–711. Cambridge University Press (2003)
4. Rigo, M.: Formal Languages, Automata and Numeration Systems, Part 1. Wiley (2014)
5. Rigo, M.: Formal Languages, Automata and Numeration Systems, Part 2. Wiley (2014)

6. TERESE. Term Rewriting Systems. Cambridge University Press (2003)
7. Zantema, H.: Normalization of infinite terms. In: Voronkov, A. (ed.) RTA 2008. LNCS, vol. 5117, pp. 441–455. Springer, Heidelberg (2008). https://doi.org/10.1007/978-3-540-70590-1_30
8. Zantema, H.: Well-definedness of streams by termination. In: Treinen, R. (ed.) RTA 2009. LNCS, vol. 5595, pp. 164–178. Springer, Heidelberg (2009). https://doi.org/10.1007/978-3-642-02348-4_12
9. Zantema, H.: Turtle graphics of morphic sequences. Fractals **24**(1) (2016). Preversion http://www.win.tue.nl/~hzantema/turtle.pdf
10. Zantema, H.: Characterizing morphic sequences. Extended version (2023). https://arxiv.org/abs/2309.10562
11. Zantema, H.: Spelen met oneindigheid: verrassende figuren en patronen. Noordboek 2023, p. 240 (2023). (in Dutch)
12. Zantema, H.: Playing with Infinity: Turtles, Patterns and Pictures. Taylor & Francis (2024 to appear)
13. Zantema, H., Endrullis, J.: Proving equality of streams automatically. In: Schmidt-Schlauss, M. (ed.) Proceedings of the 22nd International Conference on Rewriting Techniques and Applications. Leibniz International Proceedings in Informatics (LIPIcs), Dagstuhl, Germany, vol. 10, pp. 393–408. Schloss Dagstuhl–Leibniz-Zentrum fuer Informatik (2011)
14. Zantema, H., Raffelsieper, M.: Proving productivity in infinite data structures. In: Lynch, C. (ed.) Proceedings of the 21st International Conference on Rewriting Techniques and Applications. Leibniz International Proceedings in Informatics (LIPIcs), Dagstuhl, Germany, vol. 6, pp. 401–416. Schloss Dagstuhl–Leibniz-Zentrum fuer Informatik (2010)

Author Index

V. Capretta et al. (Eds.): Logics and Type Systems in Theory and Practice, LNCS 14560, p. 273, 2024.
https://doi.org/10.1007/978-3-031-61716-4

Printed in the United States
by Baker & Taylor Publisher Services